Communications and Control Engineering

Series Editors

E.D. Sontag • M. Thoma • A. Isidori • J.H. van Schuppen

Published titles include:

Stability and Stabilization of Infinite Dimensional Systems with Applications
Zheng-Hua Luo, Bao-Zhu Guo and Omer Morgul

Nonsmooth Mechanics (Second edition)
Bernard Brogliato

Nonlinear Control Systems II
Alberto Isidori

L_2-Gain and Passivity Techniques in Nonlinear Control
Arjan van der Schaft

Control of Linear Systems with Regulation and Input Constraints
Ali Saberi, Anton A. Stoorvogel and Peddapullaiah Sannuti

Robust and H_∞ Control
Ben M. Chen

Computer Controlled Systems
Efim N. Rosenwasser and Bernhard P. Lampe

Control of Complex and Uncertain Systems
Stanislav V. Emelyanov and Sergey K. Korovin

Robust Control Design Using H_∞ Methods
Ian R. Petersen, Valery A. Ugrinovski and Andrey V. Savkin

Model Reduction for Control System Design
Goro Obinata and Brian D.O. Anderson

Control Theory for Linear Systems
Harry L. Trentelman, Anton Stoorvogel and Malo Hautus

Functional Adaptive Control
Simon G. Fabri and Visakan Kadirkamanathan

Positive 1D and 2D Systems
Tadeusz Kaczorek

Identification and Control Using Volterra Models
Francis J. Doyle III, Ronald K. Pearson and Bobatunde A. Ogunnaike

Non-linear Control for Underactuated Mechanical Systems
Isabelle Fantoni and Rogelio Lozano

Robust Control (Second edition)
Jürgen Ackermann

Flow Control by Feedback
Ole Morten Aamo and Miroslav Krstić

Learning and Generalization (Second edition)
Mathukumalli Vidyasagar

Constrained Control and Estimation
Graham C. Goodwin, María M. Seron and José A. De Doná

Randomized Algorithms for Analysis and Control of Uncertain Systems
Roberto Tempo, Giuseppe Calafiore and Fabrizio Dabbene

Switched Linear Systems
Zhendong Sun and Shuzhi S. Ge

Subspace Methods for System Identification
Tohru Katayama

Digital Control Systems
Ioan D. Landau and Gianluca Zito

Multivariable Computer-controlled Systems
Efim N. Rosenwasser and Bernhard P. Lampe

Dissipative Systems Analysis and Control (2nd Edition)
Bernard Brogliato, Rogelio Lozano, Bernhard Maschke and Olav Egeland

Algebraic Methods for Nonlinear Control Systems
Giuseppe Conte, Claude H. Moog and Anna M. Perdon

Polynomial and Rational Matrices
Tadeusz Kaczorek

Simulation-based Algorithms for Markov Decision Processes
Hyeong Soo Chang, Michael C. Fu, Jiaqiao Hu and Steven I. Marcus

Iterative Learning Control
Hyo-Sung Ahn, Kevin L. Moore and YangQuan Chen

Wei Ren and Randal W. Beard

Distributed Consensus in Multi-vehicle Cooperative Control

Theory and Applications

Springer

Wei Ren, PhD
Department of Electrical and Computer Engineering
Utah State University
Logan, UT 84322-4120
USA

Randal W. Beard, PhD
Department of Electrical and Computer Engineering
Brigham Young University
Provo, UT 84602
USA

ISBN 978-1-84800-014-8 ISBN 978-1-84800-015-5 (eBook)

DOI 10.1007/978-1-84800-015-5

Communications and Control Engineering Series ISSN 0178-5354

British Library Cataloguing in Publication Data
Ren, Wei
Distributed consensus in multi-vehicle cooperative control :
theory and applications. - (Communications and control engineering)
1. Digital control systems 2. Automatic control -
Mathematics 3. Algorithms
I. Title II. Beard, Randal W.
629.8'9'015181

ISBN 978-1-84800-014-8

Library of Congress Control Number: 2007934265

Cover design: LE-TEX Jelonek, Schmidt & Vöckler GbR, Leipzig, Germany

Printed on acid-free paper

9 8 7 6 5 4 3 2 1

springer.com

This work is dedicated to

My wife, Fei Cheng, and
My parents, Hongyi Ren and Liying Wang
- Wei Ren

My wife, Andrea
- Randal W. Beard

Preface

Recent advances in miniaturizing of computing, communication, sensing, and actuation have made it feasible to envision large numbers of autonomous vehicles (air, ground, and water) working cooperatively to accomplish an objective. Cooperative control of multiple vehicle systems has potential impact in numerous civilian, homeland security, and military applications. Potential civilian applications include monitoring forest fires, oil fields, pipelines, and tracking wildlife. Potential homeland security applications include border patrol and monitoring the perimeter of nuclear power plants. For the military, applications include surveillance, reconnaissance, and battle damage assessment. However, for all of these applications, communication bandwidth and power constraints will preclude centralized command and control.

This book addresses the problem of information consensus, where a team of vehicles must communicate with its neighbors to agree on key pieces of information that enable them to work together in a coordinated fashion. The problem is particularly challenging because communication channels have limited range and experience fading and dropout. The study of information flow and sharing among multiple vehicles in a group plays an important role in understanding the coordinated movements of these vehicles. As a result, a critical problem for cooperative control is to design appropriate distributed algorithms such that the group of vehicles can reach consensus on the shared information in the presence of limited and unreliable information exchange and dynamically changing interaction topologies.

Our interest in distributed consensus algorithms and their applications was motivated by our research efforts in cooperative control of multiple vehicle systems and, in particular, teams of unmanned air vehicles. Air vehicles are constantly moving and consequently their ability to communicate is dynamically changing. In addition, in current military scenarios involving unmanned air vehicles, large assets like the Predator may have two-way communication capabilities, but micro air vehicles may have only the ability to receive commands. Therefore, we were motivated to study distributed coordination problems where the communication network is noisy, limited, time-varying, and possibly unidirectional.

Of course, coming into consensus, or agreement, is not the only issue. Each member of the team must act to achieve the team objective using the best available information. The interplay between communications/consensus and control introduces significant challenges that are only beginning to be understood. In much of the current research on cooperative control, either the consensus problem is studied in the absence of an application, or the cooperative control problem is studied under the assumption of full and reliable communication.

Our objective in writing this research monograph is to summarize our work in cooperative control using distributed consensus algorithms. The monograph is roughly divided into two parts. In the first half of the book (Chapters 1–7), we describe theoretical results on distributed consensus algorithms where the dynamics of the information state evolve according to first- and second-order dynamics and according to rigid body attitude dynamics. The consensus algorithms require only neighbor-to-neighbor interaction, which minimizes power consumption, increases stealth, and improves the scalability and robustness of the team. The second half of the book (Chapters 8–14) describes our attempts to apply the theory to a variety of applications in cooperative control, including formation keeping for wheeled mobile robots and spacecraft and cooperative perimeter tracking and timing for a team of unmanned air vehicles. We maintain a website `http://www.engineering.usu.edu/ece/faculty/wren/book/consensus` at which can be found sample simulation and experimental videos and other useful materials associated with the book.

The results in this book and particularly the results in Chapters 8–14 would not have been possible without the efforts and support of our colleagues and students. In particular, we are indebted to Professor Tim McLain at Brigham Young University for his leadership in the area of cooperative control for unmanned air vehicles and for countless discussions on consensus and other applications of cooperative control. We are also indebted to Professor Ella Atkins at the University of Michigan and Professors YangQuan Chen and Mac McKee at Utah State University for many fruitful discussions on research ideas. We also acknowledge the efforts of Nathan Sorensen, Yongcan Cao, Haiyang Chao, William Bourgeous, and Larry Ballard at Utah State University, and Derek Kingston, Jonathan Lawton, Brett Young, David Casbeer, Ryan Holt, Derek Nelson, Blake Barber, Stephen Griffiths, David Johansen, and Andrew Eldridge at Brigham Young University. We are thankful to our editor Oliver Jackson for his interest in our project and his professionalism. In addition, we acknowledge IEEE, John Wiley & Sons, Elsevier, AIAA, and Taylor & Francis for granting us the permission to reuse materials from our publications copyrighted by these publishers in this book. The last section of each chapter gives a detailed list of the references used in the chapter. Finally, we gratefully acknowledge the support of our research on consensus algorithms and cooperative control by the Utah Water Research Laboratory and Community/University Research Initiative as well as National Science Foundation under Information Technology Research Grant CCR-0313056, NASA under STTR Contract No. NNA04AA19C, Air Force Office of Scientific Research under Award No. F49550-04-0209, F49620-01-1-0091, and F49620-02-C-0094, and Defense Agency Research Projects Agency under Grant NBCH1020013.

Utah State University, Logan, Utah — *Wei Ren*
Brigham Young University, Provo, Utah — *Randal W. Beard*
September 2007

Contents

Part I

Overview of Consensus Algorithms in Cooperative Control

1

Overview of Consensus Algorithms in Cooperative Control

This chapter overviews consensus algorithms in cooperative control. The motivation for information consensus in cooperative control is given. A literature review on consensus algorithms is provided. Theoretical results regarding consensus-seeking under both time-invariant and dynamically changing communication topologies are summarized. A few specific applications of consensus algorithms to multivehicle cooperative control are described. The organization of the monograph is also introduced.

1.1 Introduction

The abundance of embedded computational resources in autonomous vehicles enables enhanced operational effectiveness through cooperative teamwork in civilian and military applications. Compared to autonomous vehicles that perform solo missions, greater efficiency and operational capability can be realized from teams of autonomous vehicles operating in a coordinated fashion. Potential applications for multivehicle systems include space-based interferometers; combat, surveillance, and reconnaissance systems; hazardous material handling; and distributed reconfigurable sensor networks. To enable these applications, various cooperative control capabilities need to be developed, including formation control, rendezvous, attitude alignment, flocking, foraging, task and role assignment, payload transport, air traffic control, and cooperative search.

Cooperative control of multiple autonomous vehicles poses significant theoretical and practical challenges. First, the research objective is to develop a system of subsystems rather than a single system. Second, the communication bandwidth and connectivity of the team are often limited, and the information exchange among vehicles may be unreliable. It is also difficult to decide what to communicate and when and with whom the communication takes place. Third, arbitration between team goals and individual goals needs to be negotiated. Fourth, the computational resources of each individual vehicle will always be limited.

Recent years have seen significant interest and research activity in the area of coordinated and cooperative control of multiple autonomous vehicles (*e.g.*, [11, 15, 22, 24, 25, 28, 37, 49, 51, 62–65, 70, 72, 80, 82, 95, 100–102, 121, 123, 133, 142, 151, 153, 189, 200, 203, 225–228, 234, 244]). Much of this work assumes the availability of global team knowledge, the ability to plan group actions in a centralized manner, and/or perfect and unlimited communication among the vehicles.

A centralized coordination scheme relies on the assumption that each member of the team has the ability to communicate to a central location or share information *via* a fully connected network. As a result, the centralized scheme does not scale well with the number of vehicles. The centralized scheme may result in a catastrophic failure of the overall system due to its single point of failure. Also, real-world communication topologies are usually not fully connected. In many cases, they depend on the relative positions of the vehicles and on other environmental factors and are therefore dynamically changing in time. In addition, wireless communication channels are subject to multipath, fading and drop-out. Therefore, cooperative control in the presence of real-world communication constraints becomes a significant challenge.

To understand the fundamental issues inherent in all cooperative control problems, we offer the following, intuitively appealing, fundamental axiom:

Axiom 1.1 *Shared information is a necessary condition for cooperation.*

Information necessary for cooperation may be shared in various ways. For example, relative position sensors may enable vehicles to construct state information for other vehicles [37], knowledge may be communicated among vehicles using a wireless network [69], or joint knowledge might be preprogrammed into the vehicles before a mission begins [16]. Under this axiom, information exchange becomes a central issue in cooperative control. In the following, we will refer to the information that is necessary for cooperation as *coordination information* or *coordination variable* [139, 193]. Suppose that a particular cooperation strategy has been devised and shown to work if the team has global access to the coordination information. Cooperation will occur if each member in the team has access to consistent, accurate, and complete coordination information. However, in the presence of an unreliable, dynamically changing communication topology and dynamically changing local situational awareness of each vehicle, it is not possible for all of the vehicles to have access to consistent, accurate, or complete coordination information, that is, the vehicles may have different instantiations of the coordination variable. Therefore, distributed algorithms need to be developed to ensure that the team is converging to a consistent view of the coordination information.

As an example, consider a meet-for-dinner problem. In this problem, a group of friends decides to meet for dinner at a particular restaurant but fail to specify a precise time to meet. On the afternoon of the dinner appointment, all of the individuals realize that they are uncertain about the time that the group will meet. A centralized solution to this problem is for the group to have a conference call, to poll all individuals regarding their preferred time for dinner, and to average the answers to arrive at a time when the group will meet. However, this centralized solution requires that a conference line be available and that the time of the conference call be known to the group. Because whatever algorithm was used to convey the time of the conference call to the group could also have been used to convey the time to meet for dinner, the central problem remains.

The coordination variable in this example is the time when the group will meet for dinner. The particular time is not what is important, but rather that each individual in the group has a consistent understanding of that information. A distributed solution to the problem would be for each individual to call, one at a time, a subset of the group. Given his or her current estimate of the meeting time, *i.e.*, his or her instantiation of the coordination variable, the individual might update his or her estimate of the meeting time to be a weighted average of his or her current meeting time and that of the person with whom he or she is conversing. The question is to determine under what conditions this strategy will enable the entire team to converge to a consistent meeting time.

To illustrate the meet-for-dinner example, suppose that there are 10 agents who communicate with exactly one other individual, chosen randomly from the group, for a random length of time. After the communication has expired, the process is repeated. Figure 1.1 shows the evolution of the dinner times with the above mentioned distributed approach, where the initial state is uniformly assigned. Note that the entire team converges to a consistent meeting time under switching communication topologies.

For cooperative control strategies to be effective, a team of vehicles must be able to respond to unanticipated situations or changes in the environment that are sensed as a cooperative task is carried out. For some applications (*e.g.*, cooperative observation on the same phenomenon or target), as the environment changes, the vehicles in the team must agree as to what changes took place. For some other applications (*e.g.*, accurate formation geometry maintenance), the vehicles need to maintain relative states between each other or achieve different group behaviors. A direct consequence of Axiom 1.1 is that cooperation requires that the group of vehicles agrees on the instantiations of the coordination variable or that differences between the instantiations of the coordination variable converge to prespecified values. A critical problem for cooperative control is to determine algorithms so that a team of vehicles can agree on the instantiations of the coordination variable or that differences between the instantiations of the coordination variable converge to prespecified values in the presence of (i) imperfect sensors, (ii) communication dropout, (iii) sparse communication topologies, and (iv) noisy and unreliable communication links.

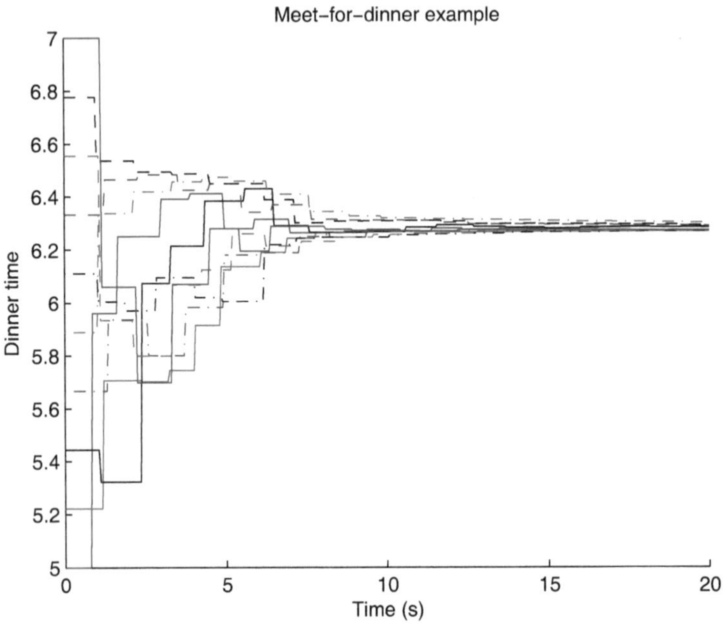

Fig. 1.1. Discrete-time meet-for-dinner simulation

1.2 Literature Review: Consensus Algorithms

When multiple vehicles agree on the value of a variable of interest, they are said to have reached *consensus*. Information consensus guarantees that vehicles sharing information over a network topology have a consistent view of information that is critical to the coordination task. To achieve consensus, there must be a shared variable of interest, called the *information state*, as well as appropriate algorithmic methods for negotiating to reach consensus on the value of that variable, called the *consensus algorithms*. The information state represents an instantiation of the coordination variable for the team. Examples include a local representation of the center and shape of a formation, the rendezvous time, the length of a perimeter being monitored, the direction of motion for a multivehicle swarm, and the probability that a military target has been destroyed. By necessity, consensus algorithms are designed to be distributed, assuming only neighbor-to-neighbor interaction between vehicles. Vehicles update the value of their information states based on the information states of their neighbors. The goal is to design an update law so that the information states of all of the vehicles in the network converge to a common value.

Consensus algorithms have a historical perspective in [32, 41, 55, 81, 131, 236], to name a few, and have recently been studied extensively in the context of cooperative control of multiple autonomous vehicles [69, 97, 126, 145, 158, 190]. Some results in consensus algorithms can be understood in the context of connective stability [215]. Consensus algorithms have applications in rendezvous [26, 58, 124, 125, 135, 216, 220, 221], formation control [69,115,118,127,134,165,174], flocking [50,59,120,147,155,169,232,238], attitude alignment [117,176,179,188], perimeter monitoring [38], decentralized task assignment [6, 143], and sensor networks [76, 154, 159, 223, 260].

1.2.1 Fundamental Consensus Algorithms

The basic idea of a consensus algorithm is to impose similar dynamics on the information states of each vehicle. If the communication network among vehicles allows continuous communication or if the communication bandwidth is sufficiently large, then the information state update of each vehicle is modeled using a differential equation. On the other hand, if the communication data arrive in discrete packets, then the information state update is modeled using a difference equation. This section overviews fundamental consensus algorithms in which a scalar information state is updated by each vehicle using, respectively, a first-order differential equation and a first-order difference equation.

Suppose that there are n vehicles in the team. The team's communication topology can be represented by directed graph $\mathcal{G}_n \triangleq (\mathcal{V}_n, \mathcal{E}_n)$, where $\mathcal{V}_n = \{1, \ldots, n\}$ is the node set and $\mathcal{E}_n \subseteq \mathcal{V}_n \times \mathcal{V}_n$ is the edge set (see Appendix B for graph theory notations). For example, Fig. 1.2 shows three different communication topologies for three vehicles. The communication topology may be time varying due to vehicle motion or communication dropouts. For example, communication dropouts might occur when an unmanned air vehicle (UAV) banks away from its neighbor or flies through an urban canyon. The most common continuous-time consensus algorithm [69, 97, 126, 158, 190] is given by

$$\dot{x}_i(t) = -\sum_{j=1}^{n} a_{ij}(t)[x_i(t) - x_j(t)], \quad i = 1, \ldots, n, \tag{1.1}$$

where $a_{ij}(t)$ is the (i, j) entry of adjacency matrix $\mathcal{A}_n \in \mathbb{R}^{n \times n}$ associated with $\mathcal{G}_n$ at time t and x_i is the information state of the ith vehicle. Setting $a_{ij} = 0$ denotes the fact that vehicle i cannot receive information from vehicle j. A consequence of (1.1) is that the information state $x_i(t)$ of vehicle i is driven toward the information states of its neighbors. The critical convergence question is, when do the information states of all of the vehicles converge to a common value?

Although (1.1) ensures that the information states of the team agree, it does not dictate a specified common value. For example, consider a cooperative rendezvous problem where a team of vehicles is tasked to arrive simultaneously

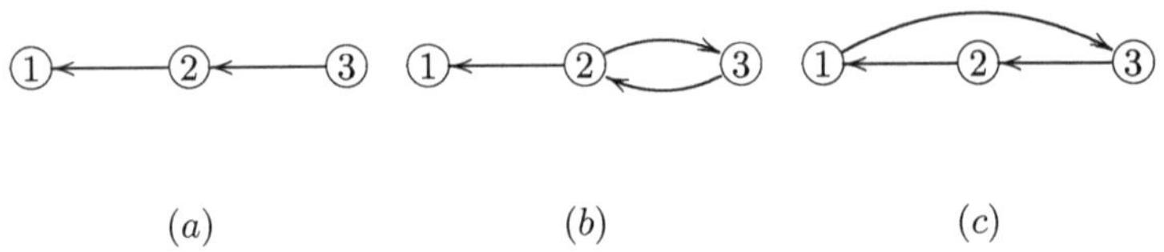

Fig. 1.2. Three different communication topologies for three vehicles. Subplot (c) is strongly connected because there is a directed path between every pair of nodes. However, (a) and (b) are not strongly connected.

at a specified location known to all vehicles. Because the rendezvous time is not given and may need to be adjusted in response to pop-up threats or other environmental disturbances, the team needs to reach consensus on the rendezvous time. To do this, each vehicle first creates an information state x_i that represents the ith vehicle's understanding of the rendezvous time. To initialize its information state, each vehicle determines a time at which it is able to rendezvous with the team and sets $x_i(0)$ to this value. Each team member then communicates with its neighbors and negotiates a team arrival time using consensus algorithm (1.1). Onboard controllers then maneuver each vehicle to rendezvous at the negotiated arrival time. When environmental conditions change, individual vehicles may reset their information state and thus cause the negotiation process to resume.

Note that (1.1) does not permit specifying a desired information state. We show in Section 1.2.2 that if the communication topology is fixed and the gains a_{ij} are time invariant, then the common asymptotic value is a linear combination of the initial information states. In general, it is possible to guarantee only that the common value is a convex combination of the initial information states.

Consensus algorithm (1.1) is written in matrix form as

$$\dot{x}(t) = -\mathcal{L}_n(t)x(t),$$

where $x = [x_1, \ldots, x_n]^T$ is the information state and $\mathcal{L}_n(t) = [\ell_{ij}(t)] \in \mathbb{R}^{n \times n}$ is the nonsymmetrical Laplacian matrix associated with $\mathcal{G}_n$ (see Appendix B). Consensus is *achieved* or *reached* by a team of vehicles if, for all $x_i(0)$ and all $i, j = 1, \ldots, n$, $|x_i(t) - x_j(t)| \to 0$, as $t \to \infty$.

When communication among vehicles occurs at discrete instants, the information state is updated using a difference equation. The most common discrete-time consensus algorithm has the form [97,145,190,236]

$$x_i[k+1] = \sum_{j=1}^{n} d_{ij}[k]x_j[k], \quad i = 1, \ldots, n, \tag{1.2}$$

where k denotes a communication event; $d_{ij}[k]$ is the (i,j) entry of a row-stochastic matrix $\mathcal{D} = [d_{ij}] \in \mathbb{R}^{n \times n}$ (see Appendix C for matrix theory notations) at the discrete-time index k with the additional assumption that

$d_{ii}[k] > 0$ for all $i = 1, \ldots, n$ and $d_{ij}[k] > 0$ for all $i \neq j$ if information flows from vehicle j to vehicle i and $d_{ij}[k] = 0$ otherwise. Intuitively, the information state of each vehicle is updated as the weighted average of its current state and the current states of its neighbors. Note that a vehicle maintains its current information state if it does not exchange information with other vehicles at that instant. Discrete-time consensus algorithm (1.2) is written in matrix form as

$$x[k+1] = \mathcal{D}[k]x[k].$$

Similar to the continuous-time case, consensus is *achieved* or *reached* if, for all $x_i[0]$ and for all $i, j = 1, \ldots, n$, $|x_i[k] - x_j[k]| \to 0$, as $k \to \infty$.

1.2.2 Convergence Analysis of Consensus Algorithms

Convergence Analysis for Time-invariant Communication Topologies

In this section, we investigate conditions under which the information states of consensus algorithm (1.1) converge when the communication topology is time invariant and the gains a_{ij} are constant, that is, the nonsymmetrical Laplacian matrix $\mathcal{L}_n$ is constant. As noted in Appendix B, zero is always an eigenvalue of $-\mathcal{L}_n$, and all nonzero eigenvalues of $-\mathcal{L}_n$ have negative real parts. As also noted in Appendix B, the column vector $\mathbf{1}_n$ of ones is an eigenvector associated with the zero eigenvalue, which implies that $\text{span}\{\mathbf{1}_n\}$ is contained in the kernel of $\mathcal{L}_n$. It follows that if zero is a simple eigenvalue of $\mathcal{L}_n$, then $x(t) \to \bar{x}\mathbf{1}_n$, where $\bar{x}$ is a scalar constant, which implies that $|x_i(t) - x_j(t)| \to 0$, as $t \to \infty$, for all $i, j = 1, \ldots, n$. Convergence analysis therefore focuses on conditions to ensure that zero is a simple eigenvalue of $\mathcal{L}_n$. Otherwise the kernel of $\mathcal{L}_n$ includes elements that are not in $\text{span}\{\mathbf{1}_n\}$, in which case consensus is not guaranteed.

If the directed graph of $\mathcal{L}_n$ is strongly connected (see Appendix B), then zero is a simple eigenvalue of $\mathcal{L}_n$ [69, Proposition 3]. However, this condition is not necessary. For example, consider the nonsymmetrical Laplacian matrices

$$\mathcal{L}_{3(1)} = \begin{bmatrix} 1 & -1 & 0 \\ 0 & 1.5 & -1.5 \\ 0 & 0 & 0 \end{bmatrix}, \quad \mathcal{L}_{3(2)} = \begin{bmatrix} 1 & -1 & 0 \\ 0 & 1.5 & -1.5 \\ 0 & -2 & 2 \end{bmatrix}, \quad \mathcal{L}_{3(3)} = \begin{bmatrix} 1 & -1 & 0 \\ 0 & 1.5 & -1.5 \\ -2 & 0 & 2 \end{bmatrix}, \tag{1.3}$$

of the directed graphs shown in Fig. 1.2. Although all of the nonsymmetrical Laplacian matrices in (1.3) have simple zero eigenvalues, Figs. 1.2a and 1.2b are not strongly connected. The common feature is that the directed graphs of $\mathcal{L}_{3(1)}$, $\mathcal{L}_{3(2)}$, and $\mathcal{L}_{3(3)}$ all have directed spanning trees. As shown in [115,127, 193,255], zero is a simple eigenvalue of $\mathcal{L}_n$ if and only if the associated directed graph has a directed spanning tree. This result implies that (1.1) achieves consensus if and only if the directed communication topology has a directed spanning tree or the undirected communication topology is connected.

For discrete-time consensus algorithm (1.2), Theorem C.1 implies that all eigenvalues of $\mathcal{D}$ are either in the open unit disk or at 1. As shown in [190], if 1 is a simple eigenvalue of $\mathcal{D}$, then $\lim_{k\to\infty} \mathcal{D}^k \to \mathbf{1}_n \nu^T$, as $k \to \infty$, where ν is an $n \times 1$ nonnegative left eigenvector of $\mathcal{D}$ associated with the eigenvalue 1 and satisfies $\nu^T \mathbf{1}_n = 1$. As a result, $x[k] = \mathcal{D}^k x[0] \to \mathbf{1}_n \nu^T x[0]$, as $k \to \infty$, which implies that, for all i, $x_i[k] \to \nu^T x[0]$, as $k \to \infty$, and thus $|x_i[k] - x_j[k]| \to 0$, as $k \to \infty$.

Theorem C.5 implies that $1 = \rho(A)$ is a simple eigenvalue of row-stochastic matrix A if directed graph $\Gamma(A)$ is strongly connected, or equivalently, if A is irreducible. As in the continuous-time case, this condition is sufficient but not necessary. Furthermore, for row-stochastic matrix $\mathcal{D}$, $\Gamma(\mathcal{D})$ has a directed spanning tree if and only if $\lambda = 1$ is a simple eigenvalue of $\mathcal{D}$ and is the only eigenvalue of modulus one [190]. As a result, under a time-invariant communication topology with constant gains a_{ij}, (1.2) achieves consensus if and only if either the directed communication topology has a directed spanning tree or the undirected communication topology is connected [190].

Equilibrium State under a Time-invariant Communication Topology

We now investigate the consensus equilibrium for the special case in which the communication topology is time invariant and the gains a_{ij} are constant (*i.e.*, constant $\mathcal{L}_n$). When the directed communication topology has a directed spanning tree, it follows from [193] that $\lim_{t\to\infty} e^{-\mathcal{L}_n t} \to \mathbf{1}_n \nu^T$, where $\nu = [\nu_1, \ldots, \nu_n]^T$ is an $n \times 1$ nonnegative left eigenvector of $\mathcal{L}_n$ associated with the simple zero eigenvalue and satisfies $\sum_{j=1}^n \nu_j = 1$. As a result, for each $i = 1, \ldots, n$, $x_i(t) \to \sum_{j=1}^n \nu_j x_j(0)$, as $t \to \infty$, that is, the equilibrium state is a weighted average of the initial information states in the network. However, some of the components of ν may be zero, implying that the information states of some vehicles do not contribute to the equilibrium.

To illustrate this phenomenon, consider the nonsymmetrical Laplacian matrices given in (1.3). It can be verified that, for $\mathcal{L}_{3(1)}$, $x(t) \to x_3(0)\mathbf{1}_3$, for $\mathcal{L}_{3(2)}$, $x(t) \to [0.5714x_2(0) + 0.4286x_3(0)]\mathbf{1}_3$, and, for $\mathcal{L}_{3(3)}$, $x(t) \to [0.4615x_1(0) + 0.3077x_2(0) + 0.2308x_3(0)]\mathbf{1}_3$. Note that with $\mathcal{L}_{3(1)}$, the initial information states of vehicles 1 and 2 do not affect the equilibrium. With $\mathcal{L}_{3(2)}$, the initial information state of vehicle 1 does not affect the equilibrium. However, with $\mathcal{L}_{3(3)}$, all of the vehicle's initial information states affect the equilibrium. Observing the directed graphs shown in Fig. 1.2, we can see that, for $\mathcal{L}_{3(1)}$, vehicle 3 is the only vehicle that can pass information to all of the other vehicles in the team, either directly or indirectly. Similarly, for $\mathcal{L}_{3(2)}$, both vehicles 2 and 3 can pass information to the entire team, whereas, for $\mathcal{L}_{3(3)}$, all vehicles can pass information to the entire team.

Next, define the nonnegative matrix $M = \max_i \ell_{ii} I_n - \mathcal{L}_n$. Because ν is the nonnegative left eigenvector of $\mathcal{L}_n$ corresponding to the zero eigenvalue, ν is also the nonnegative left eigenvector of M corresponding to the eigenvalue

$\max_i \ell_{ii}$ of M. From Theorem C.1, it follows that $\rho(M) = \max_i \ell_{ii}$. If the directed communication graph is strongly connected, so is the directed graph of M, which also implies that M is irreducible (see Appendix C). By Theorem C.5, if M is irreducible, then ν is positive. Therefore, when the directed communication topology is strongly connected, all of the initial information states contribute to the consensus equilibrium because $\nu_i \neq 0$ for all i. Furthermore, if $\nu_i = 1/n$ for all i, then the consensus equilibrium is the average of the initial information states, a condition called *average consensus* [158]. If the directed communication topology is both strongly connected and balanced, then $\mathbf{1}_n$ is a left eigenvector of $\mathcal{L}_n$ associated with the simple zero eigenvalue. Therefore, as shown in [158], average consensus is achieved if and only if the directed communication topology is both strongly connected and balanced. It can be shown that, in the case of undirected communication, average consensus is achieved if and only if the topology is connected [158].

To illustrate these ideas, Fig. 1.3 shows time histories of the information states for two different updates strategies. Figure 1.3a shows the information states for $\dot{x} = -\mathcal{L}_{3(3)}x$, where $\mathcal{L}_{3(3)}$ is given in (1.3). Because the directed graph of $\mathcal{L}_{3(3)}$ is strongly connected, all of the vehicle's initial conditions contribute to the equilibrium state. However, the equilibrium is not an average consensus because the directed graph is not balanced. In contrast, Fig. 1.3b shows the time histories of the information states for $\dot{x} = -\text{diag}(w)\mathcal{L}_{3(3)}x$, where w is the positive column left eigenvector of $\mathcal{L}_{3(3)}$ corresponding to the zero eigenvalue satisfying $w^T\mathbf{1}_3 = 1$ and $\text{diag}(w)$ is the diagonal matrix whose diagonal entries are given by w. It can be shown that directed graph $\Gamma\left[\text{diag}(w)\mathcal{L}_{3(3)}\right]$ is strongly connected and balanced, resulting in average consensus.

In contrast, when the directed communication topology has a directed spanning tree, the consensus equilibrium is equal to the weighted average of the initial conditions of those vehicles that have a directed path to all other vehicles [193]. Requiring a directed spanning tree is less stringent than requiring a strongly connected and balanced graph. However, as shown above, the consensus equilibrium is a function only of the initial information states of those vehicles that have a directed path to all other vehicles.

Convergence Analysis for Dynamic Communication Topologies

Communication topologies are often dynamic. For example, communication links among vehicles might be unreliable due to multipath effects and other disturbances. Alternatively, if information is exchanged by means of line-of-sight sensors, the neighbors visible to a vehicle might change over time, *e.g.*, when a UAV banks away from its neighbor. Therefore, in this section, we investigate conditions under which consensus algorithms converge under random switching of the communication topologies.

One approach to analyzing switching topologies is to use algebraic graph theory, which associates each graph topology with an algebraic structure of

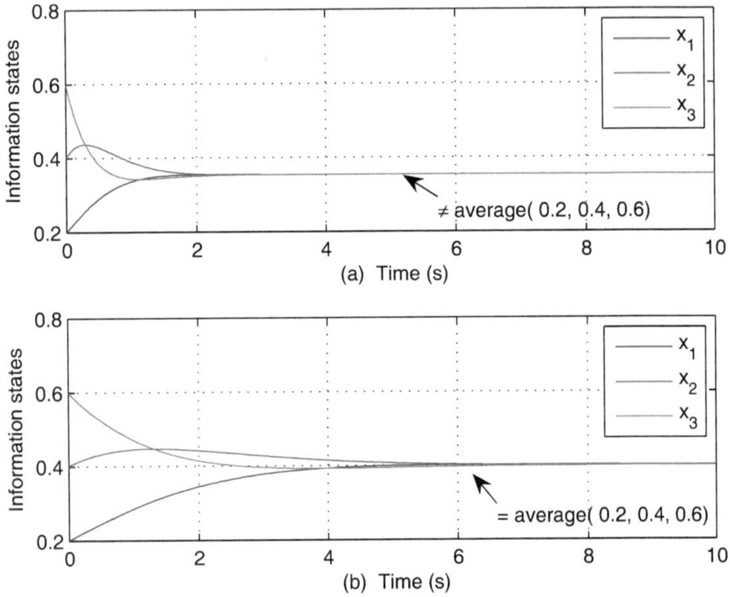

Fig. 1.3. Consensus for three vehicles. Subplots (a) and (b) correspond to $\dot{x} = -\mathcal{L}_{3(3)}x$ and $\dot{x} = -\text{diag}(w)\mathcal{L}_{3(3)}x$, respectively. Because 0.4 is the average of the initial states $(0.2, 0.4, 0.6)$, average consensus is achieved in (b), where the directed graph is strongly connected and balanced, but not in (a), where the directed graph is only strongly connected.

corresponding matrices. Because (1.1) is linear, its solution can be written as $x(t) = \Phi(t,0)x(0)$, where $\Phi(t,0)$ is the transition matrix corresponding to $-\mathcal{L}_n(t)$. $\Phi(t,0)$ is a row-stochastic matrix with positive diagonal entries for all $t \geq 0$ [192]. Consensus is achieved if $\lim_{t\to\infty} \Phi(t,0) \to \mathbf{1}_n\mu^T$, where μ is a column vector. It is typical to assume that the communication topology is piecewise constant over finite lengths of time, called *dwell times*, and that dwell times are bounded below by a positive constant [97]. In this case, $\mathcal{A}_n(t)$ and hence $\mathcal{L}_n(t)$ are piecewise constant with dwell times $\tau_j = t_{j+1} - t_j$, where $t_1, t_2, \ldots$ are the switching instants, and thus consensus is achieved if $\lim_{j\to\infty} e^{-\mathcal{L}_n(t_j)\tau_j} e^{-\mathcal{L}_n(t_{j-1})\tau_{j-1}} \cdots e^{-\mathcal{L}_n(t_0)\tau_0} = \mathbf{1}_n\mu^T$. Because $e^{-\mathcal{L}_n(t_j)(t-t_j)}$ is a row-stochastic matrix, convergence analysis involves the study of infinite products of stochastic matrices.

A classical result given in [253] (see also [192]) demonstrates the convergence property of infinite products of SIA matrices (see Appendix C). Specifically, let $\mathcal{S} = \{S_1, S_2, \ldots, S_k\}$ be a finite set of SIA matrices with the property that every finite product $S_{i_j} S_{i_{j-1}} \cdots S_{i_1}$ is SIA. Then, for each infinite sequence $S_{i_1}, S_{i_2}, \ldots$ there exists a column vector ν such that

$\lim_{j\to\infty} S_{i_j} S_{i_{j-1}} \cdots S_{i_1} = \mathbf{1}\nu^T$. Because the number of potential communication topologies is finite, the set of matrices $\{S_j \triangleq e^{-\mathcal{L}_n(t_j)(t_{j+1}-t_j)}\}_{j=1}^{\infty}$ is finite if the allowable dwell times $\tau_j = t_{j+1} - t_j$ are drawn from a finite set. Reference [97] shows that these matrices are SIA and uses this result to show that the heading angles of a swarm of vehicles achieve consensus using the nearest neighbor rules of [239]. This is a special case of discrete-time consensus algorithm (1.2), if there exists an infinite sequence of contiguous, uniformly bounded time intervals, having one of a finite number of different lengths, with the property that across each interval, the union (see Appendix B) of the undirected communication topologies is connected. See [78, 126, 158, 208] for extensions.

Consider, on the other hand, the more realistic assumption that the dwell times are drawn from an infinite but bounded set or $\mathcal{A}_n(t)$ is piecewise continuous[1] and its nonzero and hence positive entries are uniformly lower and upper bounded. In this case, let $\mathcal{S} = \{S_1, S_2, \dots\}$ be an infinite set of $n \times n$ SIA matrices, let N_t be the number of different types (see Appendix B) of all of the $n \times n$ SIA matrices, and define the matrix function $\chi(P) = 1 - \min_{i_1,i_2} \sum_j \min(p_{i_1 j}, p_{i_2 j})$. Then, $\lim_{j\to\infty} S_{i_j} S_{i_{j-1}} \cdots S_{i_1} = \mathbf{1}\nu^T$ if there exists a constant $d \in [0,1)$ such that, for every $W \triangleq S_{k_1} S_{k_2} \cdots S_{k_{N_t+1}}$, it follows that $\chi(W) \le d$ [253]. It can be shown that this condition is satisfied if there exists an infinite sequence of contiguous, uniformly bounded time intervals, with the property that across each interval, the union of the directed communication topologies has a directed spanning tree [190, 192]. Reference [168] considers a similar problem by studying the products of row-stochastic matrices with a lower triangular structure. In addition, a lower bound on the convergence rate of consensus algorithm (1.2) under directed switching communication topologies is derived in [36].

Lyapunov Analysis of Consensus Algorithms

Nonlinear analysis can also be used to study consensus algorithms [71, 128, 145]. For discrete-time consensus algorithm (1.2), a set-valued Lyapunov function V is defined as $V(x_1, \dots, x_n) = (\text{conv}\{x_1, \dots, x_n\})^n$, where $\text{conv}\{x_1, \dots, x_n\}$ denotes the convex hull of $\{x_1, \dots, x_n\}$, and $X^n \triangleq X \times \cdots \times X$. It is shown in [145] that $V(t_2) \subseteq V(t_1)$ for all $t_2 \ge t_1$, and that $x(t)$ approaches an element of the set $\text{span}\{\mathbf{1}_n\}$, which implies that consensus is reached. Using set-valued Lyapunov theory, [145] shows that discrete-time consensus algorithm (1.2) is uniformly globally attractive with respect to the collection of equilibrium solutions $\text{span}\{\mathbf{1}_n\}$ if and only if there exists $K \ge 0$ such that the union of the directed communication topologies has a directed spanning tree across each interval of length Kh, where h is the sample time.

[1] Accordingly, $\mathcal{L}_n(t)$ is piecewise continuous.

For continuous-time consensus algorithm (1.1), [144] considers the Lyapunov candidate $V(x) = \max\{x_1, \ldots, x_n\} - \min\{x_1, \ldots, x_n\}$. It is shown in [144] that the equilibrium set span$\{\mathbf{1}_n\}$ is uniformly exponentially stable if there is an interval length $T > 0$ such that, for all t, the directed graph of $-\int_t^{t+T} \mathcal{L}_n(s)ds$ has a directed spanning tree.

As an alternative analytic method, [219,246,247] applies nonlinear contraction theory to synchronization and schooling applications, which are related to information consensus. In particular, (1.1) is analyzed under undirected switching communication topologies, and a convergence result identical to the result given in [97] is derived. In addition, [13] uses passivity as a design tool for consensus algorithms over an undirected communication topology.

Information consensus is also studied from a stochastic point of view in [87,88,256], which consider a random network, where the existence of an information channel between a pair of vehicles at each time is probabilistic and independent of other channels, resulting in a time-varying undirected communication topology. For example, adjacency matrix $\mathcal{A}_n = [a_{ij}] \in \mathbb{R}^{n\times n}$ for an undirected random graph is defined as $a_{ii}(p) = 0$, $a_{ij}(p) = 1$ with probability p, and $a_{ij} = 0$ with probability $1 - p$ for all $i \neq j$. In [88], consensus over an undirected random network is addressed by notions from stochastic stability.

Communication Delays and Asynchronous Consensus

When information is exchanged among vehicles through communication, time delays from both message transmission and processing after receipt must be considered. Let σ_{ij} denote the time delay for information communicated from vehicle j to reach vehicle i. In this case, (1.1) is modified as

$$\dot{x}_i = -\sum_{j=1}^{n} a_{ij}(t)[x_i(t-\sigma_{ij}) - x_j(t-\sigma_{ij})].$$

In the simplest case, where $\sigma_{ij} = \sigma$ and the communication topology is time-invariant, undirected, and connected, average consensus is achieved if and only if $0 \leq \sigma < \frac{\pi}{2\lambda_{max}(\mathcal{L}_n)}$ [158], where $\mathcal{L}_n$ is the Laplacian matrix of the undirected communication topology and $\lambda_{max}(\cdot)$ denotes the largest eigenvalue of a matrix. See [30, 119] for extensions.

Alternatively, consider the case in which the time delay affects only the information state that is being transmitted so that (1.1) is modified as

$$\dot{x}_i = -\sum_{j=1}^{n} a_{ij}(t)[x_i(t) - x_j(t-\sigma_{ij})].$$

When $\sigma_{ij} = \sigma$ and the communication topology is directed and switching, the consensus result for switching topologies remains valid for an arbitrary time delay σ [144].

For discrete-time consensus algorithm (1.2), it is shown in [230] that if consensus is reached under a time-invariant undirected communication topology, then the presence of communication delays does not affect consensus. In addition, the result in [145] is extended to take into account bounded time delays in [12]. Furthermore, [258] shows sufficient conditions for consensus under dynamically changing communication topologies and bounded time-varying communication delays.

More generally, in an asynchronous consensus framework [31,35,67,68,140], each vehicle exchanges information asynchronously and updates its state with possibly outdated information from its local neighbors. As a result, heterogenous vehicles, time-varying communication delays, and packet dropout must be taken into account in the same asynchronous consensus framework. Reference [68] categorizes several consensus results in the literature according to synchronism, connectivity, and direction of information flow.

1.2.3 Synthesis and Extensions of Consensus Algorithms

Consensus Synthesis

In some applications, consensus algorithms must satisfy given requirements or optimize performance criteria. For example, when a UAV or micro air vehicle (MAV) swarm consists of hundreds or thousands of vehicles, it might be desirable to solve the fastest distributed linear averaging (FDLA) problem, which is defined as follows [259]. Let $W = [w_{ij}] \in \mathbb{R}^{n\times n}$ be such that $w_{ij} = 0$ if information is not exchanged between vehicle i and vehicle j. Given $x[k+1] = Wx[k]$, find W to minimize

$$r_{asym}(W) = \sup_{x[0]\neq \bar{x}} \lim_{k\to\infty} \left(\frac{\|x[k]-\bar{x}\|}{\|x[0]-\bar{x}\|} \right)^{1/k},$$

subject to the condition that $\lim_{t\to\infty} W^k = \frac{1}{n}\mathbf{1}_n\mathbf{1}_n^T$, where $\bar{x} = \frac{1}{n}\mathbf{1}_n\mathbf{1}_n^T x[0]$. In other words, the FDLA problem is to find the weight matrix W that guarantees the fastest convergence to the average consensus value. In contrast to discrete-time consensus algorithm (1.2), weights w_{ij} can be negative [259]. With the additional constraint $w_{ij} = w_{ji}$, the FDLA problem reduces to a numerically solvable semidefinite program [259]. A related problem is considered in [106], where an iterative, semidefinite-programming-based approach is developed to maximize the algebraic connectivity of the Laplacian matrix of undirected graphs (see Appendix B) with the motivation that the algebraic connectivity of the Laplacian matrix characterizes the convergence rate of the consensus algorithm.

Another problem is considered in [202], which focuses on designing consensus algorithms in which the information state is updated according to $\dot{x}_i = u_i$, and the information available to the ith vehicle is given by $y_i = G_i x$, where $x = [x_1, \ldots, x_n]^T$, $y_i \in \mathbb{R}^{m_i}$, and $G_i \in \mathbb{R}^{m_i\times n}$. The information control input is designed in the form of $u_i = k_i y_i + z_i$, where k_i is a row vector with m_i components and z_i is a scalar.

More generally, consider an interconnected network of n vehicles whose information states are updated according to $\dot{x}_i = \sum_{j=1}^{n} A_{ij} x_j + B_{1i} w_i + B_{2i} u_i$, $i = 1, \ldots, n$, where $x_i \in \mathbb{R}^n$ denotes the information state, $w_i \in \mathbb{R}^m$ denotes disturbances, and $u_i \in \mathbb{R}^r$ denotes the information control input with $i = 1, \ldots, n$. Letting x, w, and u be column vectors with components x_i, w_i, and u_i, respectively, the dynamics of x are denoted by $\dot{x} = Ax + B_1 w + B_2 u$. Reference [52] focuses on synthesizing a decentralized state feedback control law that guarantees consensus for the closed-loop system without disturbances as well as synthesizing a state-feedback controller that achieves not only consensus but optimal $\mathcal{H}_2$ performance for disturbance attenuation.

Extensions of Consensus Algorithms

Consensus algorithm (1.1) is extended in various ways in the literature. For example, [17, 47] generalize the consensus equilibrium to a weighted power mean or arbitrary functions of the initial information states. In [104], quantized consensus problems are studied, where the information state at each node is an integer. In [224], an external input is incorporated in (1.1) so that the information state tracks a time-varying input. Consensus with a constant reference state is addressed in [99, 143], and consensus is addressed with a time-varying reference state in [90, 180]. In [229], necessary and sufficient conditions are derived so that a collection of systems is controlled by a team leader. An approach based on nonsmooth gradient flows is developed in [48] to guarantee that average consensus is reached in finite time.

The single-integrator consensus algorithm given by (1.1) is also extended to double-integrator dynamics in [89, 186, 261] to model more naturally the evolution of physical phenomena, such as a coaxial rotorcraft MAV that can be controlled through gentle maneuvers with a decoupled double-integrator model. For double-integrator dynamics, the consensus algorithm is given by

$$\ddot{x}_i = -\sum_{j=1}^{n} a_{ij}(t)[(x_i - x_j) + \gamma(\dot{x}_i - \dot{x}_j)],$$

where $\gamma > 0$ denotes the coupling strength between the information state derivatives and both x_i and $\dot{x}_i$ are transmitted between team members. It is shown in [186] that both the communication topology and coupling strength γ affect consensus-seeking in the general case of directed information exchange. To achieve consensus, the directed communication topology must have a directed spanning tree and γ must be sufficiently large. See [196] for extensions to higher-order dynamics.

Related to consensus algorithms are synchronization phenomena arising in systems of coupled nonlinear oscillators. The classical Kuramoto model [114] consists of n coupled oscillators with dynamics given by

$$\dot{\theta}_i = \omega_i + \frac{k}{n}\sum_{j=1}^{n}\sin(\theta_j - \theta_i), \tag{1.4}$$

where θ_i and ω_i are, respectively, the phase and natural frequency of the ith oscillator; and k is the coupling strength. Note that model (1.4) assumes full connectivity of the network. Model (1.4) is generalized in [98] to nearest neighbor information exchange as

$$\dot{\theta}_i = \omega_i + \frac{k}{n}\sum_{j=1}^{n}a_{ij}(t)\sin(\theta_j - \theta_i).$$

Connections between phase models of coupled oscillators and kinematic models of self-propelled particle groups are studied in [213]. Analysis and design tools are developed to stabilize the collective motions. The stability of the generalized Kuramoto coupled nonlinear oscillator model is studied in [98], where it is proven that, for couplings above a critical value, all oscillators synchronize given identical and uncertain natural frequencies. Extensions of [98] to a tighter lower bound on the coupling strength are given in [46] for the traditional Kuramoto model with full connectivity. The result in [98] is also extended to account for heterogenous time delays and switching topologies in [162].

Synchronization of coupled oscillators with other nonlinear dynamics is also studied in the literature. As an example, consider a network of n vehicles with information dynamics given by

$$\dot{x}_i = f(x_i, t) + \sum_{j=1}^{n}a_{ij}(t)(x_j - x_i), \tag{1.5}$$

where $x = [x_1, \ldots, x_n]^T$. In [219], partial contraction theory is applied to derive conditions under which consensus is reached for vehicles with dynamics (1.5). As another example, [166] studies a dynamic network of n nonlinear oscillators, where the state equation for each oscillator is given by

$$\dot{x}_i = f(x_i) + \gamma\sum_{j=1}^{n}a_{ij}(t)(x_j - x_i),$$

where $x_i \in \mathbb{R}^m$ and $\gamma > 0$ denotes the global coupling strength parameter. It is shown in [166] that the algebraic connectivity of the network Laplacian matrix plays a central role in synchronization.

1.2.4 Design of Coordination Strategies *via* Consensus Algorithms

In this section, we briefly describe a few applications of consensus algorithms to multivehicle coordination problems.

Rendezvous Problem

The rendezvous problem requires that a group of vehicles in a network rendezvous at a time or a location determined through team negotiation. Consensus algorithms can be used to perform the negotiation in a way that is robust to environmental disturbances such as nonuniform wind for a team of UAVs. The rendezvous problem for a group of mobile autonomous vehicles is studied in [124, 125], where synchronous and asynchronous cases are considered. In [124, 125], vehicles execute a sequence of stop-and-go maneuvers to rendezvous in a distributed manner without communication between neighbors. A stop-and-go maneuver takes place within a time interval consisting of a sensing period during which neighbors' positions are determined, as well as a maneuvering period during which vehicles move in response to the positions of their neighbors.

Figure 1.4 shows a simple coordination framework for multivehicle rendezvous, where a consensus manager applies distributed consensus algorithms to guarantee that all vehicles reach consensus on a rendezvous objective such as a rendezvous time or rendezvous location. Based on the output of the consensus manager, each vehicle uses a local control law to drive itself to achieve the rendezvous time and/or location. An application of Fig. 1.4 is described in [110], where multiple UAVs are controlled to converge simultaneously on the boundary of a radar detection area to maximize the element of surprise. Teamwide consensus is reached on time-over-target, requiring each vehicle to adjust its velocity to ensure synchronous arrival.

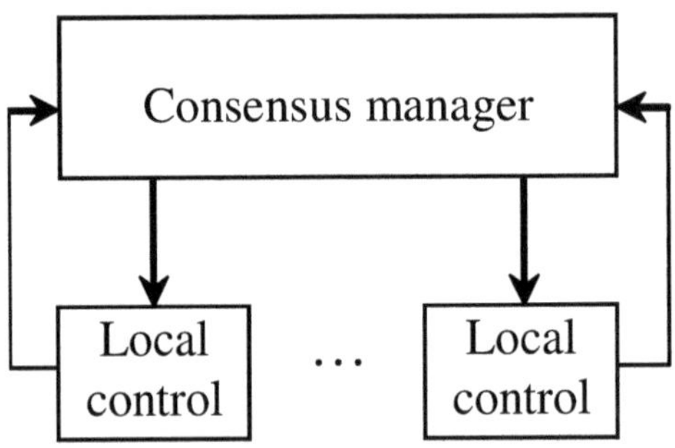

Fig. 1.4. A simple coordination framework for multivehicle rendezvous. The consensus manager applies distributed consensus algorithms to guarantee that the team reaches consensus on a rendezvous objective. Based on the output of the consensus manager, each vehicle applies local control laws to achieve the rendezvous objective.

Formation Stabilization

The formation stabilization problem requires that vehicles collectively maintain a prescribed geometric shape. This problem is relatively straightforward in the centralized case, where all team members know the desired shape and location of the formation. On the other hand, in the decentralized formation stabilization problem, each vehicle knows the desired formation shape, but the location of the formation needs to be negotiated among team members. The information state for this problem includes the center of the formation. Each vehicle initializes its information state by proposing a formation center that does not require it to maneuver into formation. The consensus algorithm is then employed by the team of vehicles to negotiate a formation center known to all members of the team.

In [69], an information flow filter is used to improve stability margins and formation accuracy through propagation of the formation center to all vehicles. Formation stabilization for multiple unicycles is studied in [127] using a consensus algorithm to achieve point, line, and general formation patterns. In addition, the simplified pursuit strategy for wheeled-vehicle formations in [133] can be considered a special case of continuous-time consensus algorithm (1.1), where the communication topology is a unidirectional ring. Furthermore, feedback control laws are derived in [115] using relative information between neighboring vehicles to stabilize vehicle formations.

Formation Maneuvering and Flocking

Consensus algorithms can be applied to execute decentralized formation maneuvers. For example, in [118], a class of formation maneuvers is studied where the desired position of each robot, $h_i^d(t)$, is either communicated to the team by a centralized entity or is preprogrammed on each robot. The robots are to maintain a prespecified formation shape even during transients and in response to environmental disturbances. In other words, when one robot slows down or maneuvers to avoid an obstacle, the other robots must maneuver to maintain the formation shape. The intervehicle communication network is limited and requires a decentralized approach to maintain the formation. The mobile robot dynamic model is feedback linearized as the double-integrator system $\ddot{h}_i = u_i$, where h_i denotes the location of a point on the ith robot that is not on the wheel axis and u_i denotes the control input. The decentralized formation control law is given in [118] as

$$u_i = -K_g \tilde{h}_i - D_g \dot{h}_i - K_f \sum_{j=1}^{n} a_{ij}(\tilde{h}_i - \tilde{h}_j) - D_f \sum_{j=1}^{n} a_{ij}(\dot{h}_i - \dot{h}_j), \tag{1.6}$$

where K_g and K_f are symmetrical positive-definite matrices, D_g and D_f are symmetrical positive-semidefinite matrices, and $\tilde{h}_i \triangleq h_i - h_i^d$. In control law (1.6), the first two terms guarantee that h_i approaches h_i^d, whereas the second two terms guarantee that the pairs $\tilde{h}_i$, $\tilde{h}_j$ and $\dot{h}_i$, $\dot{h}_j$ reach consensus. If consensus can be reached for each $\tilde{h}_j$, the preservation of the desired formation shape is guaranteed during maneuvers.

A similar approach can be applied to the rigid body attitude dynamics

$$\dot{\hat{q}}_i = -\frac{1}{2}\omega_i \times \hat{q}_i + \frac{1}{2}\bar{q}_i\omega_i, \quad \dot{\bar{q}}_i = -\frac{1}{2}\omega_i \cdot \hat{q}_i,$$
$$J_i\dot{\omega}_i = -\omega_i \times (J_i\omega_i) + T_i,$$

where, for the ith rigid body, $\hat{q}_i \in \mathbb{R}^3$, $\bar{q}_i \in \mathbb{R}$, and $q_i = [\hat{q}_i^T, \bar{q}_i]^T \in \mathbb{R}^4$ is the unit quaternion, that is, the Euler parameters (see Appendix D), $\omega_i \in \mathbb{R}^3$ is the angular velocity, and $J_i \in \mathbb{R}^{3\times 3}$ and $T_i \in \mathbb{R}^3$ are, respectively, the inertia tensor and the control torque. Defining $\text{vec}\left([\hat{q}, \bar{q}]^T\right) = \hat{q}$ as the operator that extracts the vector part of a quaternion, the control torque is given by [117, 176, 188]

$$T_i = -k_G \text{vec}\left(q_i^{d*} q_i\right) - D_G\omega_i - k_S \sum_{j=1}^{n} a_{ij}\text{vec}\left(q_j^* q_i\right) - D_S \sum_{j=1}^{n} a_{ij}(\omega_i - \omega_j), \quad (1.7)$$

where $k_G > 0$ and $k_S \geq 0$ are scalars, D_G is a symmetrical positive-definite matrix, D_S is a symmetrical positive-semidefinite matrix, q^* is the quaternion conjugate, and q^d is the centrally commanded quaternion. The first two terms in (1.7) align the rigid body with the prespecified desired attitude q_i^d. The second two terms in (1.7) are consensus terms that cause the team to maintain attitude alignment during the transients and in response to environmental disturbances [176].

Using biologically observed motions of flocks of birds, [198] defines three rules of flocking and applies them to generate realistic computer animations. The three rules of flocking are collision avoidance, velocity matching, and flock centering. Together these rules maintain the flock in close proximity without collision. Reference [198] motivates the use of similar rules for multivehicle robotic systems [155, 232]. As an example, consider the vehicle dynamics

$$\dot{r}_i = v_i, \qquad \dot{v}_i = u_i,$$

where r_i and v_i are the position and velocity of vehicle i, respectively, and u_i denotes its input. In [155], the control input u_i is defined as

$$u_i = -\frac{\partial V(r)}{\partial r_i} + \sum_{j=1}^{n} a_{ij}(r)(v_j - v_i) + f_i^\gamma, \quad (1.8)$$

where the first term is the gradient of a collective potential function $V(r)$, the second term drives the system toward velocity consensus, and the third term incorporates navigational feedback. In (1.8), the first term guarantees flock centering and collision avoidance among the vehicles, the second term guarantees velocity matching among the vehicles, and the third term achieves a group objective. Equation (1.8) has been validated for flocking with undirected communication topologies.

1.3 Monograph Overview

The subject of this monograph is distributed coordination of multiple autonomous vehicles. The objective of distributed coordination is to have multiple autonomous vehicles work together efficiently to achieve collective group behavior *via* local interaction. This monograph introduces distributed consensus algorithms and their applications in cooperative control of multiple autonomous vehicles. The consensus algorithms require only neighbor-to-neighbor information exchange, which minimizes the power consumption, increases the stealth, and improves the scalability and robustness of the team. In addition, the consensus algorithms allow the interaction topologies among the vehicles to be dynamically changing, sparse, or intermittent. This feature is particularly useful for real-world application scenarios where communication topologies are usually not fully connected, communication links are often noisy and unreliable, and vehicles have only limited communication range and bandwidth. This monograph includes both theoretical and experimental results in distributed coordination of multiple ground robots, spacecraft, and UAVs. The theoretical results address distributed consensus algorithms and their extensions for single-integrator, double-integrator, and rigid body attitude dynamics and show convergence analysis results in the presence of directed, limited, and unreliable information exchange among vehicles. Those results extend many existing results in the area of cooperative control. In the application chapters of the book, we apply the distributed consensus algorithms to several multivehicle cooperative control applications, including formation keeping for wheeled mobile robots and spacecraft and cooperative perimeter tracking and timing for a team of UAVs. The application results demonstrate issues and challenges in multivehicle cooperative control.

This monograph consists of six parts and six appendices. The first part overviews consensus algorithms in cooperative control of multiple autonomous vehicles (Chapter 1). The second part introduces consensus algorithms for single-integrator dynamics (Chapter 2) and consensus algorithms with a reference state (Chapter 3). The third part introduces consensus algorithms for double-integrator dynamics (Chapter 4) and their extensions to a reference model (Chapter 5). The fourth part focuses on attitude consensus for rigid body attitude dynamics (Chapter 6) and relative attitude maintenance and reference attitude tracking (Chapter 7). The fifth part introduces consensus-based design methodologies for distributed multivehicle cooperative control (Chapter 8). The sixth part applies consensus algorithms to several multivehicle cooperative control applications: rendezvous and axial alignment with multiple wheeled mobile robots (Chapter 9), distributed formation control of multiple wheeled mobile robots with a virtual leader (Chapter 10), a decentralized behavioral approach to multiple robot formation maneuvers (Chapter 11), deep space spacecraft formation flying (Chapter 12), cooperative fire monitoring with multiple UAVs (Chapter 13), and cooperative surveillance with multiple UAVs (Chapter 14). In addition, Appendices A–F review, respectively, selected notations and abbreviations, graph theory notations, matrix theory notations, rigid body attitude dynamics, linear system theory background, and nonlinear stability theory background.

1.4 Notes

The results in this chapter are based mainly on [191]. For further literature review on consensus algorithms, see [156].

Acknowledgment is given to

Part II

Consensus Algorithms for Single-integrator Dynamics

2

Consensus Algorithms for Single-integrator Dynamics

This chapter introduces consensus algorithms for single-integrator dynamics. Both continuous-time and discrete-time consensus algorithms are analyzed under both fixed and dynamically changing interaction topologies. This chapter shows, respectively, that consensus under fixed interaction topologies can be achieved asymptotically if and only if the directed interaction topology has a directed spanning tree and consensus under dynamically changing interaction topologies can be achieved asymptotically if there exist infinitely many uniformly bounded, consecutive time intervals such that the union of the directed interaction topologies across each such interval has a directed spanning tree.

2.1 Fundamental Algorithms

Suppose that there are n vehicles in a team. We use directed graph $\mathcal{G}_n \triangleq (\mathcal{V}_n, \mathcal{E}_n)$, where $\mathcal{V}_n = \{1, \ldots, n\}$ is the node set and $\mathcal{E}_n \subseteq \mathcal{V}_n \times \mathcal{V}_n$ is the edge set, to model the interaction topology among the n vehicles. Let $\mathcal{A}_n = [a_{ij}] \in \mathbb{R}^{n \times n}$ and $\mathcal{L}_n = [\ell_{ij}] \in \mathbb{R}^{n \times n}$ be, respectively, the adjacency matrix and the nonsymmetrical Laplacian matrix associated with $\mathcal{G}_n$ (see Appendix B for graph theory notations). Note that an undirected graph can be viewed as a special case of a directed graph. A directed graph takes into account the general case where information flow may be unidirectional. Unidirectional communication links are useful when, for example, some vehicles in a heterogeneous team have transceivers, whereas other less capable vehicles have only receivers. Furthermore, vehicles in a team may have nonuniform transmission strength. In addition, in the case of information exchange through local sensing, vehicles may be equipped with sensors that have only a limited field of view, which may result in directed interaction topologies.

Consider information states with single-integrator dynamics given by

$$\dot{\xi}_i = u_i, \quad i = 1, \ldots, n, \tag{2.1}$$

where $\xi_i \in \mathbb{R}^m$ is the information state and $u_i \in \mathbb{R}^m$ is the information control input of the ith vehicle. A continuous-time consensus algorithm is given by

$$u_i = -\sum_{j=1}^{n} a_{ij}(t)(\xi_i - \xi_j), \quad i = 1, \dots, n, \tag{2.2}$$

where $a_{ij}(t)$ is the (i,j) entry of the adjacency matrix $\mathcal{A}_n(t) \in \mathbb{R}^{n\times n}$ at time t. Note that $a_{ij}(t) > 0$ if $(j,i) \in \mathcal{E}_n$ at time t and $a_{ij}(t) = 0$ otherwise, $\forall j \neq i$. Intuitively, the information state of each vehicle is driven toward the information states of its neighbors. Equations (2.1) and (2.2) can be written in matrix form as

$$\dot{\xi} = -[\mathcal{L}_n(t) \otimes I_m]\xi, \tag{2.3}$$

where $\xi = [\xi_1^T, \dots, \xi_n^T]^T$, $\mathcal{L}_n(t) \in \mathbb{R}^{n\times n}$ is the nonsymmetrical Laplacian matrix at time t and $\otimes$ denotes the Kronecker product. With (2.2), consensus is *achieved* or *reached* by the team of vehicles if, for all $\xi_i(0)$ and all $i,j = 1, \dots, n$, $\|\xi_i(t) - \xi_j(t)\| \to 0$, as $t \to \infty$.

When interaction among vehicles occurs at discrete instants, the information state is updated using a difference equation. A discrete-time consensus algorithm is given by

$$\xi_i[k+1] = \sum_{j=1}^{n} d_{ij}[k]\xi_j[k], \quad i = 1, \dots, n, \tag{2.4}$$

where $k \in \{0, 1, \dots\}$ is the discrete-time index; $d_{ij}[k]$ is the (i,j) entry of a row-stochastic matrix $\mathcal{D}_n = [d_{ij}] \in \mathbb{R}^{n\times n}$ (see Appendix C) at the discrete-time index k with the additional assumption that $d_{ii}[k] > 0$ for all $i = 1, \dots, n$ and $d_{ij}[k] > 0$, $\forall i \neq j$, if $(j,i) \in \mathcal{E}_n$ and $d_{ij}[k] = 0$ otherwise. Intuitively, the information state of each vehicle is updated as the weighted average of its current state and the current states of its neighbors. Equation (2.4) can be written in matrix form as

$$\xi[k+1] = (\mathcal{D}_n[k] \otimes I_m)\xi[k]. \tag{2.5}$$

With (2.4), consensus is *achieved* or *reached* by the team of vehicles if, for all $\xi_i[0]$ and all $i,j = 1, \dots, n$, $\|\xi_i[k] - \xi_j[k]\| \to 0$, as $k \to \infty$.

Note that consensus algorithms (2.2) and (2.4) are distributed in the sense that each vehicle needs only information from its neighbors. Also note that the interaction topology may be changing dynamically due to unreliable transmission or a limited communication/sensing range. We use $\mathcal{G}_n(t)$ and $\mathcal{G}_n[k]$ to denote the dynamically changing interaction topologies corresponding to (2.2) and (2.4), respectively. For a given interaction topology, the weights $a_{ij}(t)$ in (2.2) and $d_{ij}[k]$ in (2.4) may be time-varying to represent possibly the time-varying relative confidence of each vehicle's information state or relative reliabilities of different interaction links among vehicles. As a result, both $\mathcal{L}_n(t)$ in (2.3) and $\mathcal{D}_n[k]$ in (2.5) may be dynamically changing over time.

Remark 2.1. The continuous-time update rule in [97] can be viewed as a special case of continuous-time consensus algorithm (2.2) by letting $a_{ij}(t) = \frac{1}{n}$ if $(j,i) \in \mathcal{E}_n$ at time t and $a_{ij}(t) = 0$ otherwise, $\forall j \neq i$. The Vicsek model [239] can be viewed as a special case of discrete-time consensus algorithm (2.4) by letting $d_{ij}[k] = \frac{1}{1+|\mathcal{N}_i|}$ if $(j,i) \in \mathcal{E}_n$ at the discrete-time index k and $d_{ij}[k] = 0$ otherwise, $\forall j \neq i$, and $d_{ii}[k] = \frac{1}{1+|\mathcal{N}_i|}$, where $|\mathcal{N}_i|$ denotes the number of neighbors of vehicle i, that is, each vehicle simply averages its own information state with those that are communicated to it. The simplified Vicsek model used in [97] can be viewed as a special case of (2.4) by letting $d_{ij}[k] = \frac{1}{g}$ if $(j,i) \in \mathcal{E}_n$ at the discrete-time index k and $d_{ij}[k] = 0$ otherwise, $\forall j \neq i$, and $d_{ii}[k] = 1 - \sum_{j \neq i} d_{ij}[k]$, where $g > n$ is a constant. □

We have the following preliminary results:

Lemma 2.2. *Continuous-time algorithm* (2.2) *achieves consensus asymptotically if and only if*

$$\begin{aligned}\Phi(t,0) =& I_n + \int_0^t -\mathcal{L}_n(\sigma_1)\, d\sigma_1 + \int_0^t -\mathcal{L}_n(\sigma_1) \int_0^{\sigma_1} -\mathcal{L}_n(\sigma_2)\, d\sigma_2\, d\sigma_1 \\ &+ \cdots \to \mathbf{1}_n c^T, \end{aligned} \tag{2.6}$$

as $t \to \infty$, *where* I_n *is the* $n \times n$ *identity matrix,* $\mathbf{1}_n$ *denotes the* $n \times 1$ *column vector with all of the entries equal to 1, and* c *is an* n *vector of constant coefficients.*

Proof: Note that the team of vehicles reaches consensus asymptotically if and only if the set

$$\mathcal{S} = \{\xi \in \mathbb{R}^n : \xi_1 = \xi_2 = \cdots = \xi_n\},$$

is attractive and positively invariant.

Because $\xi(t) = [\Phi(t,0) \otimes I_m]\xi(0)$ (see Appendix E), (2.6) implies that

$$\lim_{t\to\infty} \xi(t) = (\mathbf{1}_n c^T \otimes I_m)\xi(0) = \begin{bmatrix} (c^T \otimes I_m)\xi(0) \\ \vdots \\ (c^T \otimes I_m)\xi(0) \end{bmatrix},$$

which implies that $\mathcal{S}$ is attractive and positively invariant.

Conversely, if $\mathcal{S}$ is attractive and positively invariant, then

$$\lim_{t\to\infty} \xi(t) = \lim_{t\to\infty} [\Phi(t,0) \otimes I_m]\xi(0) = \mathbf{1}_n \otimes \alpha,$$

where α is an $m \times 1$ vector of constant coefficients, which in turn implies that

$$\lim_{t\to\infty} \Phi(t,0) = \mathbf{1}_n c^T.$$

■

Lemma 2.3. *Discrete-time algorithm* (2.4) *achieves consensus asymptotically if and only if*

$$\mathcal{D}_n[k-1]\mathcal{D}_n[k-2]\cdots\mathcal{D}_n[2]\mathcal{D}_n[1]\mathcal{D}_n[0] \to \mathbf{1}_n c^T, \tag{2.7}$$

as $k \to \infty$, *where* c *is an* n *vector of constant coefficients.*

Proof: Note that $\xi[k] = (\mathcal{D}_n[k-1]\mathcal{D}_n[k-2]\cdots\mathcal{D}_n[1]\mathcal{D}_n[0] \otimes I_m)\xi[0]$. The proof is similar to that of Lemma 2.2. ■

2.2 Consensus Under Fixed Interaction Topologies

In this section, we assume that the interaction topology is time-invariant and the weights a_{ij} in (2.2) and d_{ij} in (2.4) are constant, that is, $\mathcal{L}_n$ in (2.3) and $\mathcal{D}_n$ in (2.5) are constant. We will derive necessary and sufficient conditions for consensus of information using both continuous-time and discrete-time consensus algorithms.

2.2.1 Consensus Using a Continuous-time Algorithm

In the following, we first consider the case where the information state is inherently constant. We then consider the case where the information state is dynamically evolving in time. This is the case, for example, in formation control problems where the information state is the dynamic state of a virtual leader.

Static Consensus

We need the following lemma to derive our main results.

Lemma 2.4. *Given a matrix* $A = [a_{ij}] \in \mathbb{R}^{n\times n}$, *where* $a_{ii} \le 0$, $a_{ij} \ge 0$, $\forall i \ne j$, *and* $\sum_{j=1}^n a_{ij} = 0$ *for each* i, *then* A *has at least one zero eigenvalue with an associated eigenvector* $\mathbf{1}_n$, *and all nonzero eigenvalues are in the open left half plane. Furthermore,* A *has exactly one zero eigenvalue if and only if the directed graph of* A, *denoted by* $\Gamma(A)$, *has a directed spanning tree.*

Proof: For the first statement, note that A has zero row sums, which implies that A has at least one zero eigenvalue with an associated eigenvector $\mathbf{1}_n$. Also note that A is diagonally dominant and has nonpositive diagonal entries. It follows from Theorem C.1 that all nonzero eigenvalues of A are in the open left half plane.

The second statement of the lemma will be shown using an induction argument.

(Sufficiency.) *Step 1.* The first step is to show that A has exactly one zero eigenvalue if $\Gamma(A)$ is itself a directed spanning tree.

Noting that $\Gamma(A)$ is itself a directed spanning tree, renumber the vehicles consecutively by depth in the directed spanning tree, with the root numbered as 1. In other words, children of 1 are numbered 2 to q_1, children of 2 to q_1 are labeled $q_1 + 1$ to q_2, and so on. Note that the associated matrix A is lower diagonal with only one diagonal entry equal to zero.

Step 2. Let $Q = [q_{ij}] \in \mathbb{R}^{n\times n}$, where $q_{ii} \leq 0$, $q_{ij} \geq 0$, $\forall i \neq j$, and $\sum_{j=1}^n q_{ij} = 0$ for each i. Let $S = [s_{ij}] \in \mathbb{R}^{n\times n}$ satisfy the same properties as those of Q. Also let $\Gamma(Q)$ and $\Gamma(S)$ be directed graphs of Q and S, respectively. We assume that $s_{\ell\ell} = q_{\ell\ell} - k_{\ell m}$, $s_{\ell m} = q_{\ell m} + k_{\ell m}$, and $s_{ij} = q_{ij}$ otherwise, where $k_{\ell m} > 0$ denotes the weight for the edge (m, ℓ), $m \neq \ell$, that is, $\Gamma(S)$ corresponds to a graph where one more edge from node m to node ℓ is added to $\Gamma(Q)$, where $m \neq \ell$. Denote $p_Q(t) \triangleq \det(tI - Q)$ and $p_S(t) \triangleq \det(tI - S)$ as the characteristic polynomial of Q and S, respectively. Let $Q_t \triangleq tI - Q$ and $S_t \triangleq tI - S$. Given any matrix M, denote $M([i, j])$ as the submatrix of M formed by deleting the ith row and jth column.

Next, we will show that if Q has exactly one zero eigenvalue, then so does S. Without loss of generality, we assume that the new edge added to $\Gamma(Q)$ is from node m to node 1, where $m \neq 1$, for simplicity, because we can always renumber node ℓ as node 1.

Obviously, S has at least one zero eigenvalue, and all other nonzero eigenvalues are in the open left half plane following the first statement of this lemma. Below, we will show that S has only one zero eigenvalue.

Assume that $Q_t = [q_{tij}]$, and $S_t = [s_{tij}]$, $i, j = 1, \ldots, n$. Accordingly, it can be seen that $s_{t11} = t - s_{11} = t - q_{11} + k_{1m} = q_{t11} + k_{1m}$, $s_{t1m} = -s_{1m} = -q_{1m} - k_{1m} = q_{t1m} - k_{1m}$, and $s_{tij} = q_{tij}$ otherwise. Also note that $\det S_t([1, j]) = \det Q_t([1, j])$, $j = 1, \ldots, n$. Then we know that

$$
\begin{aligned}
\det S_t =& \sum_{j=1}^n (-1)^{1+j} s_{t1j} \det S_t([1, j]) \\
=& \sum_{j=1}^n (-1)^{1+j} q_{t1j} \det S_t([1, j]) \\
&+ k_{1m} \det S_t([1, 1]) - (-1)^{1+m} k_{1m} \det S_t([1, m]) \\
=& \det Q_t + k_{1m} \{\det S_t([1, 1]) + (-1)^m \det S_t([1, m])\}.
\end{aligned}
$$

Consider a matrix $E = [e_{ij}]$, $i, j = 1, \ldots, n - 1$, given by adding $[s_{21}, s_{31}, \ldots, s_{n1}]^T$ to the $(m - 1)$st column of $S([1, 1])$. The matrix E can be denoted as

$$
E = \begin{bmatrix} s_{22} & s_{23} & \cdots & s_{2m} + s_{21} & \cdots & s_{2n} \\ s_{32} & s_{33} & \cdots & s_{3m} + s_{31} & \cdots & s_{3n} \\ \vdots & \vdots & \vdots & \ddots & \vdots & \vdots \\ s_{n2} & s_{n3} & \cdots & s_{nm} + s_{n1} & \cdots & s_{nn} \end{bmatrix}.
$$

Thus $e_{i(m-1)} = s_{(i+1)m} + s_{(i+1)1}$, $i = 1, \ldots, n-1$. Given $p \times p$ matrices $\bar{F} = \begin{bmatrix} f_{11} & \cdots & f_{1k} + h_{1k} & \cdots & f_{1p} \\ \vdots & \vdots & \vdots & \vdots & \vdots \\ f_{p1} & \cdots & f_{pk} + h_{pk} & \cdots & f_{pp} \end{bmatrix}$, $F = \begin{bmatrix} f_{11} & \cdots & f_{1k} & \cdots & f_{1p} \\ \vdots & \vdots & \vdots & \vdots & \vdots \\ f_{p1} & \cdots & f_{pk} & \cdots & f_{pp} \end{bmatrix}$, and $H = \begin{bmatrix} f_{11} & \cdots & h_{1k} & \cdots & f_{1p} \\ \vdots & \vdots & \vdots & \vdots & \vdots \\ f_{p1} & \cdots & h_{pk} & \cdots & f_{pp} \end{bmatrix}$, the properties of determinants imply that $\det\bar{F} = \det F + \det H$. Therefore, it can be verified that

$$\det(tI - E) = \det S_t([1,1]) + (-1)^m \det S_t([1,m]).$$

Obviously, E has zero row sums and nonpositive diagonal entries and is diagonally dominant. The first statement of the lemma implies that E has at least one zero eigenvalue and all nonzero eigenvalues are in the open left half plane. As a result, the Routh stability criterion implies that the characteristic polynomial of E denoted as $\det(tI - E)$ has a nonnegative coefficient in the first power of t. We also know that Q has a positive coefficient for the first power of t in its characteristic polynomial $\det Q_t$ because Q has exactly one zero eigenvalue and all other eigenvalues are in the open left half plane. Noting that $\det S_t = \det Q_t + k_{1m}\det(tI - E)$, it is obvious that $p_S(t)$ has a positive coefficient for the first power of t. Therefore, S can only have one zero eigenvalue.

Step 3. If $\Gamma(A)$ is itself a directed spanning tree, we know from Step 1 that A has exactly one zero eigenvalue. If not, graph $\Gamma(A)$ can be constructed by consecutively adding edges to the directed spanning tree. Step 2 implies that adding one additional edge to the directed spanning tree results in an associated matrix that also has exactly one zero eigenvalue. We can recursively add additional edges, where Step 2 implies that the matrix associated with each new graph has exactly one zero eigenvalue, until we obtain $\Gamma(A)$. By induction, we know that A has exactly one zero eigenvalue if $\Gamma(A)$ has a directed spanning tree.

(Necessity.) If $\Gamma(A)$ does not have a directed spanning tree, then there exists a vehicle that separates two subgroups that do not exchange information or there exist at least two vehicles that do not receive any information from their neighbors. For the first case, suppose that the first subgroup has p vehicles and the second subgroup has q vehicles, where $p + q = n - 1$. By renumbering the vehicles by their subgroup, we assume that vehicles 1 to p are in the first subgroup, vehicles $p+2$ to n are in the second subgroup, and vehicle $p+1$ separates the two subgroups. Note that A can then be written as

$$A = \begin{bmatrix} A_p & 0_{p\times 1} & 0_{p\times q} \\ \bar{a}_p & a_{p+1,p+1} & \bar{a}_q \\ 0_{q\times p} & 0_{q\times 1} & A_q \end{bmatrix},$$

where $A_p \in \mathbb{R}^{p\times p}$, $\bar{a}_p = [a_{p+1,1}, \ldots, a_{p+1,p}]$, $\bar{a}_q = [a_{p+1,p+2}, \ldots, a_{p+1,n}]$, and $A_q \in \mathbb{R}^{q\times q}$. It is straightforward to see that both A_p and A_q have zero row sums and hence have at least one zero eigenvalue. It thus follows that $\text{Rank}(A_p) \leq p - 1$ and $\text{Rank}(A_q) \leq q - 1$. Therefore, from the structure of A, we know that $\text{Rank}(A) \leq n - 2$, implying that A has at least two zero eigenvalues. For the second case, A has at least two zero rows, which implies that A has at least two zero eigenvalues. ■

Corollary 2.5. *The nonsymmetrical Laplacian matrix $\mathcal{L}_n$ of a directed graph has a simple zero eigenvalue with an associated eigenvector $\mathbf{1}_n$ and all of the other eigenvalues are in the open right half plane if and only if the directed graph has a directed spanning tree.*

Proof: If we multiply the nonsymmetrical Laplacian matrix by -1, we get a matrix satisfying the properties defined in Lemma 2.4. ■

Next, we will show that the team of vehicles reaches consensus asymptotically using continuous-time algorithm (2.2) if $-\mathcal{L}_n$ in (2.3) has exactly one zero eigenvalue and all other eigenvalues are in the open left half plane. The following result also computes the value of the information state that is reached through the consensus process.

Lemma 2.6. *If $\mathcal{L}_n$ is the $n \times n$ nonsymmetrical Laplacian matrix, then $e^{-\mathcal{L}_n t}$, $\forall t \geq 0$, is a row-stochastic matrix with positive diagonal entries. In addition,* $\text{Rank}(\mathcal{L}_n) = n - 1$ *if and only if $\mathcal{L}_n$ has a simple zero eigenvalue. Furthermore, if $\mathcal{L}_n$ has a simple zero eigenvalue and $\nu = [\nu_1, \ldots, \nu_n]^T \geq 0$ satisfy $\mathbf{1}_n^T \nu = 1$ and $\mathcal{L}_n^T \nu = 0$,[1] then $e^{-\mathcal{L}_n t} \to \mathbf{1}_n \nu^T$, as $t \to \infty$.*

Proof: For the first statement, given eigenvalues $\lambda_i \in \sigma(-\mathcal{L}_n)$ with eigenvectors z_i, $i = 1, \ldots, n$, where $\sigma(\cdot)$ represents the spectrum of a matrix, we know that $e^{\lambda_i t} \in \sigma(e^{-\mathcal{L}_n t})$ with the same eigenvectors as $-\mathcal{L}_n$. Noting that $-\mathcal{L}_n$ has at least one zero eigenvalue with an associated eigenvector $\mathbf{1}_n$ from Lemma 2.4, it follows that $e^{-\mathcal{L}_n t}$ has at least one eigenvalue equal to one with the same eigenvector $\mathbf{1}_n$. Thus we know that $e^{-\mathcal{L}_n t}\mathbf{1}_n = \mathbf{1}_n$, which implies that $e^{-\mathcal{L}_n t}$ always has row sums equal to 1. Define the nonnegative matrix $M = \beta I_n - \mathcal{L}_n$, where β is the maximum of the diagonal entries of $\mathcal{L}_n$. We can see that $e^{-\mathcal{L}_n t} = e^{-\beta t} e^{Mt}$, which is obviously nonnegative and has positive diagonal entries. As a result, $e^{-\mathcal{L}_n t}$, $\forall t \geq 0$, is a row-stochastic matrix with positive diagonal entries, which implies that $\rho(e^{-\mathcal{L}_n t}) = 1$, where $\rho(\cdot)$ denotes the spectral radius of a matrix.

For the second statement, if $\mathcal{L}_n$ has a simple zero eigenvalue, then $\text{Nullity}(\mathcal{L}_n) = 1$, which in turn implies $\text{Rank}(\mathcal{L}_n) = n - 1$. Next, we show that $\text{Rank}(\mathcal{L}_n) = n - 1$ implies that $\mathcal{L}_n$ has a simple zero eigenvalue. Given the system $\dot{x} = -\mathcal{L}_n x$, where $x \in \mathbb{R}^n$, the solution is $x(t) = e^{-\mathcal{L}_n t} x(0)$. Note that $e^{-\mathcal{L}_n t}$ is a row-stochastic matrix with positive diagonal entries from the first statement of the lemma. Then it follows that $\left\| e^{-\mathcal{L}_n t} \right\|_\infty = 1$, $\forall t \geq 0$, which in turn implies that the system $\dot{x} = -\mathcal{L}_n x$ is marginally stable. Note

[1] That is, ν is the $n \times 1$ nonnegative left eigenvector of $\mathcal{L}_n$ associated with the zero eigenvalue and satisfies $\mathbf{1}_n^T \nu = 1$.

that the linear time-invariant (LTI) system $\dot{x} = Fx$, where $F \in \mathbb{R}^{n\times n}$, is marginally stable if and only if all of the eigenvalues of F have nonpositive real parts and every eigenvalue with zero real parts should have its geometric multiplicity equal to its algebraic multiplicity. Also note that $-\mathcal{L}_n$ has at least one zero eigenvalue and all nonzero eigenvalues of $-\mathcal{L}_n$ have negative real parts from Lemma 2.4. Because the system $\dot{x} = -\mathcal{L}_n x$ is marginally stable, the geometric multiplicity of the zero eigenvalue of $\mathcal{L}_n$ equals its algebraic multiplicity. Note that $\text{Rank}(\mathcal{L}_n) = n-1$ implies that $\text{Nullity}(\mathcal{L}_n) = 1$, that is, the geometric multiplicity of the zero eigenvalue is one. It follows that the algebraic multiplicity of the eigenvalue zero is also one, which in turn implies that $\mathcal{L}_n$ has a simple zero eigenvalue.

For the third statement, note from Lemma 2.4 that all of the nonzero eigenvalues of $-\mathcal{L}_n$ are in the open left half plane. If $-\mathcal{L}_n$ has exactly one zero eigenvalue, then $e^{-\mathcal{L}_n t}$, $\forall t > 0$, has exactly one eigenvalue equal to 1 and all of the other eigenvalues have moduli less than 1. Let $J = [j_{ml}]$, $m, l = 1, \ldots, n$, be the Jordan matrix corresponding to $-\mathcal{L}_n$; then $j_{mm} = \lambda_m$. Without loss of generality, assume that $\lambda_n = 0$ and λ_m is in the open left half plane, $m = 1, \ldots, n-1$.

Let $-\mathcal{L}_n = PJP^{-1}$, where $P = [p_1, \ldots, p_n]$ is an $n \times n$ matrix. Note that p_n can correspond to an eigenvector associated with the eigenvalue $\lambda_n = 0$. Without loss of generality, choose $p_n = \mathbf{1}_n$ as the eigenvector.

We know that $e^{-\mathcal{L}_n t} = Pe^{Jt}P^{-1}$. It can be verified that

$$e^{Jt} \to \begin{bmatrix} 0 & 0 & \cdots & 0 \\ \vdots & \vdots & \ddots & \vdots \\ 0 & 0 & \cdots & 0 \\ 0 & 0 & \cdots & 1 \end{bmatrix},$$

as $t \to \infty$, from the property of $-\mathcal{L}_n$. After some manipulation, we know that $e^{-\mathcal{L}_n t} \to \mathbf{1}_n \nu^T$, as $t \to \infty$, where ν_i, $i = 1, \ldots, n$, corresponds to the last row of the matrix P^{-1}. The result $\mathbf{1}_n^T \nu = 1$ comes from the fact that $e^{-\mathcal{L}_n t}$ has row sums equal to 1 for any t.

We also need to show that $\nu \geq 0$ and $\mathcal{L}_n^T \nu = 0$. Now consider the matrix $e^{-\mathcal{L}_n k}$, $k = 0, 1, 2, \ldots$. Obviously $e^{-\mathcal{L}_n k}$ will approach $\mathbf{1}_n \nu^T$, as $k \to \infty$. From Lemma C.2, ν is an eigenvector of $(e^{-\mathcal{L}_n})^T$ associated with the simple eigenvalue 1, which in turn implies that ν is an eigenvector of $-\mathcal{L}_n^T$ associated with the simple eigenvalue 0. Equivalently, it follows that $\mathcal{L}_n^T \nu = 0$. From Theorem C.4, $(e^{-\mathcal{L}_n})^T$ has a nonnegative eigenvector $x \geq 0$ associated with the simple eigenvalue 1. Thus it can be seen that $\nu = \alpha x$ for some $\alpha \neq 0$. Because $\sum_{i=1}^n \nu_i = 1$, it must be true that $\alpha > 0$, which implies that $\nu \geq 0$. ∎

Remark 2.7. Note that if we replace $-\mathcal{L}_n$ with $-\gamma\mathcal{L}_n$ in (2.3), where $\gamma > 0$, we can increase consensus speed by increasing γ. The solution of (2.3) with this new matrix is given by $\xi = e^{-\gamma\mathcal{L}_n t}\xi(0)$, which converges faster than the original solution, if we choose $\gamma > 1$. □

The following theorem shows a necessary and sufficient condition for consensus with a time-invariant interaction topology and constant gains a_{ij}.

Theorem 2.8. *Suppose that $\mathcal{A}_n$ is constant. Algorithm (2.2) achieves consensus asymptotically if and only if directed graph $\mathcal{G}_n$ has a directed spanning tree. In particular, $\xi_i(t) \to \sum_{i=1}^n \nu_i \xi_i(0)$, as $t \to \infty$, where $\nu = [\nu_1, \ldots, \nu_n]^T \geq 0$, $\mathbf{1}_n^T \nu = 1$, and $\mathcal{L}_n^T \nu = 0$.*

Proof: (Sufficiency.) From Lemma 2.2, we need to show that $e^{-\mathcal{L}_n t} \to \mathbf{1}_n c^T$, where c is an $n \times 1$ vector. Obviously $-\mathcal{L}_n$ in (2.3) associated with $\mathcal{G}_n$ has the same properties as A in Lemma 2.4. The fact that $\mathcal{G}_n$ has a directed spanning tree implies that the directed graph of $-\mathcal{L}_n$ has a directed spanning tree. Therefore, we know from Lemma 2.4 that $-\mathcal{L}_n$ has exactly one zero eigenvalue and all other eigenvalues are in the open left half plane. As a result, we know that the continuous-time algorithm (2.2) achieves consensus asymptotically according to Lemma 2.6. With (2.2), the solution of (2.1) is given by $\xi(t) = (e^{-\mathcal{L}_n t} \otimes I_m)\xi(0)$. Therefore, it follows from Lemma 2.6 that $\xi_i(t) \to \sum_{i=1}^n \nu_i \xi_i(0)$, $i = 1, \ldots, n$, as $t \to \infty$.

(Necessity.) Suppose that algorithm (2.2) achieves consensus asymptotically but that $\mathcal{G}_n$ does not have a directed spanning tree. Then there exist at least two vehicles i and j such that there is no path in $\mathcal{G}_n$ that contains both i and j. Therefore it is impossible to bring data between these two vehicles into consensus, which implies that consensus cannot be achieved asymptotically. ∎

Consensus Equilibrium

We have shown that algorithm (2.2) achieves consensus asymptotically if and only if $\mathcal{G}_n$ has a directed spanning tree. In addition, $\xi_i(t)$ will converge to $\sum_{i=1}^n \nu_i \xi_i(0)$, as $t \to \infty$, where $\sum_{i=1}^n \nu_i = 1$ and $\nu_i \geq 0$. A natural question is whether each initial condition $\xi_i(0)$ will contribute to the final consensus equilibrium point. We assume in this subsection that $\mathcal{G}_n$ has a directed spanning tree.

Observe that if there is a node k in $\mathcal{G}_n$ without an incoming edge (there is at most one such node if $\mathcal{G}_n$ has a directed spanning tree), the linear update law corresponding to this node is given by $\dot{\xi}_k = 0$ from (2.2), which implies that $\xi_k(t) = \xi_k(0)$ for all t. Therefore, the other nodes must converge to $\xi_k(0)$, that is, $\nu_k = 1$ and $\nu_i = 0$, $\forall i \neq k$.

In general, the initial condition of a node contributes to the equilibrium value if and only if the node has a directed path to all other nodes in $\mathcal{G}_n$. Thus $\nu_i \neq 0$ for every node that has a directed path to all other nodes in $\mathcal{G}_n$ and $\nu_i = 0$ otherwise. As a special case, the initial condition of each node in a directed graph contributes to the final equilibrium point if and only if the directed graph is strongly connected. The above argument can be explained as follows. If there is no path from node j to node m in $\mathcal{G}_n$, it is impossible for $\xi_m(t)$ to be influenced by $\xi_j(0)$. On the other hand, if there is a path from node j to every other node in $\mathcal{G}_n$, then $\xi_i(t)$, $\forall i \neq j$, will be influenced by $\xi_j(0)$.

The fact that $\nu_i \geq 0$, $i = 1, \ldots, n$, can also be explained from the following perspective. Assume that $\nu_\ell < 0$ for some ℓ. Consider the case $\xi_\ell(0) > 0$ and $\xi_i(0) = 0$, $\forall i \neq \ell$. We know that $\xi_i(t)$ will converge to $\sum_{i=1}^n \nu_i \xi_i(0) = \nu_\ell \xi_\ell(0)$, which is negative. Following continuous-time algorithm (2.2), $\dot{\xi}_\ell(0) < 0$ if there exists an incoming edge to node ℓ and $\dot{\xi}_\ell(0) = 0$ otherwise. In the first case, $\xi_\ell(t)$ will decrease and $\xi_i(t)$, $\forall i \neq \ell$ cannot decrease because $\dot{\xi}_i(0) \geq 0$, which implies that $\xi_i(t)$ will be driven to a value c with $0 \leq c < \xi_\ell(0)$. In the second case, $\xi_i(t)$ will be driven to $\xi_\ell(0)$. Both cases are contradictory to the above result. Therefore, $\nu_i \geq 0$, $i = 1, \ldots, n$.

If $\xi_i(t) \to \frac{1}{n} \sum_{i=1}^n \xi_i(0)$ (*i.e.*, $\nu_i = \frac{1}{n}$ for all i in Theorem 2.8), then the consensus equilibrium is the average of the initial information states, a condition called *average consensus*.

Corollary 2.9. *[158] Suppose that $\mathcal{A}_n$ is constant. Algorithm* (2.2) *achieves average consensus asymptotically if and only if directed graph $\mathcal{G}_n$ is strongly connected and balanced or undirected graph $\mathcal{G}_n$ is connected.*

In summary of the subsections on static consensus and consensus equilibrium, we have the following two lemmas:

Lemma 2.10. *Suppose that $z = [z_1^T, \ldots, z_p^T]^T$ with $z_i \in \mathbb{R}^m$ and $\mathcal{L}_p \in \mathbb{R}^{p \times p}$ satisfies the property* (B.2). *Then the following five conditions are equivalent.*
(i) $\mathcal{L}_p$ has a simple zero eigenvalue with an associated eigenvector $\mathbf{1}_p$ and all other eigenvalues have positive real parts,
(ii) $(\mathcal{L}_p \otimes I_m)z = 0$ implies that $z_1 = \cdots = z_p$,
(iii) Consensus is reached asymptotically for the system $\dot{z} = -(\mathcal{L}_p \otimes I_m)z$,
(iv) The directed graph of $\mathcal{L}_p$ has a directed spanning tree,
(v) The rank of $\mathcal{L}_p$ is $p - 1$.

Lemma 2.11. *Suppose that z and $\mathcal{L}_p$ are defined in Lemma 2.10. Then the following four conditions are equivalent.*
(i) The directed graph of $\mathcal{L}_p$ has a directed spanning tree and vehicle k has no incoming edges,[2]
(ii) The directed graph of $\mathcal{L}_p$ has a directed spanning tree and every entry of the kth row of $\mathcal{L}_p$ is zero,
(iii) Consensus is reached asymptotically for the system $\dot{z} = -(\mathcal{L}_p \otimes I_m)z$, where $z_i(t) \to z_k(0)$, $i = 1, \ldots, p$, as $t \to \infty$,
(iv) Vehicle k is the only node that has a directed path to all other vehicles on the team.

---

[2] At most, one such vehicle can exist when the directed graph has a directed spanning tree.

Relative State Deviations

Note that (2.2) represents the fundamental form of consensus algorithms. The algorithm can be extended to achieve different convergence results. For example, (2.2) can be extended to guarantee that the differences in the information states converge to desired values, *i.e.*, $\xi_i - \xi_j \to \Delta_{ij}(t)$, where $\Delta_{ij}(t)$ denotes the desired (time-varying) deviation between ξ_i and ξ_j. We apply the following algorithm for relative state deviations:

$$u_i = \dot{\delta}_i - \sum_{j=1}^{n} a_{ij}[(\xi_i - \xi_j) - (\delta_i - \delta_j)], \quad i = 1, \ldots, n, \tag{2.8}$$

where $\Delta_{ij} \stackrel{\triangle}{=} \delta_i - \delta_j$, $\forall i \neq j$, denotes the desired deviation between the information states. Note that by appropriately choosing δ_ℓ, $\ell = 1, \ldots, n$, we can guarantee that the differences in the information states converge to desired values. Algorithm (2.8) has applications in formation control, where the team forms a certain formation shape by maintaining relative positions among vehicles. Also note that (2.2) corresponds to the case where $\Delta_{ij} \equiv 0$, $\forall i \neq j$.

We have the following corollary for relative state deviations:

Corollary 2.12. *Suppose that $\mathcal{A}_n$ is constant. With* (2.8), $\xi_i(t) - \xi_j(t) \to \Delta_{ij}(t)$, *as $t \to \infty$, if and only if directed graph $\mathcal{G}_n$ has a directed spanning tree.*

Proof: With (2.8), (2.1) can be written as

$$\dot{\hat{\xi}}_i = -\sum_{j=1}^{n} a_{ij}(\hat{\xi}_i - \hat{\xi}_j), \quad i = 1, \ldots, n,$$

where $\hat{\xi}_i \stackrel{\triangle}{=} \xi_i - \delta_i$. From Theorem 2.8, we know that $\hat{\xi}_i(t) \to \hat{\xi}_j(t)$, as $t \to \infty$, if and only if $\mathcal{G}_n$ has a directed spanning tree. The rest of the proof then follows the fact that $\hat{\xi}_i(t) \to \hat{\xi}_j(t)$, as $t \to \infty$, is equivalent to $\xi_i(t) - \xi_j(t) \to \Delta_{ij}(t)$, as $t \to \infty$. ■

Dynamic Consensus

Note that a perturbation term may exist in (2.2). The perturbation term may represent an external signal or disturbance. As a result, consensus algorithm (2.2) becomes

$$u_i = -\sum_{j=1}^{n} a_{ij}(\xi_i - \xi_j) + w_i, \quad i = 1, \ldots, n, \tag{2.9}$$

where $w_i \in \mathbb{R}^m$ denotes the perturbation term.

In the special case where the information state of each vehicle is driven by the same time-varying input $w^f(t) \in \mathbb{R}^m$, which might represent an *a priori* known feedforward signal or an exogenous input signal, the associated consensus algorithm is given by

$$u_i = -\sum_{j=1}^{n} a_{ij}(\xi_i - \xi_j) + w^f, \quad i = 1, \ldots, n. \tag{2.10}$$

Equation (2.10) can also be written in matrix form as

$$\dot{\xi} = -(\mathcal{L}_n \otimes I_m)\xi + \mathbf{1}_n \otimes w^f. \tag{2.11}$$

We have the following theorem regarding consensus of the information states ξ_i, $i = 1, \ldots, n$, using (2.10).

Theorem 2.13. *Suppose that $\mathcal{A}_n$ is constant. Algorithm* (2.10) *achieves consensus asymptotically if and only if directed graph $\mathcal{G}_n$ has a directed spanning tree. In particular, $\xi_i(t) \to \sum_{i=1}^n \nu_i \xi_i(0) + \int_0^t w^f(\tau)d\tau$, as $t \to \infty$, where $\nu = [\nu_1, \ldots, \nu_n]^T \geq 0$, $\mathbf{1}_n^T \nu = 1$, and $\mathcal{L}_n^T \nu = 0$.*

Proof: (Sufficiency.) The solution of (2.11) is given by $\xi(t) = (e^{-\mathcal{L}_n t} \otimes I_m)\xi(0) + \int_0^t [e^{-\mathcal{L}_n(t-\tau)} \otimes I_m][\mathbf{1}_n \otimes w^f(\tau)]d\tau$ (see Appendix E). Note from Lemma 2.6 that $e^{-\mathcal{L}_n t} \to \mathbf{1}_n \nu^T$ and $e^{-\mathcal{L}_n(t-\tau)}\mathbf{1}_n = \mathbf{1}_n$. This proves the sufficient part.

(Necessity.) The necessary part follows directly from Theorem 2.8. ■

When $w_i \neq w_j$ in (2.9), we have the following theorem:

Theorem 2.14. *Suppose that $\mathcal{A}_n$ is constant. Under the condition that directed graph $\mathcal{G}_n$ has a directed spanning tree, if $\|w_i - w_j\|$ is uniformly bounded, so is $\|\xi_i - \xi_j\|$, $\forall i \neq j$.*

Proof: Let $\xi_{ij} \triangleq \xi_i - \xi_j$ and $w_{ij} \triangleq w_i - w_j$. Also let $\tilde{\xi} = [\xi_{12}^T, \xi_{13}^T, \ldots, \xi_{1n}^T]^T$ and $\tilde{w} = [w_{12}^T, w_{13}^T, \ldots, w_{1n}^T]^T$. Note that $\xi_{ij} = \xi_{1j} - \xi_{1i}$. Using (2.9), (2.1) can be written in matrix form as

$$\dot{\tilde{\xi}} = (Q \otimes I_m)\tilde{\xi} + \tilde{w}, \tag{2.12}$$

where $Q \in \mathbb{R}^{(n-1)\times(n-1)}$ is constant. Note that Theorem 2.13 implies that under the condition that $w_i = w_j$, $\forall i \neq j$, and $\mathcal{G}_n$ contains a directed spanning tree, algorithm (2.9) achieves consensus asymptotically, which implies that $\tilde{\xi}(t) \to 0$, as $t \to \infty$. It thus follows that (2.12) is asymptotically stable under the same condition that $w_i = w_j$, $\forall i \neq j$, by noting that $w_i = w_j$, $\forall i \neq j$, implies $\tilde{w} = 0$. Because (2.12) is an LTI system, asymptotical stability implies bounded-input bounded-state stability (see Appendix E), that is, if $\tilde{w}$ is uniformly bounded, so is $\tilde{\xi}$. Equivalently, if $\|w_i - w_j\|$ is uniformly bounded, so is $\|\xi_i - \xi_j\|$, $\forall i \neq j$. ■

Remark 2.15. Theorem 2.14 implies that using (2.9), (2.1) is input-to-state stable (ISS) (see Appendix F) under the assumption of the theorem. □

Illustrative Example

We consider a scenario where six vehicles are to rendezvous at a position along a parameterized trajectory represented by $\{r_x[\tau(t)], r_y[s(t)]\}$. Figure 2.1 shows the corresponding communication topology among these vehicles. Note the existence of a directed spanning tree.

It is assumed that each vehicle knows the parameterized trajectory. Therefore, the parameters τ and s represent the minimum information needed to achieve the coordination objective, *i.e.*, τ and s are the information states. We will instantiate τ and s on each vehicle as τ_i and s_i, $i = 1, \ldots, 6$. Here we let $\xi_i = [\tau_i, s_i]^T$, $i = 1, \ldots, 6$.

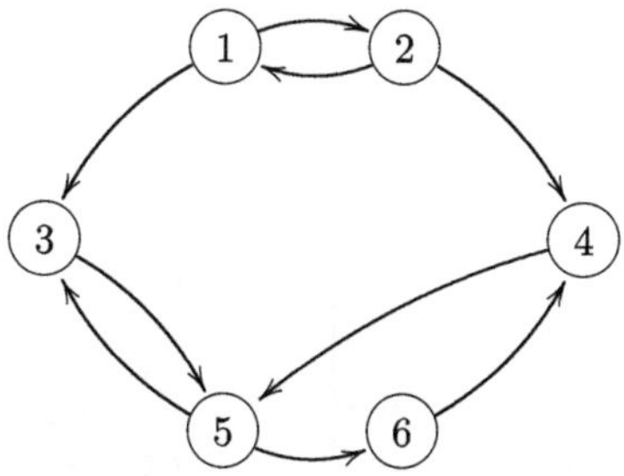

Fig. 2.1. Communication topology

Based on the communication topology shown in Fig. 2.1, the matrix $-\mathcal{L}_6$ is given by

$$-\mathcal{L}_6 = \gamma \begin{bmatrix} -1.5 & 1.5 & 0 & 0 & 0 & 0 \\ 2 & -2 & 0 & 0 & 0 & 0 \\ 0.9 & 0 & -2.8 & 0 & 1.9 & 0 \\ 0 & 1.2 & 0 & -2.5 & 0 & 1.3 \\ 0 & 0 & 1.4 & 1.8 & -3.2 & 0 \\ 0 & 0 & 0 & 0 & 0.7 & -0.7 \end{bmatrix},$$

where $\gamma > 0$ is a coefficient. The initial conditions for each instantiation of τ and s are given by $\tau_i(0) = 0.2i - 0.1$ and $s_i(0) = 0.2i$, $i = 1, \ldots, 6$.

Figure 2.2 shows the consensus scenario using the continuous-time algorithm (2.2) for $\gamma = 1$ and $\gamma = 5$, respectively. We can see that only the initial conditions of vehicle 1 and vehicle 2 affect the equilibrium value, which is consistent with the communication topology shown in Fig. 2.1, where it can be seen that only vehicle 1 and vehicle 2 have a directed path to all other vehicles. Figure 2.3 shows the same consensus scenario corresponding to the communication topology formed by deleting the edge $(2, 1)$ in Fig. 2.1. It can be seen that each instantiation of τ and s converges to $\tau_1(0)$ and $s_1(0)$, respectively. In both Figs. 2.2 and 2.3, consensus is reached faster for $\gamma = 5$ than for $\gamma = 1$.

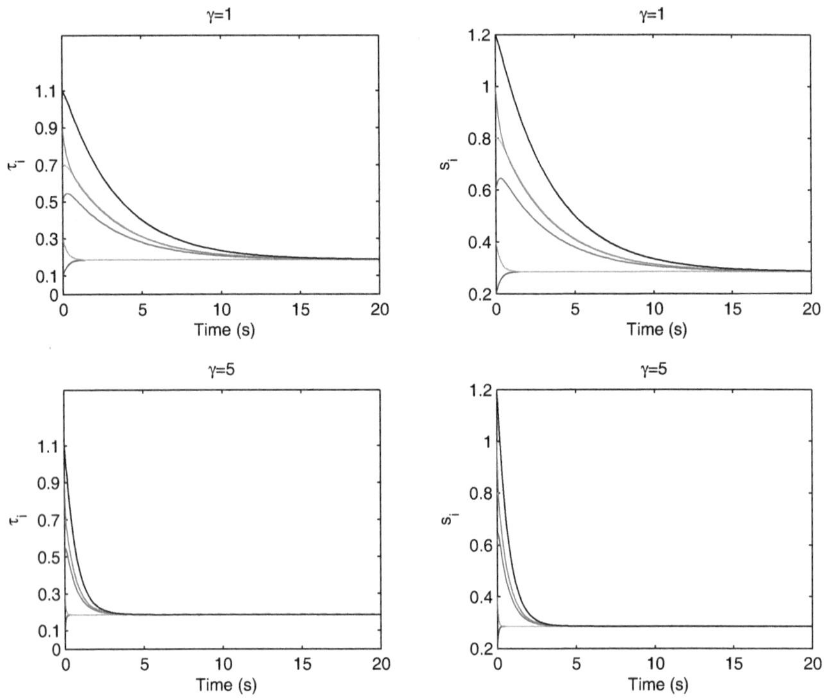

Fig. 2.2. Consensus of τ_i and s_i using continuous-time algorithm (2.2)

Figure 2.4 illustrates a dynamic consensus scenario using algorithm (2.10) for $\gamma = 1$ and $\gamma = 5$, respectively. The common predefined planning schemes for τ and s are given by $\dot{\tau} = \frac{1}{5}|\sin(t)|$ and $\dot{s} = \frac{1}{4}|\cos(t)|$, respectively. Here, we let $w^f(t) = [\frac{1}{5}|\sin(t)|, \frac{1}{4}|\cos(t)|]^T$ in (2.10). It can be seen that consensus is achieved asymptotically and that both τ_i and s_i follow their predefined planning schemes.

2.2.2 Consensus Using a Discrete-time Algorithm

We need the following lemmas to derive our main results:

Lemma 2.16. *If a nonnegative matrix $A = [a_{ij}] \in \mathbb{R}^{n\times n}$ has the same positive constant row sums given by $\mu > 0$, then μ is an eigenvalue of A with an associated eigenvector $\mathbf{1}_n$ and $\rho(A) = \mu$, where $\rho(\cdot)$ denotes the spectral radius. In addition, the eigenvalue μ of A has algebraic multiplicity equal to one, if and only if the directed graph of A, denoted by $\Gamma(A)$, has a directed spanning tree. Furthermore, if $a_{ii} > 0$, $i = 1, \ldots, n$, then $|\lambda| < \mu$ for every eigenvalue $\lambda \neq \mu$. Moreover, if $\Gamma(A)$ has a directed spanning tree and $a_{ii} > 0$, $i = 1, \ldots, n$, then μ is the unique eigenvalue of maximum modulus.*

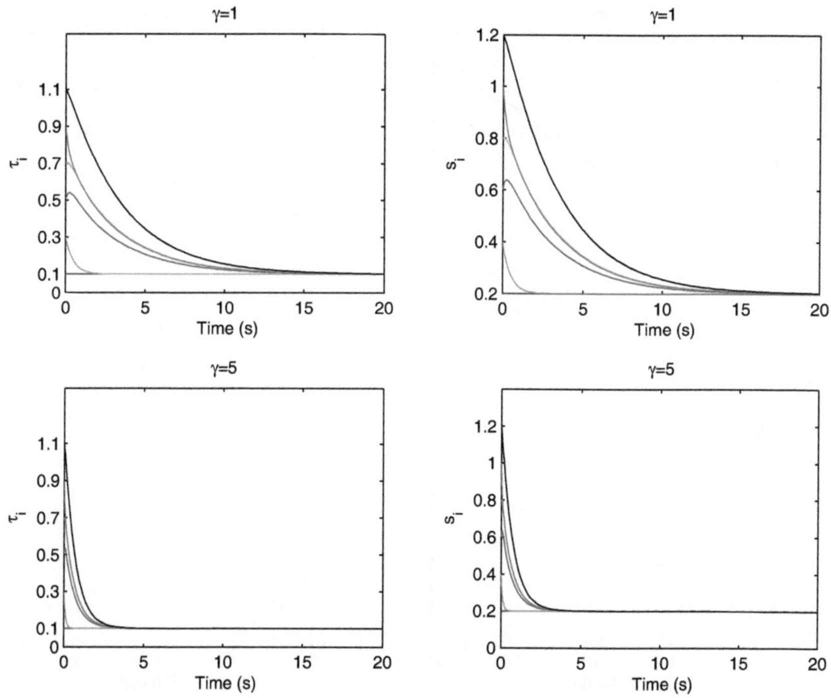

Fig. 2.3. Consensus of τ_i and s_i without edge $(2, 1)$ using continuous-time algorithm (2.2)

Proof: For the first statement, note that A has the same positive constant row sums given by $\mu > 0$, which implies that $A\mathbf{1}_n = \mu\mathbf{1}_n$. Theorem C.1 implies that all eigenvalues of A are located in the union of the n discs given by

$$\bigcup_{i=1}^{n}\{z \in \mathbb{C} : |z - a_{ii}| \leq \sum_{j \neq i} |a_{ij}|\}, \tag{2.13}$$

where $\mathbb{C}$ is the set of complex numbers. It thus follows that $\rho(A) = \mu$.

For the second statement, we need to show both necessary and sufficient conditions.

(Sufficiency.) If $\Gamma(A)$ has a directed spanning tree, then the directed graph of $B = A - \mu I_n$ also has a directed spanning tree. We know that $\lambda_i(A) = \lambda_i(B) + \mu$, where $i = 1, \ldots, n$, and $\lambda_i(\cdot)$ represents the ith eigenvalue. Noting that B satisfies the conditions in Lemma 2.4, we know that zero is an eigenvalue of B with algebraic multiplicity equal to one, which implies that μ is an eigenvalue of A with algebraic multiplicity equal to one.

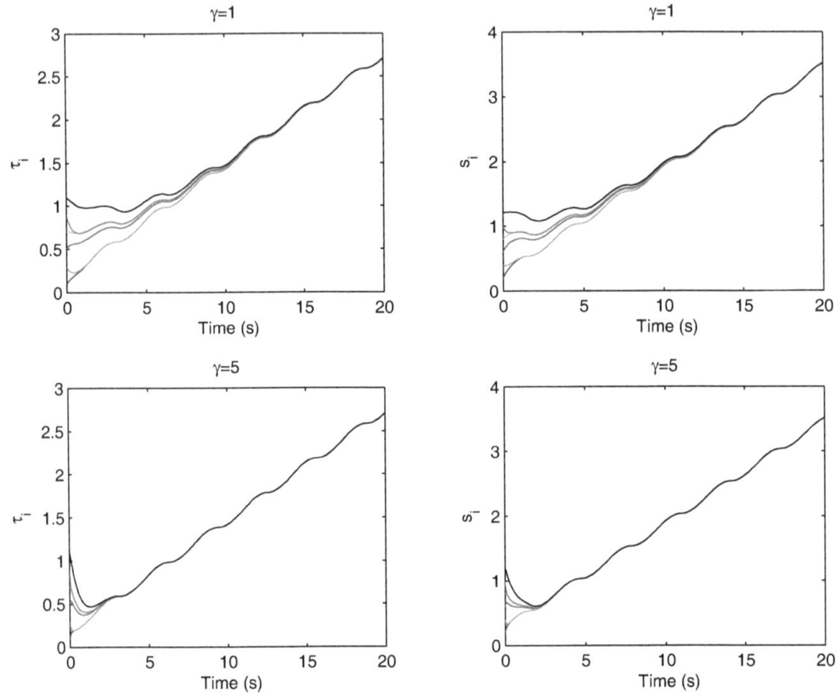

Fig. 2.4. Consensus and evolution of τ_i and s_i using algorithm (2.10)

(Necessity.) If $\Gamma(A)$ does not have a directed spanning tree, we know from Lemma 2.4 that $B = A - \mu I_n$ has more than one zero eigenvalue, which in turn implies that A has more than one eigenvalue equal to μ.

For the third statement, noting that $a_{ii} > 0$, it is easy to see that union (2.13) is included in a circle given by $\{z \in \mathbb{C} : |z| = \mu\}$ and the circular boundaries of the union of the n discs have only one intersection with the circle at $z = \mu$. Thus we know that $|\lambda| < \mu$ for every eigenvalue of A satisfying $\lambda \neq \mu$.

Combining the second and third statement, we know that μ is the unique eigenvalue of maximum modulus. ∎

Remark 2.17. Theorem C.5 states that if a nonnegative matrix A is irreducible, that is, the directed graph of A is strongly connected, then the spectral radius of A is a simple eigenvalue. We show that the irreducibility condition is too stringent for nonnegative matrices with the same positive row sums. Lemma 2.16 explicitly shows that for a nonnegative matrix A with identical positive row sums, the spectral radius of A (the row sum in this case) is a simple eigenvalue if and only if the directed graph of A has a directed spanning tree. In other words, A may be reducible but retains its spectral radius as a simple eigenvalue. Furthermore, if A has a directed spanning tree and positive diagonal entries, we know that the spectral radius of A is the unique eigenvalue of maximum modulus. □

Corollary 2.18. *A row-stochastic matrix $A = [a_{ij}] \in \mathbb{R}^{n\times n}$ has algebraic multiplicity equal to one for its eigenvalue $\rho(A) = 1$ if and only if $\Gamma(A)$ has a directed spanning tree. Furthermore, if $a_{ii} > 0$, $i = 1, \ldots, n$, then $|\lambda| < 1$ for every eigenvalue not equal to one. In addition, if $\Gamma(A)$ has a directed spanning tree and $a_{ii} > 0$, $i = 1, \ldots, n$, then 1 is the unique eigenvalue of maximum modulus.*

Lemma 2.19. *Let $A = [a_{ij}] \in \mathbb{R}^{n\times n}$ be a row-stochastic matrix. If A has an eigenvalue $\lambda = 1$ with algebraic multiplicity equal to one and all other eigenvalues satisfy $|\lambda| < 1$, then A is SIA. In particular, $\lim_{m\to\infty} A^m \to \mathbf{1}_n\nu^T$, where ν satisfies $A^T\nu = \nu$ and $\mathbf{1}_n^T\nu = 1$. Furthermore, each element of ν is nonnegative.*

Proof: The first statement of the lemma follows Lemma C.2. For the second statement, it is obvious that A^T is also nonnegative and has $\rho(A^T) = 1$ as an eigenvalue with algebraic multiplicity equal to one. Thus Theorem C.4 implies that the eigenspace of A^T associated with the eigenvalue $\lambda = 1$ is given by cx, where $c \in \mathbb{C}$, $c \neq 0$, and x is a nonnegative eigenvector. Because ν is also an eigenvector of A^T associated with the eigenvalue $\lambda = 1$ and satisfies $\mathbf{1}_n^T\nu = 1$, it follows that each element of ν must be nonnegative. ■

Next, we show a necessary and sufficient condition for consensus of information using discrete-time algorithm (2.4).

Theorem 2.20. *Suppose that $\mathcal{A}_n$ is constant. Discrete-time algorithm* (2.4) *achieves consensus asymptotically if and only if directed graph $\mathcal{G}_n$ has a directed spanning tree. In particular, $\xi_i[k] \to \sum_{i=1}^n \nu_i\xi_i[0]$, where $\nu = [\nu_1, \ldots, \nu_n]^T \geq 0$ satisfies $\mathcal{D}_n^T\nu = \nu$ and $\mathbf{1}_n^T\nu = 1$.*

Proof: From Lemma 2.3, we need to show that $\mathcal{D}_n^k \to \mathbf{1}_n c^T$, where c is a constant column vector.

(Sufficiency.) Obviously $\mathcal{D}_n$ is a row-stochastic matrix with positive diagonal entries. The fact that $\mathcal{G}_n$ has a directed spanning tree also implies that the directed graph of $\mathcal{D}_n$ has a directed spanning tree. Combining Corollary 2.18 and Lemma 2.19, we know that $\lim_{k\to\infty} \mathcal{D}_n^k \to \mathbf{1}_n\nu^T$, where ν satisfies the properties defined in the lemma.

(Necessity.) If $\mathcal{G}_n$ does not have a directed spanning tree, neither does the directed graph of $\mathcal{D}_n$, which implies, by Corollary 2.18, that the algebraic multiplicity of the eigenvalue $\rho(\mathcal{D}_n) = 1$ of $\mathcal{D}_n$ is $p > 1$. Therefore, the Jordan decomposition of $\mathcal{D}_n^k$ has the form $\mathcal{D}_n^k = MJ^kM^{-1}$, where M is full rank and J^k is lower triangular with p diagonal elements equal to one. Therefore, the rank of $\lim_{k\to\infty} \mathcal{D}_n^k$ is at least $p > 1$ which implies, by Lemma 2.3, that the team of vehicles cannot reach consensus asymptotically. ■

Using discrete-time consensus algorithm (2.4), we have similar results for consensus equilibrium analysis and dynamics consensus.

2.3 Consensus Under Dynamically Changing Interaction Topologies

We need the following lemmas.

Lemma 2.21. *Given $n \times n$ nonnegative matrices A and B, if $A \geq \alpha B$, for some $\alpha > 0$, then $\Gamma(B)$ is a subgraph of $\Gamma(A)$.*

Proof: Trivial. ∎

Lemma 2.22. *Given $n \times n$ nonnegative matrices P, Q, R, and S, if $P \sim R$ and $Q \sim S$, then $(P+Q) \sim (R+S)$ and $PQ \sim RS$, where $\sim$ denotes that two nonnegative matrices are of the same type (see Appendix C for its definition). Moreover, if an $n \times n$ time-varying nonnegative matrix $M(t)$ with continuous entries is of a fixed type for $t \in [t_1, t_2]$, where $t_1 < t_2$, then $M(t) \sim \int_{t_1}^{t_2} M(t)dt$.*

Proof: Trivial. ∎

Lemma 2.23. *Let $C(t) = [c_{ij}(t)] \in \mathbb{R}^{n \times n}$ be piecewise continuous, where $c_{ij} \geq 0$, $\forall i \neq j$, and $\sum_j c_{ij} = 0$. Let $\Phi_C(t, t_0)$ be the corresponding transition matrix. Then $\Phi_C(t, t_0)$ is a row-stochastic matrix with positive diagonal entries for any $t \geq t_0$.*

Proof: Note that

$$\Phi_C(t, t_0) = I_n + \int_{t_0}^{t} C(\sigma_1)\, d\sigma_1 + \int_{t_0}^{t} C(\sigma_1) \int_{t_0}^{\sigma_1} C(\sigma_2)\, d\sigma_2\, d\sigma_1 + \cdots. \quad (2.14)$$

Noting that $C(t)\mathbf{1}_n = 0$, we can verify that $\Phi_C(t, t_0)\mathbf{1}_n = \mathbf{1}_n$.

Let $B(t) = C(t) + \mu I_n$, where μ is a constant greater than $\max_{\tau \in [t_0, t]} \max_i |c_{ii}(\tau)|$. Note that $B(t)$ is a nonnegative matrix with positive diagonal entries. It is straightforward to see that

$$\frac{d}{dt}\Phi_C(t, t_0) = C(t)\Phi_C(t, t_0)$$

and

$$\begin{aligned}
&\frac{d}{dt}[\Phi_B(t, t_0)e^{-\mu(t-t_0)}] \\
&= B(t)\Phi_B(t, t_0)e^{-\mu(t-t_0)} - \mu\Phi_B(t, t_0)e^{-\mu(t-t_0)} \\
&= (B(t) - \mu I_n)\Phi_B(t, t_0)e^{-\mu(t-t_0)} \\
&= C(t)\Phi_B(t, t_0)e^{-\mu(t-t_0)},
\end{aligned}$$

and that $\Phi_C(t_0, t_0) = \Phi_B(t_0, t_0)e^{-\mu(t_0-t_0)} = I_n$. Therefore, it follows that $\Phi_C(t, t_0) = \Phi_B(t, t_0)e^{-\mu(t-t_0)}$. Using (2.14) to compute $\Phi_B(t, t_0)$, it is straightforward to see that $\Phi_B(t, t_0)$ is nonnegative and has positive diagonal entries. Therefore, it follows that $\Phi_C(t, t_0)$ is nonnegative and has positive diagonal entries. Combining these arguments, it follows that the transition matrix $\Phi_C(t, t_0)$ is a row-stochastic matrix with positive diagonal entries. ∎

Lemma 2.24. *Let $C(t) = [c_{ij}(t)] \in \mathbb{R}^{n\times n}$ and $\tilde{C} = [\tilde{c}_{ij}(t)] \in \mathbb{R}^{n\times n}$ be continuous on $t \in [\tau, s]$, where $s > \tau$, such that $c_{ij}(t) \geq 0$ and $\tilde{c}_{ij}(t) \geq 0$, $\forall i \neq j$, and $\sum_{j=1}^{n} c_{ij}(t) = \sum_{j=1}^{n} \tilde{c}_{ij}(t) = 0$. Let $\Phi_C(s,\tau)$ and $\Phi_{\tilde{C}}(s,\tau)$ be the corresponding transition matrices. Suppose that the directed graphs of $C(t)$ and $\tilde{C}(t)$ are the same and fixed for $t \in [\tau, s]$. Then the directed graph of $C(t)$ is a subgraph of the directed graph of $\Phi_C(s,\tau)$ and $\Phi_C(s,\tau) \sim \Phi_{\tilde{C}}(s,\tau)$.*

Proof: Let $B(t) = C(t) + \mu I_n$, where μ is a constant greater than $\max_{\tau\in[t_0,t]} \max_i |c_{ii}(\tau)|$. Following Lemma 2.23, we know that $\Phi_C(s,\tau) = \Phi_B(s,\tau)e^{-\mu(s-\tau)}$. Note that the directed graphs of $C(t)$ and $B(t)$ are the same, so are the directed graphs of $\Phi_C(s,\tau)$ and $\Phi_B(s,\tau)$. Therefore, using (2.14) to compute $\Phi_B(s,\tau)$, we can see that $\Phi_B(s,\tau) \geq \int_\tau^s B(\sigma_1)d\sigma_1$, where $\int_\tau^s B(\sigma_1)d\sigma_1 \sim B(t)$ for $t \in [\tau, s]$, or in other words, the directed graph of $B(t)$ for $t \in [\tau, s]$ is a subgraph of the directed graph of $\Phi_B(s,\tau)$. Therefore, the directed graph of $C(t)$ for $t \in [\tau, s]$ is a subset of the directed graph of $\Phi_C(s,\tau)$.

Note that $\Phi_{\tilde{C}}(s,\tau) = \Phi_{\tilde{B}}(s,\tau)e^{-\tilde{\mu}(s-\tau)}$, where $\tilde{C} = \tilde{B} - \tilde{\mu} I_n$. To show that Φ_C is of the same type as $\Phi_{\tilde{C}}$, we need to show that Φ_B is of the same type as $\Phi_{\tilde{B}}$. Note that B and $\tilde{B}$ are of the same type because their directed graphs are the same. By writing Φ_B and $\Phi_{\tilde{B}}$ as in (2.14) and comparing each term, Lemma 2.22 implies that each corresponding term is of the same type, which in turn implies that $\Phi_B(s,\tau)$ and $\Phi_{\tilde{B}}(\tilde{s},\tilde{\tau})$ are of the same type. ∎

Lemma 2.25. *[97, Lemma 2] Let $p \geq 2$ be a positive integer and let $P_1, P_2, \ldots, P_p$ be $n \times n$ nonnegative matrices with positive diagonal entries, then*

$$P_1 P_2 \cdots P_p \geq \gamma(P_1 + P_2 + \cdots + P_p),$$

where $\gamma > 0$ can be specified from P_i, $i = 1, \ldots, p$.

Lemma 2.26. *Let $S_A = \{A_1, A_2, \ldots, A_\ell\}$ be a set of row-stochastic matrices with positive diagonal entries. If the directed graph of A_i has a directed spanning tree, then A_i is SIA. If the union of the directed graphs of A_i, $i = 1, \ldots, \ell$, has a directed spanning tree, then the matrix product $\prod_{i=1}^{\ell} A_i$ is SIA.*

Proof: The first statement follows from Corollary 2.18 and Lemma 2.19. For the second statement, note that the product of row-stochastic matrices is still a row-stochastic matrix. Also note that $\prod_{i=1}^{\ell} A_i \geq \gamma \sum_{i=1}^{\ell} A_i$ for some $\gamma > 0$ according to Lemma 2.25. Because the union of the directed graphs of the matrices in S_A has a directed spanning tree, it is obvious that the directed graph of $\sum_{i=1}^{\ell} A_i$ has a directed spanning tree. Therefore, it can be seen that the directed graph of $\prod_{i=1}^{\ell} A_i$ has a directed spanning tree according to Lemma 2.21, which in turn implies, from the first statement of the lemma, that $\prod_{i=1}^{\ell} A_i$ is SIA. ∎

Lemma 2.27. *Let $\overline{\mathcal{G}_n}$ denote the set of all possible directed graphs for the n vehicles.[3] If the union of directed graphs $\{\mathcal{G}_n(t_1), \mathcal{G}_n(t_2), \ldots, \mathcal{G}_n(t_p)\}$, where*

[3] Note that $\overline{\mathcal{G}_n}$ has a finite number of elements.

$\mathcal{G}_n(t_i) \in \overline{\mathcal{G}_n}$, has a directed spanning tree and $\mathcal{L}_n(t_i)$ is the nonsymmetrical Laplacian matrix associated with each directed graph $\mathcal{G}_n(t_i)$, then the matrix product

$$e^{-\mathcal{L}_n(t_p)\Delta t_p} \cdots e^{-\mathcal{L}_n(t_2)\Delta t_2} e^{-\mathcal{L}_n(t_1)\Delta t_1}$$

is SIA, where $\Delta t_i > 0$ are bounded below.

Proof: From (2.3), each matrix $-\mathcal{L}_n(t_i)$ satisfies the properties defined in Lemma 2.4. Thus each $-\mathcal{L}_n(t_i)$ can be written as the sum of a nonnegative matrix M_{t_i} and $-\eta_{t_i} I_n$, where η_{t_i} is the maximum of the diagonal entries of $\mathcal{L}_n(t_i)$, $i = 1, \ldots, p$. Note that $e^{-\mathcal{L}_n(t_i)\Delta t_i} = e^{-\eta_{t_i}\Delta t_i} e^{M_{t_i}\Delta t_i} \geq \alpha_i M_{t_i}$ for some $\alpha_i > 0$. Since the union of directed graphs $\{\mathcal{G}_n(t_1), \mathcal{G}_n(t_2), \ldots, \mathcal{G}_n(t_p)\}$ has a directed spanning tree, we know that the union of the directed graphs of M_{t_i} has a directed spanning tree, which in turn implies from Lemma 2.21 that the union of the directed graphs of $e^{-\mathcal{L}_n(t_i)\Delta t_i}$ has a directed spanning tree. From Lemma 2.25, we know that $e^{-\mathcal{L}_n(t_p)\Delta t_p} \cdots e^{-\mathcal{L}_n(t_2)\Delta t_2} e^{-\mathcal{L}_n(t_1)\Delta t_1} \geq \gamma \sum_{i=1}^{p} e^{-\mathcal{L}_n(t_i)\Delta t_i}$ for some $\gamma > 0$, which implies that the above matrix product also has a directed spanning tree.

From Lemma 2.6, each matrix $e^{-\mathcal{L}_n(t_i)\Delta t_i}$ is a row-stochastic matrix with positive diagonal entries, which implies that the above matrix product is also a row-stochastic matrix with positive diagonal entries. Therefore, from Lemma 2.26, we know that the above matrix product is SIA. ■

Lemma 2.28. *Let $C(t) = [c_{ij}(t)] \in \mathbb{R}^{n\times n}$ be piecewise continuous for $t \in [\tau, s]$, where $s - \tau > 0$ is bounded, $c_{ij} \geq 0$, $\forall i \neq j$, and $\sum_j c_{ij} = 0$. If the union of the directed graphs of $C(t)$ for $t \in [\tau, s]$ has a directed spanning tree, then the transition matrix $\Phi_C(s, \tau)$ is SIA.*

Proof: Note that $\Phi_C(t_{\ell+1}, t_0) = \Phi_C(t_{\ell+1}, t_\ell)\Phi_C(t_\ell, t_{\ell-1}) \cdots \Phi_C(t_1, t_0)$, where $t_0 \stackrel{\triangle}{=} \tau$, $t_{\ell+1} \stackrel{\triangle}{=} s$, and t_j, $j = 1, \ldots, \ell$, denotes the times when $C(t)$ is discontinuous. From Lemma 2.24, we know that the directed graph of $C(t)$ for each $t \in [t_{i-1}, t_i]$ is a subgraph of the directed graph of $\Phi_C(t_i, t_{i-1})$, $i = 1, \ldots, \ell+1$. In other words, if the union of the directed graphs of $C(t)$ has a directed spanning tree, so does the union of the directed graphs of the corresponding transition matrices. Also note from Lemma 2.24 that each $\Phi_C(t_i, t_{i-1})$ is a row-stochastic matrix with positive diagonal entries. The proof then follows from Lemma 2.26. ■

Lemma 2.29. *If the union of a set of directed graphs $\{\mathcal{G}_n[k_1], \mathcal{G}_n[k_2], \ldots, \mathcal{G}_n[k_p]\}$, where $\mathcal{G}_n[k_j] \in \overline{\mathcal{G}_n}$, has a directed spanning tree, then the matrix product $\mathcal{D}_n[k_p] \cdots \mathcal{D}_n[k_2]\mathcal{D}_n[k_1]$ is SIA, where $\mathcal{D}_n[k_j]$ is the row-stochastic matrix in (2.5) associated with each directed graph $\mathcal{G}_n[k_j]$.*

Proof: From Lemma 2.25, we know that $\mathcal{D}_n[k_p] \cdots \mathcal{D}_n[k_2]\mathcal{D}_n[k_1] \geq \gamma \sum_{j=1}^{p} \mathcal{D}_n[k_j]$ for some $\gamma > 0$. Because the union of $\{\mathcal{G}_n[k_1], \mathcal{G}_n[k_2], \ldots, \mathcal{G}_n[k_p]\}$ has a directed spanning tree, we know that the directed graph of $\sum_{j=1}^{p} \mathcal{D}_n[k_j]$ has a directed spanning tree, which in turn implies that the directed graph of the matrix product $\mathcal{D}_n[k_p]\cdots\mathcal{D}_n[k_2]\mathcal{D}_n[k_1]$ has a directed spanning tree. Also the matrix product $\mathcal{D}_n[k_p]\cdots\mathcal{D}_n[k_2]\mathcal{D}_n[k_1]$ is a row-stochastic matrix with positive diagonal entries because row-stochastic matrices with positive diagonal entries are closed under matrix multiplication. From Lemma 2.26, we know that the matrix product $\mathcal{D}_n[k_1]\mathcal{D}_n[k_2]\cdots\mathcal{D}_n[k_p]$ is SIA. ■

Lemma 2.30. *[253] Let $\mathcal{S} = \{S_1, S_2, \ldots, S_k\}$ be a finite set of SIA matrices with the property that for each sequence $S_{i_1}, S_{i_2}, \ldots, S_{i_j}$ of positive length, the matrix product $S_{i_j}S_{i_{j-1}}\cdots S_{i_1}$ is SIA. Then for each infinite sequence $S_{i_1}, S_{i_2}, \ldots$ there exists a column vector ν such that*

$$\lim_{j\to\infty} S_{i_j}S_{i_{j-1}}\cdots S_{i_1} = \mathbf{1}_n\nu^T. \tag{2.15}$$

In addition, when $\mathcal{S}$ is an infinite set, $\chi(W) < 1$, where $W = S_{k_1}S_{k_2}\cdots S_{k_{N_t+1}}$ and N_t is the number of different types of all $n \times n$ SIA matrices (see Appendix C for definitions). Furthermore, if there exists a constant $0 \le d < 1$ satisfying $\chi(W) \le d$, then (2.15) also holds.

2.3.1 Consensus Using a Continuous-time Algorithm

In this subsection, we will focus on demonstrating that under certain conditions, the existence of a directed spanning tree jointly is sufficient for consensus under dynamically changing interaction topologies using continuous-time algorithm (2.2).

To make the problem mathematically tractable, we use the notion of dwell time introduced in [97, 146]. The idea is that there is a finite, lower bound on the time between $\mathcal{A}_n(t)$ switches, and that between the switching times, $\mathcal{A}_n$ is constant. This implies that both $\mathcal{A}_n(t)$ and $\mathcal{L}_n(t)$ are piecewise constant. Equation (2.3) can therefore be rewritten as

$$\dot{\xi}(t) = -[\mathcal{L}_n(t_i) \otimes I_m]\xi(t), \quad t \in [t_i, t_i + \tau_i), \tag{2.16}$$

where $\tau_i > 0$ is the dwell time and $t_0, t_1, t_2, \ldots$ is the infinite time sequence such that $t_{i+1} - t_i = \tau_i$.

Let $\bar{\Theta}$ be a finite set of arbitrary positive numbers. Let Υ be an infinite set generated from the set $\bar{\Theta}$, which is closed under addition and multiplications by positive integers. We assume that $\tau_i \in \Upsilon$, $i = 0, 1, \ldots$. By choosing the set $\bar{\Theta}$ properly, the dwell time can be chosen from an infinite set Υ.

We have the following theorem for algorithm (2.2) under directed switching interaction topologies.

Theorem 2.31. *Let $t_1, t_2, \ldots$ be the infinite time sequence such that $\tau_i = t_{i+1} - t_i \in \Upsilon$, $i = 0, 1, \ldots$. Let $\mathcal{G}_n(t_i) \in \overline{\mathcal{G}_n}$ be the directed graph at time $t = t_i$ and $a_{ij}(t_i) \in \bar{\Psi}$, where $\bar{\Psi}$ is a finite set of arbitrary nonnegative numbers. Continuous-time algorithm (2.2) achieves consensus asymptotically if there exists an infinite sequence of contiguous, nonempty, uniformly bounded, time* *intervals* $[t_{i_j}, t_{i_{j+1}})$, $j = 1, 2, \ldots$, *starting at* $t_{i_1} = t_0$, *with the property that the union of the directed graphs across each such interval has a directed spanning tree. Furthermore, if the union of the directed graphs after some finite time does not have a directed spanning tree, then consensus cannot be achieved asymptotically.*

Proof: The set of all possible matrices $e^{-\mathcal{L}_n(t_i)\tau_i}$, where $\tau_i \in \Upsilon$, under dynamically changing interaction topologies and weights $a_{ij}(t)$, can be chosen or constructed by matrix multiplications from the set $\bar{E} = \{e^{-\mathcal{L}_n(t_i)\tau_i}, \tau_i \in \bar{\Theta}\}$. Clearly $\bar{E}$ is finite because $\overline{\mathcal{G}_n}$, $\bar{\Psi}$, and $\bar{\Theta}$ are finite.

Note that the union of the directed graphs across each time interval $[t_{i_j}, t_{i_{j+1}})$ has a directed spanning tree. Let $\{\mathcal{L}_n(t_{i_j}), \mathcal{L}_n(t_{i_j+1}), \ldots, \mathcal{L}_n(t_{i_{j+1}-1})\}$ be a set of nonsymmetrical Laplacian matrices corresponding to each directed graph in $[t_{i_j}, t_{i_{j+1}})$. Following Lemma 2.27, the matrix product $\Phi(t_{i_{j+1}}, t_{i_j}) = e^{-\mathcal{L}_n(t_{i_{j+1}-1})\tau_{i_{j+1}-1}} \cdots e^{-\mathcal{L}_n(t_{i_j+1})\tau_{i_j+1}} e^{-\mathcal{L}_n(t_{i_j})\tau_{i_j}}$, $j = 1, 2, \ldots$, is SIA. Because $\bar{E}$ is finite and $[t_{i_j}, t_{i_{j+1}})$ is uniformly bounded, the set of all matrices $\Phi(t_{i_{j+1}}, t_{i_j})$ is finite.

Note that the solution of (2.16) is given by

$$\xi(t) = [e^{-\mathcal{L}_n(t_{i_{k+1}})(t-t_{i_{k+1}})}\Phi(t_{i_{k+1}}, t_{i_k}) \cdots \Phi(t_{i_2}, t_{i_1}) \otimes I_m]\xi(0),$$

where i_{k+1} is the largest nonnegative integer satisfying $t_{i_{k+1}} \leq t$. It follows from the first statement of Lemma 2.30 that $\Phi(t_{i_{k+1}}, t_{i_k}) \cdots \Phi(t_{i_2}, t_{i_1}) \to \mathbf{1}_n \nu^T$, as $k \to \infty$. It also follows from Lemma 2.6 that $e^{-\mathcal{L}_n(t_{i_{k+1}})(t-t_{i_{k+1}})}$ is a row-stochastic matrix satisfying $e^{-\mathcal{L}_n(t_{i_{k+1}})(t-t_{i_{k+1}})}\mathbf{1}_n \nu^T = \mathbf{1}_n \nu^T$. Therefore, we conclude that $e^{-\mathcal{L}_n(t_{i_{k+1}})(t-t_{i_{k+1}})}\Phi(t_{i_{k+1}}, t_{i_k}) \cdots \Phi(t_{i_2}, t_{i_1}) \to \mathbf{1}_n \nu^T$, as $t \to \infty$, under the assumption of the theorem, which proves the first statement of the theorem.

For the second statement, if the union of the directed graphs after some finite time $\hat{t}$ does not have a directed spanning tree, then during the infinite time interval $[\hat{t}, \infty)$, there exist at least two vehicles such that there is no path in the union of the directed graphs that contains these two vehicles, which implies that the information states of these vehicles cannot agree. ∎

Corollary 2.32. *[97] Let* $t_1, t_2, \ldots$ *be the infinite time sequence such that* $t_{i+1} - t_i = \tau_D$, $i = 0, 1, \ldots$, *where* τ_D *is a constant. Let* $\mathcal{G}_n(t_i) \in \overline{\mathcal{G}_n}$ *be the undirected graph at time* $t = t_i$. *Suppose that* $a_{ij}(t_i) = \frac{1}{n}$ *if* $(j, i) \in \mathcal{E}_n$ *and* $a_{ij}(t_i) = 0$ *otherwise,* $\forall j \neq i$, *at time* $t = t_i$. *Continuous-time algorithm* (2.2) *achieves consensus asymptotically if there exists an infinite sequence of contiguous, nonempty, uniformly bounded, time intervals* $[t_{i_j}, t_{i_{j+1}})$, $j = 1, 2, \ldots$, *starting at* $t_{i_1} = t_0$, *with the property that the union of the undirected graphs across each such interval is connected.*

Consider, on the other hand, the more realistic assumption that $\mathcal{A}_n(t)$ and hence $\mathcal{L}_n(t)$ are piecewise continuous.

Theorem 2.33. *Suppose that* $\mathcal{A}_n(t) = [a_{ij}(t)] \in \mathbb{R}^{n\times n}$ *in* (2.3) *is piecewise continuous and its nonzero and hence positive entries are both uniformly lower and upper bounded (i.e.,* $a_{ij} \in [\underline{a}, \bar{a}]$, *where* $0 < \underline{a} < \bar{a}$, *if* $(j,i) \in \mathcal{E}_n$ *and* $a_{ij} = 0$ *otherwise). Let* $t_0, t_1, \ldots$ *be the time sequence corresponding to the times at which* $\mathcal{A}_n(t)$ *switches, where it is assumed that* $t_i - t_{i-1} \geq t_L$, $\forall i = 1, 2, \ldots$ *with* t_L *a positive constant.*[4] *Continuous-time algorithm* (2.2) *achieves consensus asymptotically if there exists an infinite sequence of contiguous, nonempty, uniformly bounded time-intervals* $[t_{i_j}, t_{i_{j+1}})$, $j = 1, 2, \ldots$, *starting at* $t_{i_1} = t_0$, *with the property that the union of directed graphs* $\mathcal{G}_n(t)$ *across each such interval has a directed spanning tree.*

Proof: With $t_0, t_1, \ldots$ the switching time sequence, each interval $[t_{i-1}, t_i)$ can be divided into a finite or infinite number of subintervals such that the length of each subinterval is greater than or equal to t_L but less than or equal to $t_M = 2t_L$, and on each subinterval the directed graph $\mathcal{G}_n(t)$ is time invariant and the weights $a_{ij}(t)$ are continuous. Relabel these subintervals as $s_0, s_1, \ldots$, where $s_0 = t_0$. Note that for $t \in [s_{j-1}, s_j]$, $-\mathcal{L}_n(t)$ is continuous and satisfies the assumption of Lemma 2.24.

With (2.2), the solution of (2.1) can be denoted as $\xi(t) = [\Phi(t, t_0) \otimes I_m]\xi(t_0) = [\Phi(t, s_j)\Phi(s_j, s_{j-1}) \cdots \Phi(s_1, s_0) \otimes I_m]\xi(s_0)$, where Φ is the transition matrix corresponding to $-\mathcal{L}_n(t)$. Note that all nonzero and hence positive entries of $\mathcal{A}_n(t)$ are both uniformly lower and upper bounded. It follows that all entries of $\mathcal{L}_n(t)$ are within a compact set. If $\mathcal{G}_n(t)$ is switching with time, there are a finite number of possible interaction topologies. For each possible interaction topology, note that $\mathcal{L}_n(t)$ has the same structure in the sense that positive, zero, and negative entries are in the same places for $t \in [s_{j-1}, s_j]$. From Lemma 2.24, each transition matrix $\Phi(s_j, s_{j-1})$ is a row-stochastic matrix, where $t_L \leq s_j - s_{j-1} \leq t_M$, and $\Phi(s_j, s_{j-1})$ is of constant type over this interval for each possible interaction topology. Combining the above arguments with the fact that $\Phi(s_j, s_{j-1})$ is a continuous function of $\ell_{ij}(t)$ for $t \in [s_{j-1}, s_j]$, we see that each nonzero and hence positive entry of $\Phi(s_j, s_{j-1})$ is uniformly lower bounded for each possible interaction topology. It is straightforward to see that there are only finitely many types for $\Phi(s_j, s_{j-1})$. We know that there exists a sequence of unions of the directed graphs across some time intervals and each union is uniformly bounded and has a directed spanning tree. Thus the transition matrix $\Phi^{(k)}$ for each union is a product of finitely many matrices $\Phi(s_{k_i}, s_{k_{i-1}})$. From Lemma 2.22, the type of $\Phi^{(k)}$ is uniquely decided by the order and type of each element in its product. Also, from Lemma 2.28, we know that each $\Phi^{(k)}$ is SIA. In addition, noting that the directed graph of each $\Phi^{(k)}$ has a directed spanning tree, we see that any finite number of products of $\Phi^{(k)}$ is also SIA according to the second statement of Lemma 2.26. Noting that $\Phi^{(k)}$ can have only finitely many types, we see that for each type of $\Phi^{(k)}$, its nonzero and hence positive entries are

[4] That is, it is assumed that the interaction topology and weights $a_{ij}(t)$ cannot switch infinitely fast.

uniformly lower bounded. Let $W = \Phi^{(j_1)}\Phi^{(j_2)}\dots\Phi^{(j_{N_t}+1)}$. From the second argument of Lemma 2.30, we know that $\chi(W) < 1$. Note that W can have only finite many types, denoted as W_t. To show that $\chi(W) \leq d < 1$, it is sufficient to show that for each type, there exists a $0 \leq d_i < 1$ such that $\chi(W) \leq d_i$. This can be verified by noting that the nonzero and hence positive entries of W are uniformly lower bounded for each type. Let $d = \max\{d_1, d_2, \dots, d_{W_t}\}$. It is obvious that $\chi(W) \leq d$. From the third statement of Lemma 2.30, we can show that $\Phi(t, t_0) \to \mathbf{1}_n\nu^T$, where ν is a column vector, under the condition of the theorem. This proves the theorem. ■

Remark 2.34. The leader-following scenario is a special case of Theorem 2.33. If there is one vehicle in the team that does not have any incoming edge, but the union of the directed graphs has a directed spanning tree, then this vehicle must be the root of the directed spanning tree, or in other words, the team leader. Because consensus is guaranteed, the information states of the other vehicles asymptotically converge to the information state of the team leader. □

We have the following theorem for algorithm (2.9) under directed switching interaction topologies.

Theorem 2.35. *Given consensus algorithm* (2.9), *under the assumption of Theorem 2.33, if* $\|w_i - w_j\|$ *is uniformly bounded, so is* $\|\xi_i - \xi_j\|$, $\forall i \neq j$.

Proof: We assume that $\xi_i \in \mathbb{R}$ and $w_i \in \mathbb{R}$ for simplicity. However, the proof holds for $\xi_i \in \mathbb{R}^m$ and $w_i \in \mathbb{R}^m$ by introducing the Kronecker product. Let $\xi_{ij} \triangleq \xi_i - \xi_j$, $w_{ij} \triangleq w_i - w_j$, $\tilde{\xi} = [\xi_{12}, \xi_{13}, \dots, \xi_{1n}]^T$, and $\tilde{w} = [w_{12}, w_{13}, \dots, w_{1n}]^T$. With (2.9), (2.1) can be written as

$$\dot{\tilde{\xi}} = Q(t)\tilde{\xi} + \tilde{w}, \tag{2.17}$$

where $Q(t) \in \mathbb{R}^{(n-1)\times(n-1)}$ is switching with time. Under the assumption of Theorem 2.33, $\tilde{\xi}(t) \to 0$ uniformly in t_0, as $t \to \infty$, when $w_i = 0$. However, unlike the proof of Theorem 2.14, for a linear time-varying system (LTV) like (2.17), asymptotical stability does not necessarily imply bounded-input bounded-state stability (see Appendix E). We will show that (2.17) is *uniformly* asymptotically stable under the condition that $w_i = 0$. With (2.2), the solution of (2.1) is given by $\xi(t) = \Phi(t, 0)\xi(0)$, $t \geq 0$, where the transition matrix $\Phi(t, 0) \in \mathbb{R}^{n\times n}$ is a row-stochastic matrix as shown in Lemma 2.23. As a result, it follows that $\xi_i(t) = \sum_{j=1}^n \beta_{ij}(t)\xi_j(0)$, where $\beta_{ij}(t) \geq 0$ and $\sum_{j=1}^n \beta_{ij}(t) = 1$. Therefore, it follows that $\max_j \xi_j(t) \leq \max_j \xi_j(0)$ and $\min_j \xi_j(t) \geq \min_j \xi_j(0)$, which in turn imply that

$$\max_{j\neq 1} |\xi_{1j}(t)| \leq \max_j \xi_j(0) - \min_j \xi_j(0). \tag{2.18}$$

Let $\underline{j} = \text{argmin}_j \xi_j(0)$ and $\bar{j} = \text{argmax}_j \xi_j(0)$. Note that

$$\max_j \xi_j(0) - \min_j \xi_j(0) = \left|\xi_{\bar{j}\underline{j}}(0)\right| = \left|\xi_{1\underline{j}}(0) - \xi_{1\bar{j}}(0)\right|$$

$$\leq \left|\xi_{1\underline{j}}(0)\right| + \left|\xi_{1\bar{j}}(0)\right| \leq 2\max_{j\neq 1}|\xi_{1j}(0)|\,. \tag{2.19}$$

Combining (2.18) and (2.19) gives $\max_{j\neq 1}|\xi_{1j}(t)| \leq 2\max_{j\neq 1}|\xi_{1j}(0)|$, which implies $\left\|\tilde{\xi}(t)\right\|_\infty \leq 2\left\|\tilde{\xi}(0)\right\|_\infty$, $\forall t \geq 0$, when $w_i = 0$. Therefore, (2.17) is uniformly stable when $w_i = 0$. Combining the above arguments, it follows that (2.17) is uniformly asymptotically stable when $w_i = 0$, which in turn implies that (2.17) is uniformly exponentially stable from Theorem E.3. According to Lemma E.6, it follows that if $\|\tilde{w}\|$ is uniformly bounded, so is $\left\|\tilde{\xi}\right\|$. Equivalently, it follows that if $\|w_i - w_j\|$ is uniformly bounded, so is $\|\xi_i - \xi_j\|$, $\forall i \neq j$. ■

Remark 2.36. Theorem 2.35 implies that with (2.9), (2.1) is ISS under the assumption of the theorem. □

2.3.2 Consensus Using a Discrete-time Algorithm

In this subsection, we will show that under certain conditions the existence of a directed spanning tree jointly is sufficient for consensus under dynamically changing interaction topologies using discrete-time algorithm (2.4).

We have the following theorem for algorithm (2.4) under directed switching interaction topologies.

Theorem 2.37. *Let $\mathcal{G}_n[k] \in \overline{\mathcal{G}_n}$ be the directed graph at discrete-time index k. Also let $d_{ij}[k] \in \bar{\Psi}$, where $\bar{\Psi}$ is a finite set of nonnegative numbers that are no larger than one. Discrete-time algorithm (2.4) achieves consensus asymptotically if there exists an infinite sequence of contiguous, nonempty, uniformly bounded time intervals $[k_j, k_{j+1})$, $j = 1, 2, \ldots$, starting at $k_1 = 0$, with the property that the union of the directed graphs across each such interval has a directed spanning tree. Furthermore, if the union of the directed graphs after some finite time does not have a directed spanning tree, then consensus cannot be achieved asymptotically.*

Proof: The set of all possible matrices $\mathcal{D}_n[k]$ under dynamically changing interaction topologies and weights $d_{ij}[k]$ is finite because both $\overline{\mathcal{G}_n}$ and $\bar{\Psi}$ are finite. Note that the union of the directed graphs across each time interval $[k_j, k_{j+1})$ has a directed spanning tree. Let $\{\mathcal{D}_n[k_j], \mathcal{D}_n[k_j+1], \ldots, \mathcal{D}_n[k_{j+1}-1]\}$ be the set of row-stochastic matrices corresponding to each directed graph in $[k_j, k_{j+1})$. Following Lemma 2.29, the matrix product $\mathcal{D}_n[k_{j+1}-1]\cdots\mathcal{D}_n[k_j+1]\mathcal{D}_n[k_j]$, $j = 1, 2, \ldots$, is SIA. Then by following a proof similar to that in Theorem 2.31, the first statement of the theorem is demonstrated.

If the union of the directed graphs after some finite k does not have a directed spanning tree, then during the infinite time interval $[k, \infty)$, there exist at least two vehicles such that there is no path in the union of the directed graphs that contains these two vehicles, which implies that the information states of these vehicles cannot agree. ∎

Corollary 2.38. *[97] Let $\mathcal{G}_n[k] \in \overline{\mathcal{G}_n}$ be undirected graph at discrete-time index k. Also let $d_{ij}[k] \in \bar{\Psi}$, where $\bar{\Psi}$ is a finite set of nonnegative numbers that are no larger than one. Discrete-time algorithm (2.4) achieves consensus asymptotically if there exists an infinite sequence of contiguous, nonempty, uniformly bounded time intervals $[k_j, k_{j+1})$, $j = 1, 2, \ldots$, starting at $k_1 = 0$, with the property that the union of the undirected graphs across each such interval is connected.*

Again, consider, on the other hand, the more realistic assumption that $\mathcal{D}_n[k]$ is piecewise constant.

Theorem 2.39. *Suppose that the row-stochastic matrix $\mathcal{D}_n[k] = d_{ij}[k] \in \mathbb{R}^{n\times n}$ in (2.5) is piecewise constant and its nonzero and hence positive entries are both uniformly lower and upper bounded (i.e., $d_{ii} \in [\underline{d}, 1]$, where $0 < \underline{d} < 1$, and for all $i \neq j$, $d_{ij} \in [\underline{c}, \bar{c}]$, where $0 < \underline{c} < \bar{c} < 1$ if $(j, i) \in \mathcal{E}_n$ and $d_{ij} = 0$ otherwise). Discrete-time algorithm (2.4) achieves consensus asymptotically if there exists an infinite sequence of contiguous, nonempty, uniformly bounded time intervals $[k_j, k_{j+1})$, $j = 1, 2, \ldots$, starting at $k_1 = 0$, with the property that the union of the directed graphs across each such interval has a directed spanning tree.*

Proof: Note that $\mathcal{D}_n[k]$ is a row-stochastic matrix with positive diagonal entries. In addition, for each possible interaction topology, $\mathcal{D}_n[k]$ is of the same type and its nonzero and hence positive entries are uniformly lower bounded. We know that there exists a sequence of unions of the directed graphs across some time intervals, and each union is uniformly bounded and has a directed spanning tree. Let $\mathcal{D}_n^{(i)}$ be the product of $\mathcal{D}_n[k]$ over the ith union. Note that each $\mathcal{D}_n^{(i)}$ is SIA from Lemma 2.26. As a result, the proof follows the same reasoning as that of Theorem 2.33 with $\mathcal{D}_n^{(i)}$ playing the role of $\Phi^{(k)}$. ∎

2.3.3 Simulation Results

In this subsection, we simulate information consensus for five vehicles under dynamically changing interaction topologies using discrete-time algorithm (2.4) and continuous-time algorithm (2.16), respectively.

For simplicity, we constrain the possible interaction topologies for these five vehicles within the set $\mathcal{G}_5^s = \{\mathcal{G}_{5(1)}, \mathcal{G}_{5(2)}, \mathcal{G}_{5(3)}, \mathcal{G}_{5(4)}, \mathcal{G}_{5(5)}\}$, as shown in Fig. 2.5, which is obviously a subset of $\overline{\mathcal{G}_5}$. For discrete-time algorithm (2.4), we assume that the interaction topology switches randomly in $\mathcal{G}_5^s$ at each discrete-time index k, where $k = 0, 1, 2, \ldots$ and the sample period is 0.5 seconds. For the continuous-time algorithm, we assume that the interaction topology switches randomly in $\mathcal{G}_5^s$ at each random time $t = t_k$, $k = 0, 1, 2, \ldots$. The nonzero weights d_{ij} in (2.4) and a_{ij} in (2.16) are chosen arbitrarily *a priori* for each directed graph in $\mathcal{G}_5^s$ to satisfy the constraints of Theorems 2.37 and 2.31, respectively.

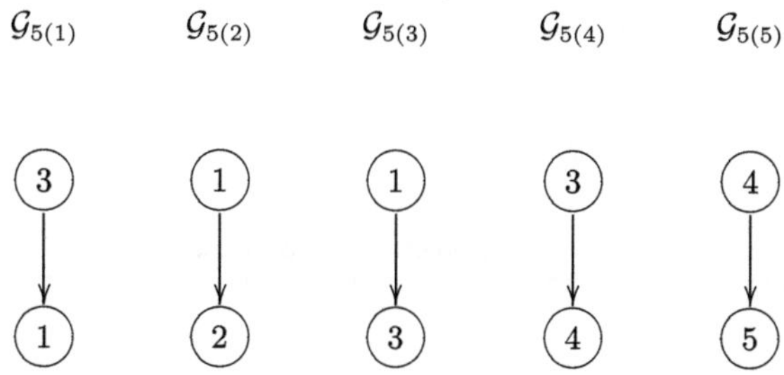

Fig. 2.5. Possible interaction topologies for five vehicles

Note that each directed graph in $\mathcal{G}_5^s$ does not have a directed spanning tree. However, as can be seen from Fig. 2.6, the union of these graphs has a directed spanning tree. Because the switching among the directed graphs in $\mathcal{G}_5^s$ is random, the condition for consensus will be generically satisfied. Simulation results show that asymptotic consensus is achieved using both discrete-time algorithm (2.4) and continuous-time algorithm (2.16).

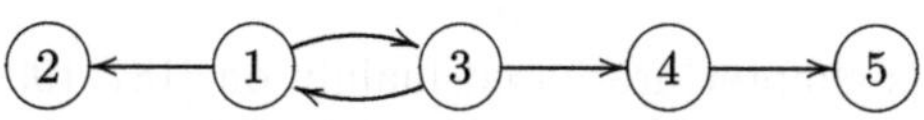

Fig. 2.6. The union of $\mathcal{G}_5^s$

The initial information state is selected arbitrarily as $0.2 * i$, $i = 1, \ldots, 5$. Figure 2.7 shows the consensus results using both discrete-time algorithm (2.4) and continuous-time algorithm (2.16). Note that consensus is reached in both cases.

Consider now a leader-following scenario where the directed graph switches in $\mathcal{G}_5^\ell \triangleq \mathcal{G}_5^s \setminus \mathcal{G}_{5(1)}$. As a result, there is no edge from node 3 to node 1. In this case, the union of the directed graphs still has a directed spanning tree. However, unlike the previous case, there is no incoming edge to node 1. Figure 2.8 shows the consensus results using both discrete-time algorithm (2.4) and continuous-time algorithm (2.16). Note that $\xi_i(t)$ (respectively, $\xi_i[k]$), $i = 2, \ldots, 5$, converges asymptotically to $\xi_1(0)$ (respectively, $\xi_1[0]$) as expected.

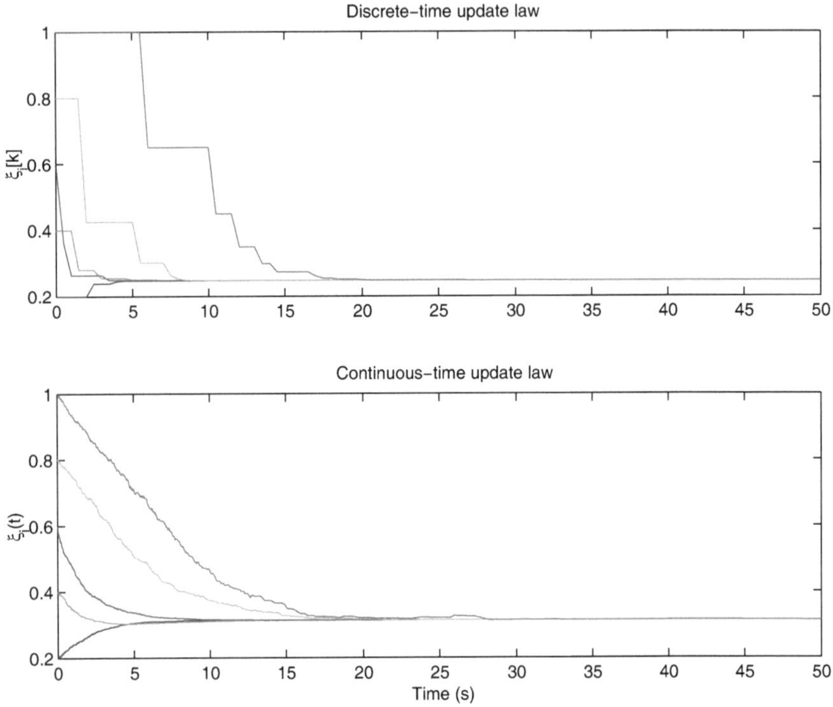

Fig. 2.7. Consensus with $\mathcal{G}_5[k]$ and $\mathcal{G}_5(t_k)$ randomly switching from $\mathcal{G}_5^s$

2.4 Notes

The results in this chapter are based mainly on [187, 190, 192, 193]. Variant forms of continuous-time consensus algorithm (2.2) are proposed in [69, 97, 126, 158, 190], and variant forms of discrete-time consensus algorithm (2.4) are proposed in [97, 145, 190, 236]. Lemma 2.25 is from [97]. Lemma 2.30 is originally from Lemma 4 and the concluding remarks in [253], but its first part is restated in [97]. Consensus over an undirected interaction topology is addressed in [97]. In particular, Corollaries 2.32 and 2.38 are from [97]. Average consensus is considered in [158]. In particular, Corollary 2.9 is from [158]. For other theoretical results and extensions of algorithms (2.2) and (2.4), see the references in Section 1.2.

Acknowledgment is given to

©2005 IEEE. Reprinted, with permission, from Wei Ren and Randal W. Beard, "Consensus seeking in multiagent systems under dynamically changing interaction topologies," *IEEE Transactions on Automatic Control*, vol. 50, no. 5, pp. 655–661, May 2005.

©2005 IEEE. Reprinted, with permission, from Wei Ren, Randal W. Beard, and Derek B. Kingston, "Multi-agent Kalman consensus with relative

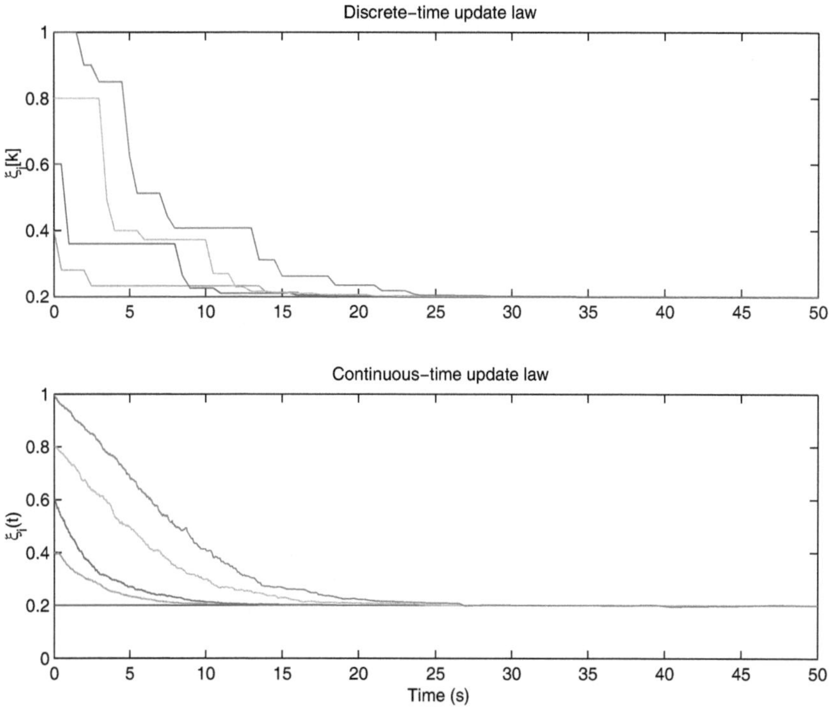

Fig. 2.8. Consensus with $\mathcal{G}_5[k]$ and $\mathcal{G}_5(t_k)$ randomly switching from $\bar{\mathcal{G}}_5^{\ell}$

uncertainty," *Proceedings of the American Control Conference*, pp. 1865–1870, Portland, OR, 2005.

3

Consensus Tracking with a Reference State

This chapter studies consensus tracking algorithms for single-integrator dynamics, where the information states of all vehicles approach a constant or time-varying consensus reference state. We first propose and analyze a consensus tracking algorithm with a constant consensus reference state. We then propose consensus tracking algorithms with a time-varying consensus reference state and show conditions under which consensus is reached on the time-varying consensus reference state in four cases, a directed fixed interaction topology, a directed switching interaction topology, bounded control inputs, and vehicle information feedback. The time-varying consensus reference state may be an exogenous signal or evolve according to a nonlinear model. These consensus tracking algorithms are also extended to achieve relative state deviations between vehicles.

3.1 Problem Statement

With algorithm (2.2), the consensus equilibrium is a weighted average of all vehicles' initial information states and hence constant. The constant consensus equilibrium, which depends on the interaction topology and weights a_{ij}, might be *a priori* unknown. The assumption of constant consensus equilibrium might not be appropriate when each vehicle's information state evolves over time, as occurs in formation control problems, where the formation evolves in two-, or three-dimensional space. In addition, algorithm (2.2) ensures only that the information states converge to a common value but does not allow specifying a particular value. Although this paradigm is useful for applications such as cooperative rendezvous where there is not a single correct value, there are many applications where there is a desired, or reference, information state. In this case, the convergence issues include both convergence to a common value, as well as convergence of the common state to its reference value.

Suppose that the team consists of the same n vehicles as in Chapter 2, together with an additional (virtual) vehicle labeled $n + 1$, which acts as the unique (virtual) leader of the team. We call vehicle $n+1$ the *team leader* and vehicles $1,\ldots,n$ the *followers*. Vehicle $n+1$ has the information state $\xi_{n+1} \triangleq \xi^r \in \mathbb{R}^m$, where ξ^r represents the *consensus reference state*. The consensus reference state satisfies

$$\dot{\xi}^r = f(t, \xi^r), \tag{3.1}$$

where $f(\cdot,\cdot)$ is bounded, piecewise continuous in t and locally Lipschitz in ξ^r.

The main objective of this chapter is to propose and analyze consensus algorithms to ensure that each vehicle in the team reaches consensus on a constant or time-varying consensus reference state that evolves with time or according to nonlinear dynamics when the consensus reference state is available only to a subgroup of the followers, called *subgroup leaders*. The *consensus tracking* problem with a consensus reference state is *solved* if $\xi_i(t) \to \xi^r(t)$, $i=1,\ldots,n$, as $t \to \infty$.

We use directed graph $\mathcal{G}_{n+1} \triangleq (\mathcal{V}_{n+1}, \mathcal{E}_{n+1})$ to model the interaction topology for the $n+1$ vehicles, where $\mathcal{V}_{n+1} = \{1,\ldots,n+1\}$ is the node set and $\mathcal{E}_{n+1} \subseteq \mathcal{V}_{n+1} \times \mathcal{V}_{n+1}$ is the edge set. Let $\mathcal{A}_{n+1} = [a_{ij}] \in \mathbb{R}^{(n+1)\times(n+1)}$ be the adjacency matrix associated with $\mathcal{G}_{n+1}$, where $a_{ij} > 0$ if $(j,i) \in \mathcal{E}_{n+1}$ and $a_{ij} = 0$ otherwise for all $i = 1,\ldots,n$ and $j = 1,\ldots,n+1$, and $a_{(n+1)j} = 0$ for all $j = 1,\ldots,n+1$. Also let $\mathcal{L}_{n+1} = [\ell_{ij}] \in \mathbb{R}^{(n+1)\times(n+1)}$ be the nonsymmetrical Laplacian matrix associated with $\mathcal{G}_{n+1}$, where $\ell_{ij} = -a_{ij}$, $i \neq j$, $\ell_{ii} = \sum_{j=1, j\neq i}^{n+1} a_{ij}$, $i,j = 1,\ldots,n+1$.

Remark 3.1. When we need to focus only on vehicles 1 to n (*i.e.*, the followers), we still use $\mathcal{G}_n \triangleq (\mathcal{V}_n, \mathcal{E}_n)$ to model the interaction topology for the n followers, as in Chapter 2. In addition, $\mathcal{A}_n \in \mathbb{R}^{n\times n}$ and $\mathcal{L}_n \in \mathbb{R}^{n\times n}$ are still, respectively, the adjacency matrix and the nonsymmetrical Laplacian matrix associated with $\mathcal{G}_n$, as in Chapter 2. □

3.2 Constant Consensus Reference State

In this section, we consider the case where the consensus reference state ξ_r is constant, that is, $f(t,\xi^r) \equiv 0$ in (3.1). For information states with single-integrator dynamics (2.1), a consensus tracking algorithm with a constant consensus reference state is given as

$$u_i = -\sum_{j=1}^{n} a_{ij}(\xi_i - \xi_j) - a_{i(n+1)}(\xi_i - \xi^r), \quad i = 1,\ldots,n, \tag{3.2}$$

where $\xi_i \in \mathbb{R}^m$ is the ith information state and a_{ij}, $i = 1,\ldots,n$, $j = 1,\ldots,n+1$, is the (i,j) entry of the adjacency matrix $\mathcal{A}_{n+1} \in \mathbb{R}^{(n+1)\times(n+1)}$. Note that algorithm (3.2) corresponds to the leader-following case in Chapter 2 (see Remark 2.34).

We have the following theorem for consensus tracking with a constant consensus reference state.

Theorem 3.2. *Suppose that $\mathcal{A}_{n+1}$ is constant. The consensus tracking problem with a constant consensus reference state is solved with algorithm* (3.2) *if and only if directed graph $\mathcal{G}_{n+1}$ has a directed spanning tree.*[1]

Proof: Noting that ξ^r is constant gives $\dot{\xi}^r \equiv 0$. With (3.2), (2.1) can be written in matrix form as

$$\dot{\underline{\xi}} = -(\mathcal{L}_{n+1} \otimes I_m)\underline{\xi},$$

where $\underline{\xi} \triangleq [\xi_1^T, \ldots, \xi_n^T, \xi^{r^T}]^T$ and $\mathcal{L}_{n+1} \in \mathbb{R}^{(n+1)\times(n+1)}$ is the nonsymmetrical Laplacian matrix. Note that all entries of the last row of $\mathcal{L}_{n+1}$ are zero and the directed graph of $\mathcal{L}_{n+1}$ has a directed spanning tree if and only if $\mathcal{G}_{n+1}$ has a directed spanning tree. Then from the statements (ii) and (iii) of Lemma 2.11 with $\mathcal{L}_{n+1}$ and $\underline{\xi}$ playing the roles of $\mathcal{L}_p$ and z, respectively, it follows that $\xi_i(t) \to \xi^r(0)$, $i = 1, \ldots, n$, as $t \to \infty$, if and only if $\mathcal{G}_{n+1}$ has a directed spanning tree. Equivalently, it follows that $\xi_i(t) \to \xi^r$, $i = 1, \ldots, n$, as $t \to \infty$, because ξ^r is constant. ∎

Example 3.3. Theorem 3.2 is illustrated by the following example. Consider a team of $n = 4$ vehicles together with the virtual team leader, vehicle 5. The virtual team leader has a constant consensus reference state $\xi^r \equiv 1$. Figures 3.1a–d show, respectively, the interaction topologies among the five vehicles for Subcases (a)–(d), where node ξ^r denotes the virtual team leader. In Subcase (a), we let $a_{15} = 1$ and $a_{j5} = 0$, $\forall j \neq 1$, which corresponds to the case where ξ^r is available to only vehicle 1. In Subcase (b), we let $a_{j5} = 1$, $j = 1, \ldots, 4$, which corresponds to the case where ξ^r is available to all followers. In Subcase (c), we let $a_{35} = a_{45} = 1$ and $a_{j5} = 0$, $\forall j \notin \{3, 4\}$, which corresponds to the case where ξ^r is available only to vehicles 3 and 4. In Subcase (d), we let $a_{45} = 1$ and $a_{j5} = 0$, $\forall j \neq 4$, which corresponds to the case where ξ^r is available only to vehicle 4.

Figures 3.2a–d show, respectively, the information states of all vehicles using (3.2). Note that ξ_i converges to ξ^r in each subcase except Subcase (d). Also note that node ξ^r has a directed path to all other nodes in Subcases (a), (b), and (c). However, there does not exist a directed path from node ξ^r to all other nodes in Subcase (d). Note that in Subcase (c), neither 3 nor 4 has a directed path to all other followers. However, as shown above, ξ_i still approaches ξ^r in this case. □

[1] Note that vehicle $n + 1$ has the state ξ^r and all entries of the last row of $\mathcal{A}_{n+1}$ are zero. From Lemma 2.11, the condition that $\mathcal{G}_{n+1}$ has a directed spanning tree is equivalent to the condition that vehicle $n + 1$ is the only vehicle that has a directed path to all other vehicles on the team.

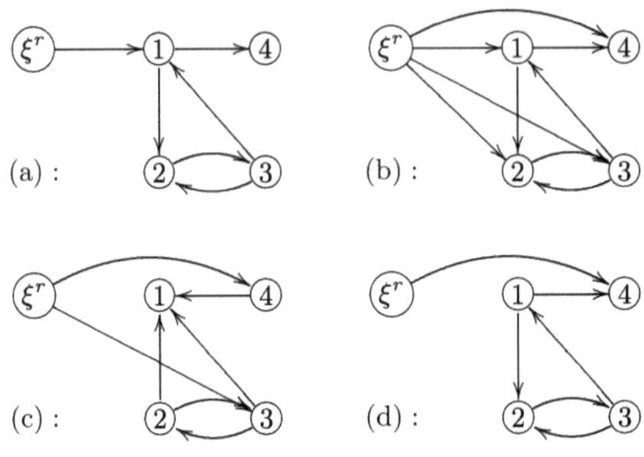

Fig. 3.1. Interaction topologies among the five vehicles, where the constant consensus reference state ξ^r may be available to one or more followers on the team

3.3 Time-varying Consensus Reference State

In this section, we assume that the consensus reference state might be a time-varying exogenous signal or evolves according to certain nonlinear dynamics.

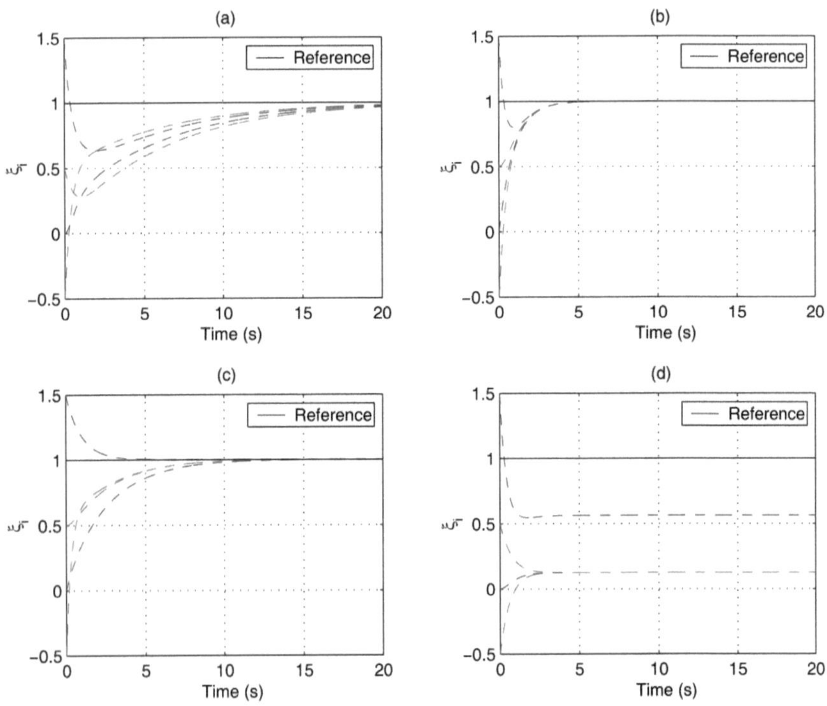

Fig. 3.2. Consensus tracking with a constant consensus reference state using (3.2)

Suppose that ξ^r satisfies the dynamics given by (3.1). We first show that algorithm (3.2) is not sufficient for consensus tracking in a time-varying consensus reference state, as illustrated by the following example.

Example 3.4. Let $\xi^r = \cos(t)$ and consider the four subcases in Example 3.3. As shown in Fig. 3.3, with algorithm (3.2), none of the subcases solves the consensus tracking problem with a time-varying consensus reference state. □

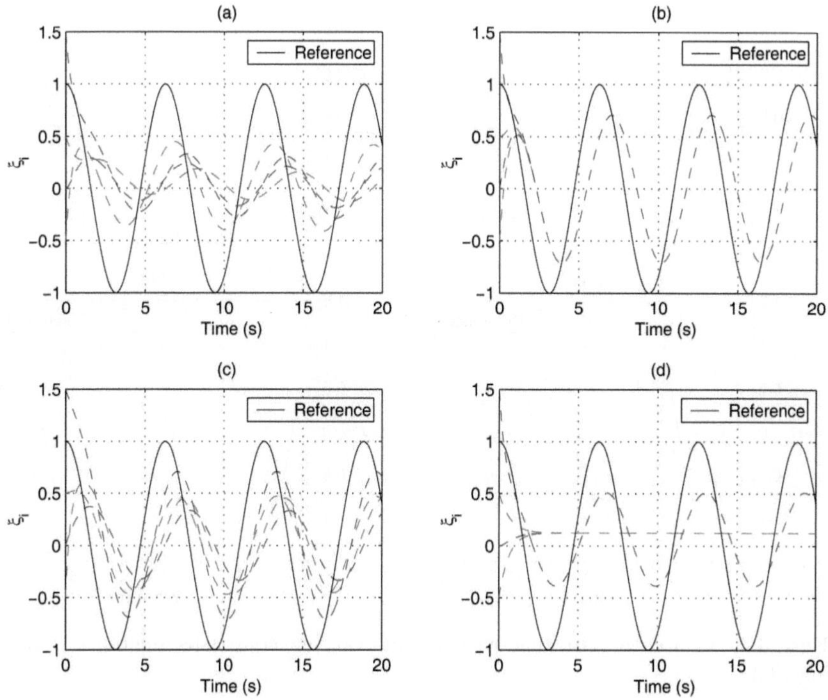

Fig. 3.3. Consensus tracking with a time-varying consensus reference state using (3.2)

We then consider the following algorithm:

$$u_i = a_{i(n+1)}\dot{\xi}^r - a_{i(n+1)}\alpha_i(\xi_i - \xi^r) - \sum_{j=1}^{n} a_{ij}(\xi_i - \xi_j), \quad i = 1, \ldots, n, \quad (3.3)$$

where α_i are positive constant scalars and a_{ij} is the (i,j) entry of the adjacency matrix $\mathcal{A}_{n+1}$ with the additional assumption that $a_{i(n+1)} = 1$ if $(n+1, i) \in \mathcal{E}_{n+1}$ and $a_{i(n+1)} = 0$ otherwise.

Example 3.5. Algorithm (3.3) is illustrated by the following example. Let $\xi^r = \cos(t)$ and consider the four subcases in Example 3.3. As shown in Fig. 3.4, with algorithm (3.3), the consensus tracking problem is solved only in Subcase (b). □

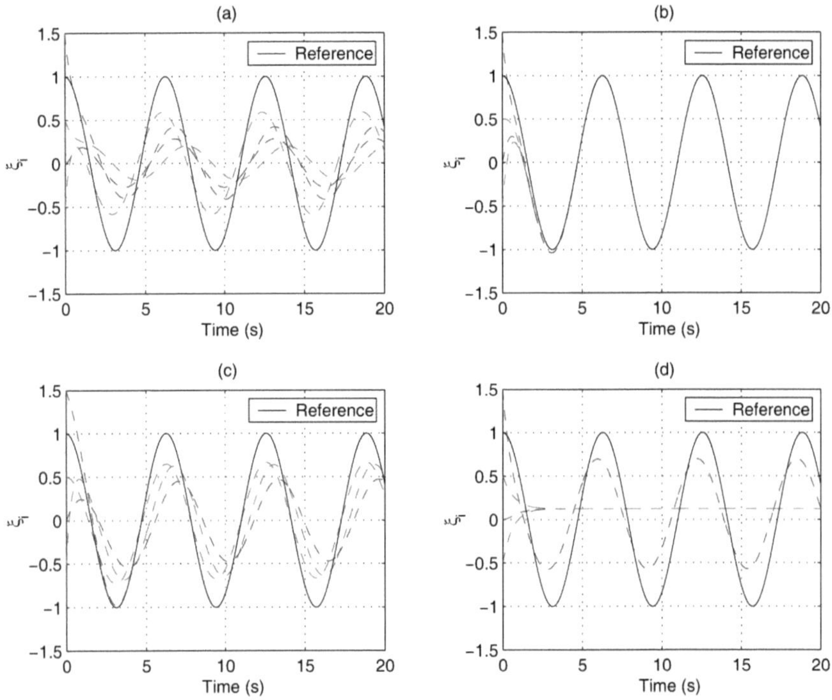

Fig. 3.4. Consensus tracking with a time-varying consensus reference state using (3.3)

We have the following theorem for consensus tracking in a time-varying consensus reference state using algorithm (3.3):

Theorem 3.6. *Suppose that $\mathcal{A}_{n+1}$ is constant. If $a_{i(n+1)} \equiv 1$, $i = 1, \ldots, n$, then the consensus tracking problem with a time-varying consensus reference state is solved with algorithm* (3.3).

Proof: With (3.3), (2.1) can be written in matrix form as

$$\dot{\tilde{\xi}} = -[(\mathcal{L}_n + Q) \otimes I_m]\tilde{\xi},$$

where $Q = \text{diag}(\alpha_1, \ldots, \alpha_n) \in \mathbb{R}^{n\times n}$, $\mathcal{L}_n = [\ell_{ij}]$ is the $n \times n$ nonsymmetrical Laplacian matrix (*i.e.*, $\ell_{ij} = -a_{ij}$, $i \neq j$, and $\ell_{ii} = \sum_{j=1, j\neq i}^{n} a_{ij}$), and $\tilde{\xi} \triangleq$

$[\tilde{\xi}_1^T, \ldots, \tilde{\xi}_n^T]^T$ with $\tilde{\xi}_i \triangleq \xi_i - \xi^r$. Note that $\mathcal{L}_n$ satisfies the property (B.2) with $p = n$. From Theorem C.1, it is straightforward to see that all eigenvalues of $-(\mathcal{L}_n + Q)$ have negative real parts. Therefore, it follows that $\tilde{\xi}(t) \to 0$, that is, $\xi_i(t) \to \xi^r(t)$, $i = 1, \ldots, n$, as $t \to \infty$. ■

Remark 3.7. Note that Theorem 3.6 does not rely on the interaction topology among the followers. Even if there is no information exchange among the followers (*i.e.*, $\mathcal{A}_n = 0_{n\times n}$), algorithm (3.3) still solves the consensus tracking problem with a time-varying consensus reference state, as long as ξ^r and $\dot{\xi}^r$ are available to all followers. However, this requirement is rather restricted. □

3.3.1 Fundamental Consensus Tracking Algorithm

In this subsection, we consider the case where ξ^r and $\dot{\xi}^r$ are available only to a subgroup of the followers. We propose a consensus tracking algorithm as

$$\begin{aligned} u_i = {} & \frac{1}{\eta_i(t)} \sum_{j=1}^{n} a_{ij}(t)[\dot{\xi}_j - \gamma(\xi_i - \xi_j)] \\ & + \frac{1}{\eta_i(t)} a_{i(n+1)}(t)[\dot{\xi}^r - \gamma(\xi_i - \xi^r)], \quad i = 1, \ldots, n, \end{aligned} \tag{3.4}$$

where $a_{ij}(t)$, $i = 1, \ldots, n$, $j = 1, \ldots, n+1$, is the (i, j) entry of adjacency matrix $\mathcal{A}_{n+1}(t)$ at time t, γ is a positive constant scalar, and $\eta_i(t) \triangleq \sum_{j=1}^{n+1} a_{ij}(t)$. Note that each vehicle's information control input depends on both its neighbors' information states and their derivatives[2] because of the tracking nature of (3.4). In practical implementation, the derivatives of the neighbors' information states can be calculated by numerical differentiation. For example, in the simplest case, $\dot{\xi}_j$ on the right side of (3.4) can be approximated by $\frac{\xi_j[k]-\xi_j[k-1]}{T}$, where k is the discrete-time index and T is the sample period.

Next, we will show conditions under which a unique solution exists for each control input and conditions under which consensus tracking is achieved over a directed fixed interaction topology and directed switching interaction topologies, respectively.

We have the following theorem for consensus tracking over a directed fixed interaction topology:

Theorem 3.8. *Suppose that $\mathcal{A}_{n+1}$ is constant. With (3.4), a unique solution for u_i, $i = 1, \ldots, n$, exists and the consensus tracking problem is solved if and only if directed graph $\mathcal{G}_{n+1}$ has a directed spanning tree.*

---

[2] The derivatives of the information states are equivalent to the information control inputs by noting from (2.1) that $\dot{\xi}_j = u_j$.

Proof: Noting that all entries of the last row of $\mathcal{A}_{n+1}$ are zero and $\mathcal{G}_{n+1}$ has a directed spanning tree, it follows that no other row of $\mathcal{A}_{n+1}$ can have all zero entries. It thus follows that $\eta_i = \sum_{j=1}^{n+1} a_{ij} \neq 0$, $i = 1, \ldots, n$. Thus (3.4) is well defined.

Noting that $\dot{\xi}_j = u_j$, we rewrite (3.4) as

$$(\sum_{j=1}^{n+1} a_{ij})u_i - \sum_{j=1}^{n} a_{ij}u_j - a_{i(n+1)}\dot{\xi}^r = -\gamma[\sum_{j=1}^{n} a_{ij}(\xi_i - \xi_j) + a_{i(n+1)}(\xi_i - \xi^r)],$$

which can be written in matrix form as

$$[\mathcal{L}_{n\times(n+1)} \otimes I_m]\underline{u} = -\gamma[\mathcal{L}_{n\times(n+1)} \otimes I_m]\underline{\xi}, \tag{3.5}$$

where $\underline{u} \triangleq [u_1^T, \ldots, u_n^T, \dot{\xi}^{r^T}]^T$, $\underline{\xi} \triangleq [\xi_1^T, \ldots, \xi_n^T, \xi^{r^T}]^T$, and $\mathcal{L}_{n\times(n+1)} = [\ell_{ij}] \in \mathbb{R}^{n\times(n+1)}$ is defined as $\ell_{ij} = -a_{ij}$, $i \neq j$, $\ell_{ii} = \sum_{j=1, j\neq i}^{n+1} a_{ij}$, $i, j = 1, \ldots, n$, and $\ell_{i(n+1)} = -a_{i(n+1)}$, $i = 1, \ldots, n$.

Note that $\mathcal{L}_{n+1} = \begin{bmatrix} \mathcal{L}_{n\times(n+1)} \\ 0_{1\times(n+1)} \end{bmatrix}$ and the directed graph of $\mathcal{L}_{n+1}$ has a directed spanning tree if and only if $\mathcal{G}_{n+1}$ has a directed spanning tree. Also note that $\mathcal{L}_{n+1}$ satisfies the property (B.2) with $p = n + 1$. From statements (iv) and (v) of Lemma 2.10, it follows that $\text{Rank}(\mathcal{L}_{n+1}) = n$ if and only if $\mathcal{G}_{n+1}$ has a directed spanning tree. This in turn implies that $\text{Rank}(\mathcal{L}_{n\times(n+1)}) = n$ if and only if $\mathcal{G}_{n+1}$ has a directed spanning tree because all of the entries in the last row of $\mathcal{L}_{n+1}$ are zero. Rewrite $\mathcal{L}_{n\times(n+1)} = [M|b]$, where $M = [m_{ij}] \in \mathbb{R}^{n\times n}$ is given as $m_{ij} = -a_{ij}$, $i \neq j$, $m_{ii} = \sum_{j=1, j\neq i}^{n+1} a_{ij}$, and $b \in \mathbb{R}^n$ is given as $b = [-a_{1(n+1)}, \ldots, -a_{n(n+1)}]^T$. Noting that $\mathcal{L}_{n\times(n+1)}$ has $n+1$ columns and each of its row sums is zero, it follows that the last column of $\mathcal{L}_{n\times(n+1)}$ depends on its first n columns, where $b = -M\mathbf{1}_n$. As a result, it follows that $\text{Rank}(M) = \text{Rank}([M|b]) = n$ if and only if $\mathcal{G}_{n+1}$ has a directed spanning tree.

Note that (3.5) can be written as

$$(M \otimes I_m)u + (b \otimes I_m)\dot{\xi}^r = -\gamma[(M \otimes I_m)\xi + (b \otimes I_m)\xi^r],$$

where $u = [u_1^T, \ldots, u_n^T]^T$ and $\xi = [\xi_1^T, \ldots, \xi_n^T]^T$. Because M has full rank and therefore is invertible if and only if $\mathcal{G}_{n+1}$ has a directed spanning tree, it follows that u can be solved uniquely under the same necessary and sufficient condition. The unique solution is given by

$$u = (M \otimes I_m)^{-1}[-\gamma(M \otimes I_m)\xi - \gamma(b \otimes I_m)\xi^r - (b \otimes I_m)\dot{\xi}^r],$$

which can be further written as

$$u = -\gamma\xi + \gamma(\mathbf{1}_n \otimes \xi^r) + \mathbf{1}_n \otimes \dot{\xi}^r. \tag{3.6}$$

By noting that $b = -M\mathbf{1}_n$ and M is invertible implies that $-M^{-1}b = \mathbf{1}_n$ under the same necessary and sufficient condition, and $(M \otimes I_m)^{-1}(b \otimes I_m) = M^{-1}b \otimes I_m$ from Lemma C.8.

Noting that $\dot{\xi} = u$, we rewrite (3.6) as

$$\dot{\xi} - \mathbf{1}_n \otimes \dot{\xi}^r = -\gamma(\xi - \mathbf{1}_n \otimes \xi^r),$$

which implies that $\xi(t) \to \mathbf{1}_n \otimes \xi^r(t)$ as $t \to \infty$. Therefore, it follows that $\xi_i(t) \to \xi^r(t)$, $i = 1, \ldots, n$, as $t \to \infty$, if and only if $\mathcal{G}_{n+1}$ has a directed spanning tree. ∎

Remark 3.9. Note that we have essentially used distributed algorithm (3.4) to achieve the same objective as centralized algorithm (3.6) can achieve. □

Example 3.10. Theorem 3.8 is illustrated by the following example. Again, consider $n = 4$ vehicles together with the virtual team leader, vehicle 5. Figure 3.5 shows the interaction topology for the five vehicles, where $a_{35} = a_{45} = 1$ and $a_{j5} = 0$, $\forall j \notin \{3, 4\}$. We consider two subcases using (3.4). In Subcase (a), let $\xi^r = \cos(t)$. In Subcase (b), assume that ξ^r satisfies (3.1) with $f(t, \xi^r) = \sin(t)\sin(2\xi^r)$, where $\xi^r(0) = 0.5$. As shown in Fig. 3.6, the information states of all followers converge to the exogenous signal $\cos(t)$ in Subcase (a) and to the solution of (3.1) in Subcase (b). □

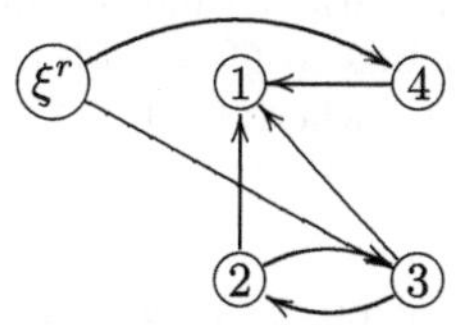

Fig. 3.5. Interaction topology among the five vehicles, where the consensus reference state ξ^r is available to vehicles 3 and 4

We have the following theorem for consensus tracking over a directed switching interaction topology:

Theorem 3.11. *Suppose that $\mathcal{A}_{n+1}(t)$ is piecewise continuous and each nonzero and hence positive entry of $\mathcal{A}_{n+1}(t)$ is chosen from a compact set $[\underline{a}, \bar{a}]$, where $\underline{a}$ an $\bar{a}$ are positive constants. Let t_0 be the initial time. Also let $t_1, t_2, \ldots$ be the switching times for $\mathcal{A}_{n+1}(t)$. With (3.4), a unique solution for u_i, $i = 1, \ldots, n$, exists if and only if directed switching graph $\mathcal{G}_{n+1}(t)$ has a directed spanning tree across each interval $[t_i, t_{i+1})$, $i = 0, 1, \ldots$. Furthermore, the consensus tracking problem is solved if directed switching graph $\mathcal{G}_{n+1}(t)$ has a directed spanning tree across each interval $[t_i, t_{i+1})$, $i = 0, 1, \ldots$.*

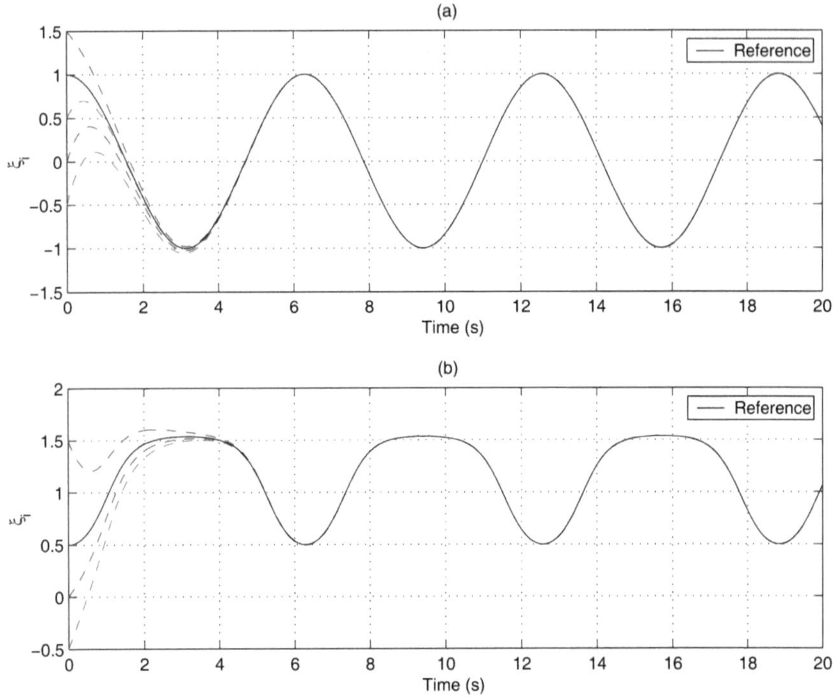

Fig. 3.6. Consensus tracking with a time-varying consensus reference state using (3.4). Subplot (a) corresponds to $\xi^r = \cos(t)$, and Subplot (b) corresponds to $f(t, \xi^r) = \sin(t)\sin(2\xi^r)$ in (3.1), where $\xi^r(0) = 0.5$.

Proof: Following the proof of Theorem 3.8, we can show that $\eta_i(t) \neq 0$, $i = 1, \ldots, n$, across each interval $[t_i, t_{i+1})$, $i = 0, 1, \ldots$ if directed switching graph $\mathcal{G}_{n+1}(t)$ has a directed spanning tree across that interval. Thus (3.8) is well defined under the condition of the theorem.

Note that (3.5) can be written as

$$[M(t) \otimes I_m]u + [b(t) \otimes I_m]\dot{\xi}^r = -\gamma\{[M(t) \otimes I_m]\xi + [b(t) \otimes I_m]\xi^r\}, \quad (3.7)$$

where $u \stackrel{\triangle}{=} [u_1^T, \ldots, u_n^T]^T$, $\xi \stackrel{\triangle}{=} [\xi_1^T, \ldots, \xi_n^T]^T$, and $M(t)$ and $b(t)$ are defined as in the proof of Theorem 3.8 except that $M(t)$ and $b(t)$ are switching in (3.7). Following the proof of Theorem 3.8, we can show that $M(t)$ is invertible across each interval $[t_i, t_{i+1})$, $i = 0, 1, \ldots$ if and only if $\mathcal{G}_{n+1}(t)$ has a directed spanning tree across that interval.

Next, we show that $M^{-1}(t)$ is bounded under the assumption of the theorem. Noting that each nonzero entry of $\mathcal{A}_{n+1}(t)$ is chosen from a compact set, it follows that each nonzero entry of $M(t)$ is also within a compact set. Two matrices are said to have the same structure if their positive, zero, and negative entries are in the same places. Under the assumption that $\mathcal{G}_{n+1}(t)$

is switching but has a directed spanning tree across each interval $[t_i, t_{i+1})$, the number of possible directed graphs for $\mathcal{G}_{n+1}(t)$ is finite. It then follows that there are a finite number of possible structures for $\mathcal{A}_{n+1}(t)$, which implies that there are a finite number of possible structures for $M(t)$. For each possible structure of $M(t)$, $M^{-1}(t)$ exists across each interval $[t_i, t_{i+1})$, which implies that $\det[M(t)] \neq 0$ across each interval $[t_i, t_{i+1})$, if $\mathcal{G}_{n+1}(t)$ has a directed spanning tree across that interval. Thus for each possible structure of $M(t)$, $\det[M(t)]$ is within a compact set that does not include zero. Also note that for each possible structure of $M(t)$, all entries of its adjoint are within a compact set. It thus follows that each entry of $M^{-1}(t)$ is within a compact set. Therefore, it follows that u can be solved uniquely across each interval $[t_i, t_{i+1})$, $i = 0, 1, \ldots$, and the unique solution is given by

$$u = [M(t) \otimes I_m]^{-1}\{-\gamma[M(t) \otimes I_m]\xi - \gamma[b(t) \otimes I_m]\xi^r - [b(t) \otimes I_m]\dot{\xi}^r\}.$$

Noting that $b(t) = -M(t)\mathbf{1}_n$ and $M(t)$ is invertible across each interval $[t_i, t_{i+1})$ under the assumption of the theorem, it follows that $-M^{-1}(t)b(t) = \mathbf{1}_n$. Thus the unique solution is given by

$$u = -\gamma\xi + \gamma(\mathbf{1}_n \otimes \xi^r) + \mathbf{1}_n \otimes \dot{\xi}^r.$$

Therefore, it follows from the proof of Theorem 3.8 that $\xi_i(t) \to \xi^r(t)$, $i = 1, \ldots, n$, as $t \to \infty$, if $\mathcal{G}_{n+1}(t)$ has a directed spanning tree across each interval $[t_i, t_{i+1})$, $i = 0, 1, \ldots$. ■

Remark 3.12. Note that the tracking nature of algorithm (3.4) requires more stringent conditions for convergence over directed switching interaction topologies than those for algorithm (2.2), where the final consensus equilibrium is a constant. In contrast to algorithm (3.3), which requires that the time-varying consensus reference state be available to each follower for consensus tracking, algorithm (3.4) allows the time-varying consensus reference state to be available only to a subgroup of the followers. □

Remark 3.13. In contrast to the leader-following strategy (*e.g.*, [244]), where the interaction topology is itself a directed spanning tree,[3] algorithm (3.4) takes into account the general case where information may flow among vehicles. We have shown that with the time-varying consensus reference state, complexity results from the information flow between neighbors. It is worthwhile to mention that the extension of consensus algorithms from a constant reference to a time-varying reference is nontrivial. It is not straightforward to apply the internal model principle of control to consensus tracking with a time-varying consensus reference state for multivehicle systems involving only local information exchange. □

[3] The leader-following strategy requires an interaction topology that is itself a directed spanning tree. Note that the condition that a graph has a directed spanning tree is not equivalent to the condition that a graph is itself a directed spanning tree. The latter condition is a special case of the former one.

Example 3.14. Theorem 3.11 is illustrated by the following example. Suppose that $\mathcal{G}_5(t)$ switches randomly from the set, $\{\mathcal{G}_{5(1)}, \ldots, \mathcal{G}_{5(4)}\}$, as shown in Fig. 3.7, with a switching time around 0.25 seconds. We choose $a_{ij}(t) = 1$ if $(j, i) \in \mathcal{E}_{n+1}$ at time t and $a_{ij}(t) = 0$ otherwise. Let $f(t, \xi^r) = \sin(t)\sin(2\xi^r)$ in (3.1) with $\xi^r(0) = 1$. Figure 3.8 shows that each vehicle tracks the consensus reference state with (3.4) under directed switching interaction topologies. □

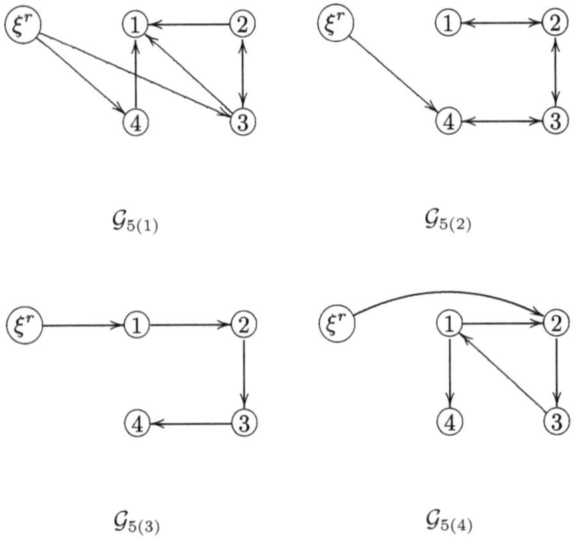

Fig. 3.7. Directed switching graphs for (3.4)

3.3.2 Consensus Tracking Algorithm with Bounded Control Inputs

In this subsection, we consider a consensus tracking algorithm that explicitly accounts for a bounded control effort. Given $x = [x_1, \ldots, x_m] \in \mathbb{R}^m$, define $\tanh(x) = [\tanh(x_1), \ldots, \tanh(x_m)]^T$. We propose a consensus tracking algorithm with bounded control inputs as

$$u_i = \frac{1}{\eta_i}[\sum_{j=1}^{n} a_{ij}\dot{\xi}_j + a_{i(n+1)}\dot{\xi}^r]$$
$$- \frac{1}{\eta_i}\Lambda_i \tanh[\sum_{j=1}^{n} a_{ij}(\xi_i - \xi_j) + a_{i(n+1)}(\xi_i - \xi^r)], \quad i = 1, \ldots, n, \quad (3.8)$$

where a_{ij}, $i = 1, \ldots, n$, $j = 1, \ldots, n+1$, is the (i, j) entry of the adjacency matrix $\mathcal{A}_{n+1}$, $\eta_i \triangleq \sum_{j=1}^{n+1} a_{ij}$, and $\Lambda_i \in \mathbb{R}^{m \times m}$ is a symmetrical positive-definite matrix.

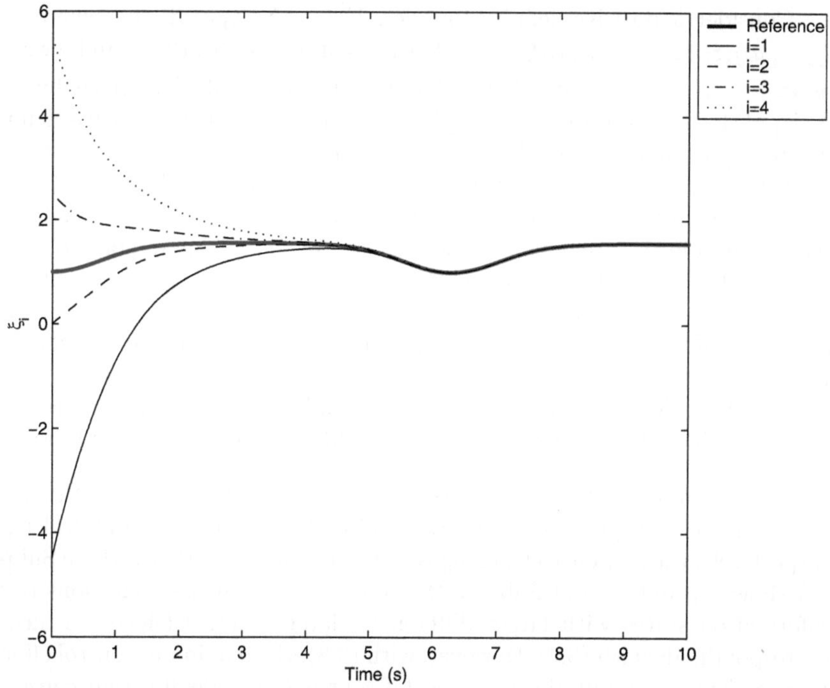

Fig. 3.8. Consensus tracking with (3.4) under directed switching interaction topologies given by Fig. 3.7

Next, we will show conditions under which a unique bounded solution exists for each control input and conditions under which consensus tracking is achieved over a directed fixed interaction topology.

Theorem 3.15. *Suppose that $\mathcal{A}_{n+1}$ is constant. With* (3.8)*, a unique solution for u_i, $i = 1, \ldots, n$, exists, $\|u_i\|_\infty$ is bounded and independent of the initial conditions of the information states, and the consensus tracking problem is solved if and only if directed graph $\mathcal{G}_{n+1}$ has a directed spanning tree.*

Proof: Following the proof of Theorem 3.8, we rewrite (3.8) as

$$(M \otimes I_m)u + (b \otimes I_m)\dot{\xi}^r = -\Lambda \tanh[(M \otimes I_m)\xi + (b \otimes I_m)\xi^r],$$

where $u \triangleq [u_1^T, \ldots, u_n^T]^T$, $\xi \triangleq [\xi_1^T, \ldots, \xi_n^T]^T$, and $\Lambda \triangleq \text{diag}(\Lambda_1, \ldots, \Lambda_n)$. As shown in the proof of Theorem 3.8, M has full rank and therefore is invertible if and only if $\mathcal{G}_{n+1}$ has a directed spanning tree. It follows that u can be solved uniquely under the same necessary and sufficient condition. In addition, it is straightforward to see that $\|u_i\|_\infty$ is bounded and independent of the initial conditions of the information states because $\dot{\xi}^r$ is bounded and $\|\tanh(\cdot)\|_\infty \leq 1$.

Consider a positive-definite function $V = \frac{1}{2}\sum_{i=1}^{n} e_i^T \Lambda_i^{-1} e_i$, where $e_i \triangleq \sum_{j=1}^{n} a_{ij}(\xi_i - \xi_j) + a_{i(n+1)}(\xi_i - \xi^r)$. Noting that with (3.8), (2.1) can be written as $\dot{e}_i = -\Lambda_i \tanh(e_i)$, it follows that the derivative of V is given by $\dot{V} = \sum_{i=1}^{n} e_i^T \Lambda_i^{-1}[-\Lambda_i \tanh(e_i)] = -\sum_{i=1}^{n} e_i^T \tanh(e_i)$, which is negative definite. Therefore, it follows from Theorem F.1 that $e_i(t) \to 0$, $i = 1, \ldots, n$, as $t \to \infty$.

Letting $e = [e_1^T, \ldots, e_n^T]^T$, then $e = (M \otimes I_m)\xi + (b \otimes I_m)\xi^r$. Thus $e_i(t) \to 0$, $i = 1, \ldots, n$, as $t \to \infty$, implies that $(M \otimes I_m)\xi(t) + (b \otimes I_m)\xi^r(t) \to 0$ as $t \to \infty$. Also note from the previous argument that $-M^{-1}b = \mathbf{1}_n$ if and only $\mathcal{G}_{n+1}$ has a directed spanning tree. It follows that $\xi(t) \to -(M \otimes I_m)^{-1}(b \otimes I_m)\xi^r(t) = -(M^{-1}b \otimes I_m)\xi^r(t) = \mathbf{1}_n \otimes \xi^r(t)$, which implies that $\xi_i(t) \to \xi^r(t)$, $i = 1, \ldots, n$, as $t \to \infty$ if and only if the directed graph of $\mathcal{G}_{n+1}$ has a directed spanning tree. ■

Note that (3.8) explicitly accounts for bounded control inputs, whereas (3.4) does not. The following example compares (3.8) with (3.4) in simulation:

Example 3.16. Let $f(t, \xi^r) = \sin(t)\sin(2\xi^r)$ in (3.1) with $\xi^r(0) = 1$. Figure 3.9 shows directed graph $\mathcal{G}_5$ used in (3.8) and (3.4). Figures 3.10 and 3.11 show, respectively, the consensus tracking results with (3.8) and (3.4), where Subplot (a) shows ξ_i and ξ^r and Subplot (b) shows u_i. For initial conditions of the information states with larger differences, larger control efforts are needed correspondingly with (3.4). However, with (3.8), the maximum control effort is independent of the initial conditions. Of course, the tradeoff is that consensus tracking with (3.8) converges more slowly than with (3.4). □

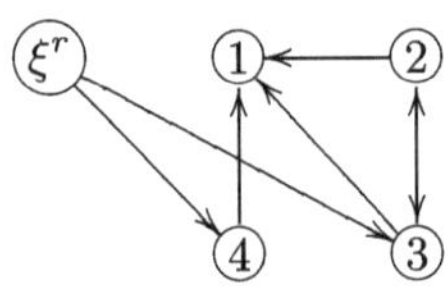

Fig. 3.9. Directed graph $\mathcal{G}_5$ used in (3.8) and (3.4)

3.3.3 Information Feedback to the Consensus Reference State

Note that with consensus tracking algorithms (3.4) and (3.8), there is no information feedback from the followers to the consensus reference state. If a vehicle cannot track the consensus reference state due to disturbance or temporal malfunction, the consensus reference state will evolve at its nominal speed without compensating for vehicle disturbance or temporal malfunction. We will propose consensus tracking algorithms that introduce feedback from the followers to the consensus reference state.

One strategy is to update ξ^r as

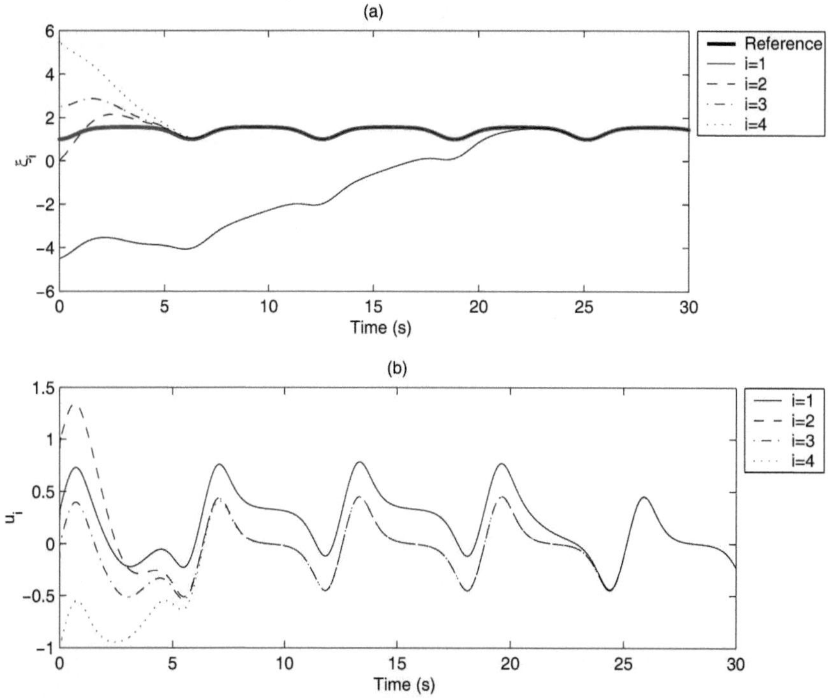

Fig. 3.10. Consensus tracking with bounded control inputs given by (3.8). Subplot (a) shows the information states, and Subplot (b) shows the information control inputs.

$$\frac{d\xi^r}{ds} = f(s, \xi^r), \qquad \frac{ds}{dt} = \frac{1}{1 + z_f}, \tag{3.9}$$

where $f(\cdot, \cdot)$ is defined as in (3.1) and z_f denotes a nonnegative consensus tracking performance function. The function z_f must satisfy the property that z_f is small if consensus is achieved well and is large otherwise. For example, we may choose a nonnegative function z_f such that $z_f = 0$ if $k_f \sum_{i \in \mathcal{V}_L} \sum_{j=1}^n a_{ij} \|\xi_i - \xi_j\|^2$ is below a threshold and z_f grows as $k_f \sum_{i \in \mathcal{V}_L} \sum_{j=1}^n a_{ij} \|\xi_i - \xi_j\|^2$ increases above the threshold, where k_f is a positive constant and $\mathcal{V}_L$ denotes the subgroup leaders (*i.e.*, the set of vehicles to which ξ^r and $\dot{\xi}^r$ are available). As a result, the consensus reference state will evolve at its nominal speed if the consensus error is small and evolves more slowly if the consensus error becomes large. Note that with the feedback strategy (3.9), Theorems 3.8, 3.11, and 3.15 remain valid.

Another strategy is to define $a_{i(n+1)}$ in (3.4) and (3.8) as

$$a_{i(n+1)} = \frac{a^*_{i(n+1)}}{1 + z_f}, \tag{3.10}$$

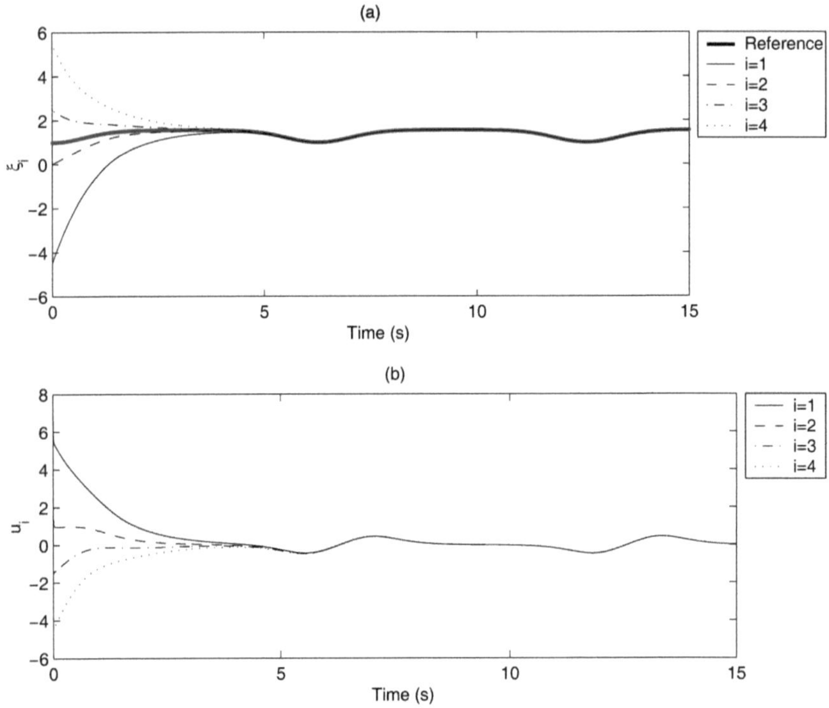

Fig. 3.11. Consensus tracking without explicitly accounting for bounded control inputs with (3.4). Subplot (a) shows the information states, and Subplot (b) shows the information control inputs.

where $a^*_{i(n+1)}$ denotes the nominal value for $a_{i(n+1)}$ when there is no feedback mechanism and z_f is a nonnegative consensus tracking performance function. The function z_f must satisfy the properties that $z_f \in [0, \bar{z}_f]$, where $\bar{z}_f$ is an upper bound for z_f, and z_f is small if consensus is achieved well and is large otherwise. With the feedback strategy (3.10), Theorem 3.11 remains valid because $a_{i(n+1)} = 0$ if $a^*_{i(n+1)} = 0$ and $a_{i(n+1)}$ belongs to a compact set when $a^*_{i(n+1)} \neq 0$ is chosen from a compact set.

Example 3.17. The following example compares consensus tracking results with and without information feedback. Suppose that $\mathcal{G}_5$ is given by $\mathcal{G}_{5(2)}$ in Fig. 3.7. Also suppose that vehicle 3 fails at $t \in [5, 7]$ seconds, where $u_3(t) \equiv 0$ at $t \in [5, 7]$ seconds. We combine the information feedback strategies (3.9), where $z_f = 50\,|\xi_4 - \xi_3|$, and (3.10), where $a_{45} = \frac{1}{1+z_f}$. Subplot (a) in Fig. 3.12 shows the consensus tracking result with the information feedback strategies (3.9) and (3.10), and Subplot (b) in Fig. 3.12 shows the consensus tracking result with (3.4) without information feedback. Note that during $t \in [5, 7]$ seconds, the information states of the vehicles stay closer and the consensus reference state evolves more slowly in Subplot (a) than in Subplot (b) due to the introduction of information feedback in Subplot (a). □

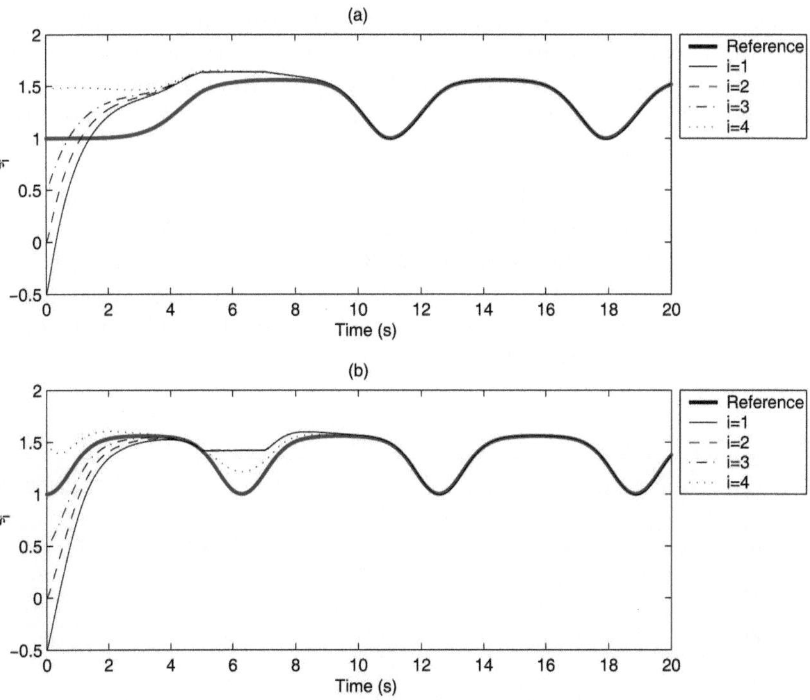

Fig. 3.12. Consensus tracking with information feedback versus without information feedback. Subplot (a) corresponds to the case with information feedback, and Subplot (b) corresponds to the case without information feedback.

3.4 Extension to Relative State Deviations

We propose the following consensus algorithm for relative state deviations with a time-varying consensus reference state

$$\begin{aligned} u_i =& \dot{\delta}_i + \frac{1}{\eta_i}\sum_{j=1}^{n} a_{ij}\{\dot{\xi}_j - \dot{\delta}_j - \gamma[(\xi_i - \xi_j) - (\delta_i - \delta_j)]\} \\ &+ \frac{1}{\eta_i} a_{i(n+1)}[\dot{\xi}^r - \gamma(\xi_i - \delta_i - \xi^r)], \end{aligned} \tag{3.11}$$

where $\Delta_{ij} \triangleq \delta_i - \delta_j$, $\forall i \neq j$, denotes the desired separation between the information states.

Theorem 3.18. *Suppose that $\mathcal{A}_{n+1}$ is constant. With algorithm* (3.11), *$\xi_i(t) \to \xi^r(t) + \delta_i(t)$ and $\xi_i(t) - \xi_j(t) \to \Delta_{ij}(t)$, as $t \to \infty$, if and only if directed graph $\mathcal{G}_{n+1}$ has a directed spanning tree.*

Proof: Define $\tilde{\xi}_i \stackrel{\triangle}{=} \xi_i - \delta_i$ and $\tilde{u}_i \stackrel{\triangle}{=} u_i - \dot{\delta}_i$. Note that $\dot{\tilde{\xi}}_i = \tilde{u}_i$. Also note that (3.11) can be rewritten in the same form as (3.4) with $\tilde{\xi}_i$ and $\tilde{u}_i$ playing the roles of ξ_i and u_i, respectively. As a result, from Theorem 3.8, $\tilde{\xi}_i(t) \to \xi^r(t)$, $i = 1, \ldots, n$, as $t \to \infty$, which implies that $\xi_i(t) \to \xi^r(t) + \delta_i(t)$ and $\xi_i(t) - \xi_j(t) \to \Delta_{ij}(t)$, as $t \to \infty$. ■

Example 3.19. Theorem 3.18 is illustrated by the following example. As in Example 3.10, consider the two subcases using (3.11). We choose $\delta_i = 1 - i$, $i = 1, \ldots, 4$. As shown in Fig. 3.13, $\xi_1(t) \to \xi^r(t)$, $\xi_2(t) \to \xi^r(t) - 1$, $\xi_3(t) \to \xi^r(t) - 2$, and $\xi_4(t) \to \xi^r(t) - 3$, as $t \to \infty$, where ξ^r is the exogenous signal $\cos(t)$ in Subcase (a) and is the solution of the nonlinear model $\dot{\xi}^r = \sin(t)\sin(2\xi^r)$ in Subcase (b). □

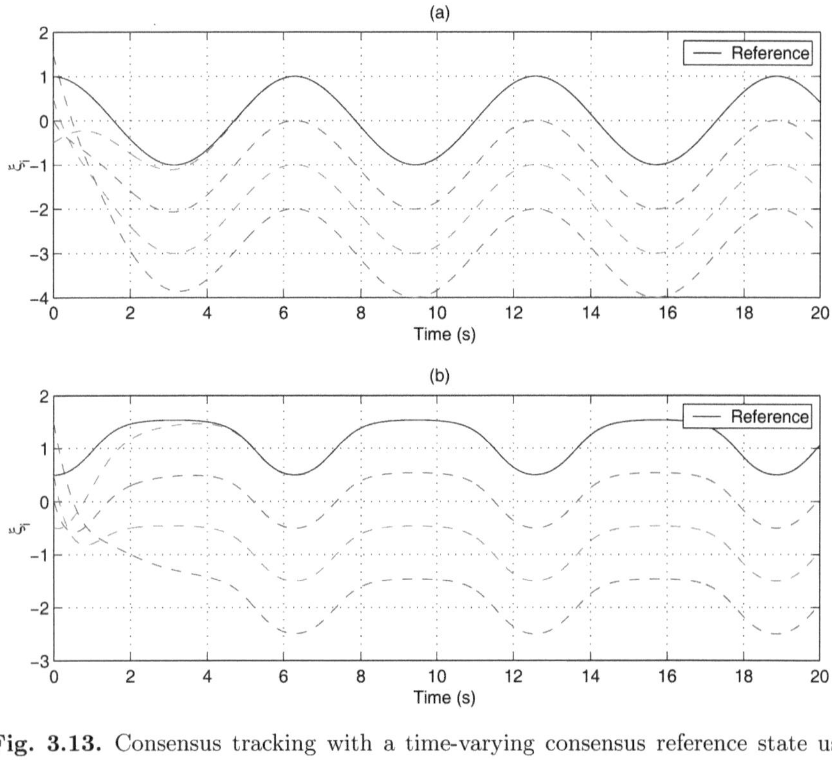

Fig. 3.13. Consensus tracking with a time-varying consensus reference state using (3.11). Subplot (a) corresponds to $\xi^r = \cos(t)$, and Subplot (b) corresponds to $f(t, \xi^r) = \sin(t)\sin(2\xi^r)$ in (3.1), where $\xi^r(0) = 0.5$.

3.5 Notes

The results in this chapter are based mainly on [173, 175, 180]. Variants of algorithm (3.2) are also considered in [99, 143] over a directed fixed interaction topology.

Acknowledgment is given to

Part III

Consensus Algorithms for Double-integrator Dynamics

4

Consensus Algorithms for Double-integrator Dynamics

This chapter introduces consensus algorithms for double-integrator dynamics that take into account motions of the information states and their derivatives, extending the algorithms for single-integrator dynamics. We first propose a fundamental consensus algorithm and derive conditions under which consensus is reached over directed fixed and switching interaction topologies, respectively. Unlike the single-integrator case, we show that having a directed spanning tree is a necessary rather than a sufficient condition for consensus-seeking with double-integrator dynamics. We focus on a formal analysis of interaction topologies that permit consensus for double-integrator dynamics. Given its importance for the stability of the coordinated system, an analysis of the consensus term control gains is also presented, specifically the strength of the information states relative to their derivatives. We then propose two other consensus algorithms that, respectively, take into account bounded control effort and remove the requirement for measurements of relative information state derivatives. We derive conditions under which consensus is reached for these two algorithms over an undirected fixed interaction topology.

4.1 Consensus Algorithm

Chapters 2 and 3 focus on consensus algorithms that take the form of single-integrator dynamics. Equations of motion of a broad class of vehicles require a double-integrator dynamic model. For example, some vehicle dynamics can be feedback linearized as double integrators, *e.g.*, mobile robot dynamic models. In the case of consensus algorithms for single-integrator dynamics, the consensus equilibrium is a constant. In contrast to the constant consensus equilibrium, it might be proper to derive consensus algorithms for double-integrator dynamics such that some information states converge to a consistent value (*e.g.*, position of the formation center) while others converge to another consistent value (*e.g.*, velocity of the formation center). Unfortunately, the extension of consensus algorithms from single-integrator dynamics to double-integrator dynamics is nontrivial.

Consider information states with double-integrator dynamics given by

$$\dot{\xi}_i = \zeta_i, \quad \dot{\zeta}_i = u_i, \quad i = 1, \ldots, n, \tag{4.1}$$

where $\xi_i \in \mathbb{R}^m$ is the information state, $\zeta_i \in \mathbb{R}^m$ is the information state derivative, and $u_i \in \mathbb{R}^m$ is the information control input associated with the ith vehicle.

Note that the interaction topologies for ξ_i and ζ_i among n vehicles are allowed to be different under some circumstances. We use directed graphs $\mathcal{G}_n^A \triangleq (\mathcal{V}_n, \mathcal{E}_n^A)$, where $\mathcal{V}_n = \{1, \ldots, n\}$ is the node set and $\mathcal{E}_n^A \subseteq \mathcal{V}_n \times \mathcal{V}_n$ is the edge set, and $\mathcal{G}_n^B \triangleq (\mathcal{V}_n, \mathcal{E}_n^B)$, where $\mathcal{V}_n$ is the same node set as above and $\mathcal{E}_n^B \subseteq \mathcal{V}_n \times \mathcal{V}_n$ is the edge set, to model the interaction topologies for ξ_i and ζ_i, respectively, among the n vehicles. Let $\mathcal{A}_n = [a_{ij}] \in \mathbb{R}^{n\times n}$ and $\mathcal{B}_n = [b_{ij}] \in \mathbb{R}^{n\times n}$ be the adjacency matrix associated with $\mathcal{G}_n^A$ and $\mathcal{G}_n^B$, respectively. Also let $\mathcal{L}_n^A = [\ell_{ij}] \in \mathbb{R}^{n\times n}$ (*i.e.*, $\ell_{ij} = -a_{ij}$, $i \neq j$, $\ell_{ii} = \sum_{j=1,j\neq i}^n a_{ij}$) and $\mathcal{L}_n^B = [\ell_{ij}] \in \mathbb{R}^{n\times n}$ (*i.e.*, $\ell_{ij} = -b_{ij}$, $i \neq j$, $\ell_{ii} = \sum_{j=1,j\neq i}^n b_{ij}$) be the nonsymmetrical Laplacian matrix associated with $\mathcal{G}_n^A$ and $\mathcal{G}_n^B$, respectively. When there is only one interaction topology associated with the n vehicles,[1] we simply use $\mathcal{G}_n \triangleq (\mathcal{V}_n, \mathcal{E}_n)$ to model the interaction topology among the n vehicles, as in Chapter 2. Similarly, $\mathcal{A}_n$ and $\mathcal{L}_n$ are, respectively, the adjacency matrix and the nonsymmetrical Laplacian matrix associated with $\mathcal{G}_n$.

We propose a consensus algorithm for (4.1) as

$$u_i = -\sum_{j=1}^n a_{ij}(t)[(\xi_i - \xi_j) + \gamma(t)(\zeta_i - \zeta_j)], \qquad i = 1, \ldots, n, \tag{4.2}$$

where $a_{ij}(t)$ is the (i,j) entry of the adjacency matrix $\mathcal{A}_n(t) \in \mathbb{R}^{n\times n}$ associated with $\mathcal{G}_n$[2] at time t and $\gamma(t)$ is a positive scalar at time t. When ξ_i and ζ_i denote, respectively, the position and velocity of the ith vehicle, (4.2) represents the acceleration of that vehicle. With (4.2), consensus is *achieved* or *reached* by the team of vehicles if, for all $\xi_i(0)$ and $\zeta_i(0)$ and all $i, j = 1, \ldots, n$, $\|\xi_i(t) - \xi_j(t)\| \to 0$ and $\|\zeta_i - \zeta_j\| \to 0$, as $t \to \infty$.

Let $\xi \triangleq [\xi_1^T, \ldots, \xi_n^T]^T$ and $\zeta \triangleq [\zeta_1^T, \ldots, \zeta_n^T]^T$. By applying algorithm (4.2), (4.1) can be written in matrix form as

$$\begin{bmatrix} \dot{\xi} \\ \dot{\zeta} \end{bmatrix} = (\Theta(t) \otimes I_m) \begin{bmatrix} \xi \\ \zeta \end{bmatrix}, \tag{4.3}$$

where

[1] For example, the interaction topologies for ξ_i and ζ_i are identical or need only ξ_i to be exchanged among vehicles.

[2] In this case, the interaction topologies for both ξ_i and ζ_i are identical.

$$\Theta = \begin{bmatrix} 0_{n\times n} & I_n \\ -\mathcal{L}_n(t) & -\gamma(t)\mathcal{L}_n(t) \end{bmatrix} \tag{4.4}$$

with $\mathcal{L}_n(t) \in \mathbb{R}^{n\times n}$ as the nonsymmetrical Laplacian matrix associated with $\mathcal{G}_n$ at time t.

4.1.1 Convergence Analysis Under Fixed Interaction Topologies

In this subsection, we focus on analyzing consensus algorithm (4.2) when $\mathcal{A}_n$ and γ are constant. Given a block matrix

$$M = \begin{bmatrix} A & B \\ C & D \end{bmatrix},$$

it is known that $\det(M) = \det(AD - CB)$ if A and C commute, where $\det(\cdot)$ denotes the determinant of a matrix.

To find the eigenvalues of Θ, we can solve the equation $\det(\lambda I_{2n} - \Theta) = 0$, where $\det(\lambda I_{2n} - \Theta)$ is the characteristic polynomial of Θ. Note that

$$\begin{aligned} \det(\lambda I_{2n} - \Theta) &= \det\left(\begin{bmatrix} \lambda I_n & -I_n \\ \mathcal{L}_n & \lambda I_n + \gamma\mathcal{L}_n \end{bmatrix}\right) \\ &= \det[\lambda^2 I_n + (1+\gamma\lambda)\mathcal{L}_n]. \end{aligned} \tag{4.5}$$

Also note that

$$\det(\lambda I_n + \mathcal{L}_n) = \prod_{i=1}^{n} (\lambda - \mu_i), \tag{4.6}$$

where μ_i is the ith eigenvalue of $-\mathcal{L}_n$.

By comparing (4.5) and (4.6), we see that

$$\det\left[\lambda^2 I_n + (1+\gamma\lambda)\mathcal{L}_n\right] = \prod_{i=1}^{n} [\lambda^2 - (1+\gamma\lambda)\mu_i],$$

which implies that the roots of (4.5) can be obtained by solving $\lambda^2 = (1 + \gamma\lambda)\mu_i$. Therefore, it is straightforward to see that the eigenvalues of Θ are given by

$$\lambda_{i\pm} = \frac{\gamma\mu_i \pm \sqrt{\gamma^2\mu_i^2 + 4\mu_i}}{2}, \tag{4.7}$$

where λ_{i+} and λ_{i-} are called the eigenvalues of Θ that are associated with μ_i.

From (4.7), we can see that Θ has $2k$ zero eigenvalues if and only if $-\mathcal{L}_n$ has k zero eigenvalues. According to Lemma 2.4, $-\mathcal{L}_n$ has at least one zero eigenvalue with an associated eigenvector $\mathbf{1}_n$, and all of the nonzero eigenvalues of $-\mathcal{L}_n$ have negative real parts. Therefore, we know that Θ has at least two zero eigenvalues. Without loss of generality, we let $\lambda_{1+} = \lambda_{1-} = 0$.

We have the following lemma regarding a necessary and sufficient condition for an information consensus using algorithm (4.2):

Lemma 4.1. *Algorithm* (4.2) *achieves consensus asymptotically if and only if Θ has exactly two zero eigenvalues and all other eigenvalues have negative real parts. Specifically,* $\xi_i(t) \to \sum_{i=1}^n p_i\xi_i(0) + t\sum_{i=1}^n p_i\zeta_i(0)$ *and* $\zeta_i(t) \to \sum_{i=1}^n p_i\zeta_i(0)$ *for large* t*, where* $p = [p_1, \ldots, p_n]^T \geq 0$, $\mathbf{1}_n^T p = 1$*, and* $\mathcal{L}_n^T p = 0$.

Proof: (Sufficiency.) We first show that the zero eigenvalue of Θ has geometric multiplicity equal to one, when Θ has exactly two zero eigenvalues. Letting $q = [q_a^T, q_b^T]^T$, where $q_a, q_b \in \mathbb{R}^n$, is an eigenvector of Θ associated with the zero eigenvalue, then we know that

$$\Theta q = \begin{bmatrix} 0_{n\times n} & I_n \\ -\mathcal{L}_n & -\gamma\mathcal{L}_n \end{bmatrix} \begin{bmatrix} q_a \\ q_b \end{bmatrix} = \begin{bmatrix} \mathbf{0}_n \\ \mathbf{0}_n \end{bmatrix},$$

which implies that $q_b = \mathbf{0}_n$ and $-\mathcal{L}_n q_a = \mathbf{0}_n$, where $\mathbf{0}_n$ denotes the $n \times 1$ vector of all zeros, that is, q_a is an eigenvector of $-\mathcal{L}_n$ associated with the zero eigenvalue of $-\mathcal{L}_n$. Because Θ has exactly two zero eigenvalues, we know from (4.7) that $-\mathcal{L}_n$ has exactly one zero eigenvalue. Therefore, we see that $-\mathcal{L}_n$ has only one linearly independent eigenvector q_a associated with the zero eigenvalue, which in turn implies that Θ has only one linearly independent eigenvector $q = [q_a^T, \mathbf{0}_n^T]^T$ associated with the zero eigenvalue, that is, the zero eigenvalue of Θ has algebraic multiplicity equal to two but geometric multiplicity equal to one.

Note that Θ can be written in Jordan canonical form as

$$\begin{aligned} \Theta &= PJP^{-1} \\ &= [w_1, \cdots, w_{2n}] \begin{bmatrix} 0 & 1 & 0_{1\times(2n-2)} \\ 0 & 0 & 0_{1\times(2n-2)} \\ 0_{(2n-2)\times 1} & 0_{(2n-2)\times 1} & J' \end{bmatrix} \begin{bmatrix} \nu_1^T \\ \vdots \\ \nu_{2n}^T \end{bmatrix}, \end{aligned} \tag{4.8}$$

where $w_j \in \mathbb{R}^{2n}$, $j = 1, \cdots, 2n$, can be chosen to be the right eigenvectors or generalized right eigenvectors of Θ, $\nu_j \in \mathbb{R}^{2n}$, $j = 1, \cdots, 2n$, can be chosen as the left eigenvectors or generalized left eigenvectors of Θ, and J' is the Jordan upper diagonal block matrix corresponding to the nonzero eigenvalues λ_{i+} and λ_{i-}, $i = 2, \cdots, n$.

Without loss of generality, we choose $w_1 = [\mathbf{1}_n^T, \mathbf{0}_n^T]^T$ and $w_2 = [\mathbf{0}_n^T, \mathbf{1}_n^T]^T$, where it can be verified that w_1 and w_2 are a right eigenvector and generalized right eigenvector of Θ associated with the zero eigenvalue, respectively. Noting that Θ has exactly two zero eigenvalues, we know that $-\mathcal{L}_n$ has a simple zero eigenvalue, which in turn implies that there exists an $n \times 1$ nonnegative vector p such that $p^T\mathcal{L}_n = 0$ and $\mathbf{1}_n^T p = 1$, as shown in Lemma 2.6. It can be verified that $\nu_1 = [p^T, \mathbf{0}_n^T]^T$ and $\nu_2 = [\mathbf{0}_n^T, p^T]^T$ are a generalized left eigenvector and left eigenvector of Θ associated with the zero eigenvalue, respectively, where $\nu_1^T w_1 = 1$ and $\nu_2^T w_2 = 1$. Noting that the eigenvalues λ_{i+} and λ_{i-}, $i = 2, \cdots, n$, have negative real parts, we see that

$$e^{\Theta t} = Pe^{Jt}P^{-1}$$
$$= P \begin{bmatrix} 1 & t & 0_{1\times(2n-2)} \\ 0 & 1 & 0_{1\times(2n-2)} \\ 0_{(2n-2)\times 1} & 0_{(2n-2)\times 1} & e^{J't} \end{bmatrix} P^{-1},$$

which converges to $\begin{bmatrix} \mathbf{1}_n p^T & t\mathbf{1}_n p^T \\ 0_{n\times n} & \mathbf{1}_n p^T \end{bmatrix}$ for large t, where we have used the fact that $e^{J't} \to 0_{(2n-2)\times(2n-2)}$ for large t.

Noting that for large t,

$$\begin{bmatrix} \xi(t) \\ \zeta(t) \end{bmatrix} \to \left(\begin{bmatrix} \mathbf{1}_n p^T & t\mathbf{1}_n p^T \\ 0_{n\times n} & \mathbf{1}_n p^T \end{bmatrix} \otimes I_m \right) \begin{bmatrix} \xi(0) \\ \zeta(0) \end{bmatrix},$$

we see that $\xi(t) \to (\mathbf{1}_n p^T \otimes I_m)\xi(0) + t(\mathbf{1}_n p^T \otimes I_m)\zeta(0)$ and $\zeta(t) \to (\mathbf{1}_n p^T \otimes I_m)\zeta(0)$ for large t. As a result, we know that $\|\xi_i(t) - \xi_j(t)\| \to 0$ and $\|\zeta_i(t) - \zeta_j(t)\| \to 0$, as $t \to \infty$, that is, consensus is achieved for the team of vehicles.

(Necessity.) Suppose that the sufficient condition that Θ has exactly two zero eigenvalues and all other eigenvalues have negative real parts does not hold. Noting that Θ has at least two zero eigenvalues, the fact that the sufficient condition does not hold implies that Θ has either more than two zero eigenvalues or it has two zero eigenvalues and at least one eigenvalue with a positive real part. Without loss of generality, assume that $\varsigma_1 = \varsigma_2 = 0$ and $\mathrm{Re}(\varsigma_3) \geq 0$, where ς_k, $k = 1, \cdots, 2n$, denotes the kth eigenvalue of Θ and $\mathrm{Re}(\cdot)$ represents the real part of a number. Letting $J = [j_{k\ell}]$ be the Jordan canonical form of Θ, we know that $j_{kk} = \varsigma_k$, $k = 1, \cdots, 2n$. Then we see that $e^{j_{kk}t} \neq 0$, $k = 1, 2, 3$, for large t, which in turn implies that the first three rows of e^{Jt} are linearly independent for large t. Therefore, we know that the rank of e^{Jt} is at least three for large t, which implies that the rank of $e^{\Theta t}$ is at least three for large t. Note that consensus is reached asymptotically if and only if $e^{\Theta t} \to \begin{bmatrix} \mathbf{1}_n p^T \\ \mathbf{1}_n q^T \end{bmatrix}$ for large t, where p and q are $n \times 1$ vectors. As a result, the rank of $e^{\Theta t}$ cannot exceed two for large t. This results in a contradiction. ∎

If all of the nonzero eigenvalues of $-\mathcal{L}_n$ are real and hence negative, it is straightforward using (4.7) to verify that all nonzero eigenvalues of Θ have negative real parts. We have the following lemma and corollary:

Lemma 4.2. *If $-\mathcal{L}_n$ has a simple zero eigenvalue and all other eigenvalues are real and hence negative, then algorithm* (4.2) *achieves consensus asymptotically for any $\gamma > 0$.*

Corollary 4.3. *Suppose that $\mathcal{G}_n$ is undirected. Then algorithm* (4.2) *achieves consensus asymptotically for any $\gamma > 0$ if and only if $\mathcal{G}_n$ is connected.*

In the general case, some nonzero eigenvalues of Θ may have positive real parts even if all of the nonzero eigenvalues of $-\mathcal{L}_n$ have negative real parts, as shown in the following five examples. In all five examples, we assume that $a_{ij} = 1$ if $(j, i) \in \mathcal{E}_n$ and $\gamma = 1$ in (4.2) unless stated otherwise. In addition, we let $\xi_i(0) = 0.2(i-1)$ and $\zeta_i(0) = 0.1(i-1)$, $i = 1, \cdots, 4$.

Example 4.4. When the interaction topology has separated subgroups, as shown in Fig. 4.1, consensus cannot be achieved for the team of vehicles because the information states and their derivatives from different separated groups do not affect each other. We also know that $-\mathcal{L}_n$ has at least two zero eigenvalues in this case from the necessity part of the proof of Lemma 2.4, which in turn implies that Θ has at least four zero eigenvalues.

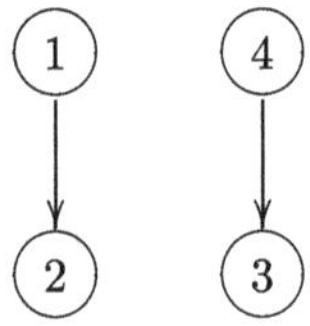

Fig. 4.1. A directed graph with separated subgroups

Figure 4.2 shows the evolution of the information states ξ_i and their derivatives ζ_i, $i = 1, \cdots, 4$, using consensus algorithm (4.2) under the interaction topology given by Fig. 4.1. Note that vehicles 1 and 2 reach consensus, and vehicles 3 and 4 also reach consensus although the whole team cannot reach consensus because the separate subgroups do not exchange information with each other. □

Example 4.5. When the interaction topology has multiple vehicles that have only outgoing edges, as shown in Fig. 4.3, where both vehicle 1 and vehicle 4 have only outgoing edges, consensus cannot be achieved for the team of vehicles because those vehicles' information states and their derivatives are not affected by any other vehicle in the team. Noting that $-\mathcal{L}_n$ has at least two rows with all zero entries in this case, we know that $-\mathcal{L}_n$ has at least two zero eigenvalues, which in turn implies that Θ has at least four zero eigenvalues.

Figure 4.4 shows the evolution of the information states ξ_i and their derivatives ζ_i, $i = 1, \cdots, 4$, using consensus algorithm (4.2) under the interaction topology given by Fig. 4.3. Note that only vehicles 1 and 2 reach consensus. □

Example 4.6. When the interaction topology is undirected and connected, as shown in Fig. 4.5, consensus is achieved asymptotically. Note that in the case of undirected graphs, $-\mathcal{L}_n$ has a simple zero eigenvalue and all other eigenvalues are negative if and only if the associated graph is connected.

Figure 4.6 shows the evolution of the information states ξ_i and their derivatives ζ_i, $i = 1, \cdots, 4$, using consensus algorithm (4.2) under the interaction topology given by Fig. 4.5. Note that consensus is reached for all vehicles. □

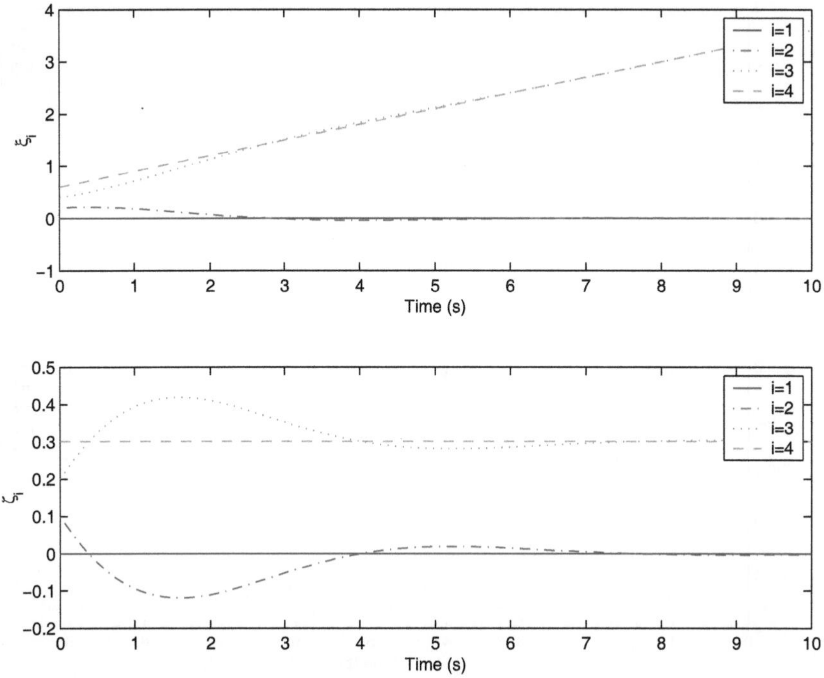

Fig. 4.2. Evolution of the information states and their derivatives under the interaction topology given by Fig. 4.1

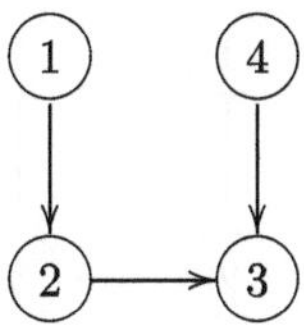

Fig. 4.3. A directed graph with multiple vehicles having only outgoing edges

Example 4.7. When the interaction topology is itself a directed spanning tree, as shown in Fig. 4.7, it is straightforward to see that $-\mathcal{L}_n$ can be written as an upper diagonal matrix by permutation transformations. As a result, we know that zero is a simple eigenvalue of $-\mathcal{L}_n$ and all other eigenvalues of $-\mathcal{L}_n$ are negative. Therefore, we know that consensus is achieved asymptotically in the case of a directed spanning tree.

Figure 4.8 shows the evolution of the information states ξ_i and their derivatives ζ_i, $i = 1, \cdots, 4$, using consensus algorithm (4.2) under the interaction topology given by Fig. 4.7. Note that consensus is reached for all vehicles. □

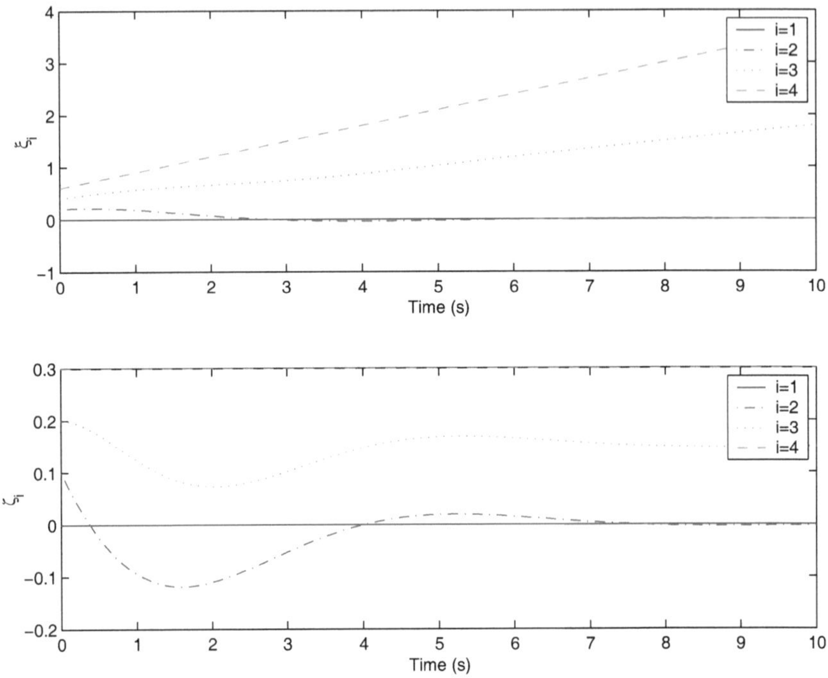

Fig. 4.4. Evolution of the information states and their derivatives under the interaction topology given by Fig. 4.3.

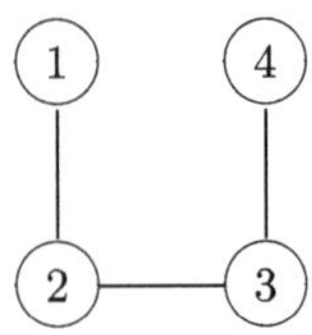

Fig. 4.5. A connected undirected graph

Example 4.8. Note that the undirected connected topology and the directed spanning tree can be considered special cases of an interaction topology that has a directed spanning tree. When the interaction topology has a directed spanning tree, as shown in Fig. 4.9, consensus may not be achieved, where the consensus algorithm is given by (2.2). However, having a directed spanning tree is a necessary condition for information consensus, as will be shown below.

Figures 4.10 and 4.11 show the evolution of the information states ξ_i and their derivatives ζ_i, $i = 1, \cdots, 4$, using consensus algorithm (4.2) under the interaction topology given by Fig. 4.9 with $\gamma = 1$ and $\gamma = 0.4$, respectively. Note that consensus is reached for $\gamma = 1$ but is not reached for $\gamma = 0.4$.

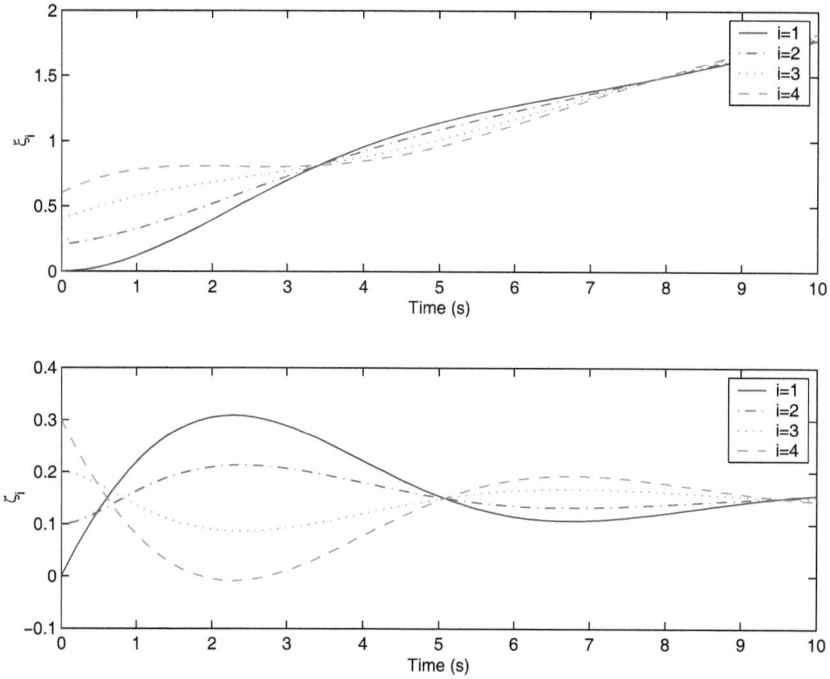

Fig. 4.6. Evolution of the information states and their derivatives under the interaction topology given by Fig. 4.5

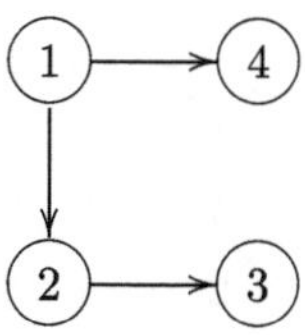

Fig. 4.7. A directed spanning tree

Unlike the previous cases where convergence does not depend on γ, consensus may not generally be reached with small γ, given interaction topologies with a directed spanning tree other than those from Examples 4.6 and 4.7.

By comparing Figs. 4.7 and 4.9, we see that more information exchange is involved in Fig. 4.9 than in Fig. 4.7 in the sense that vehicle 3 sends information to vehicle 1 in Fig. 4.9. However, although the consensus algorithm converges under the interaction topology given by Fig. 4.7 for any $\gamma > 0$, consensus algorithm (4.2) does not converge under the interaction topology given by Fig. 4.9 if γ is too small. This is somewhat contradictory to our intuition

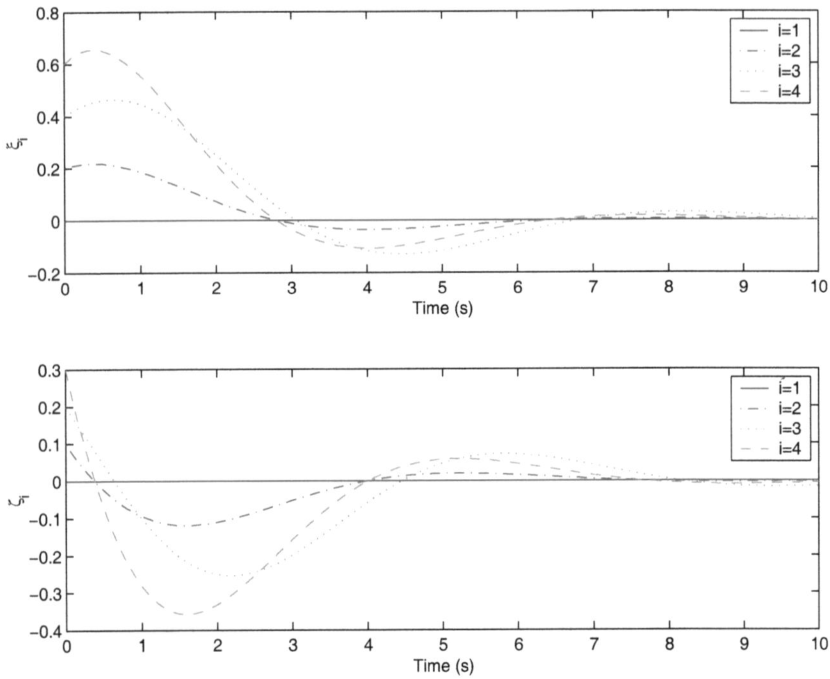

Fig. 4.8. Evolution of the information states and their derivatives under the interaction topology given by Fig. 4.7

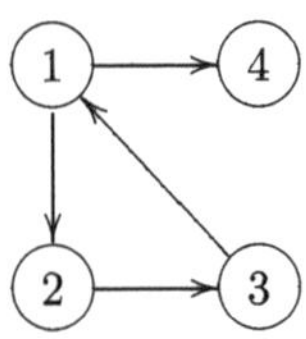

Fig. 4.9. A directed graph with a directed spanning tree

in the sense that more information exchange may lead to instability for the whole team. □

We have the following necessary condition for information consensus:

Theorem 4.9. *If algorithm* (4.2) *achieves consensus asymptotically, then directed graph* $\mathcal{G}_n$ *has a directed spanning tree.*[3]

[3] In contrast, according to Theorem 2.8, the first-order consensus algorithm (2.2) achieves consensus asymptotically if and only if the interaction topology has a directed spanning tree.

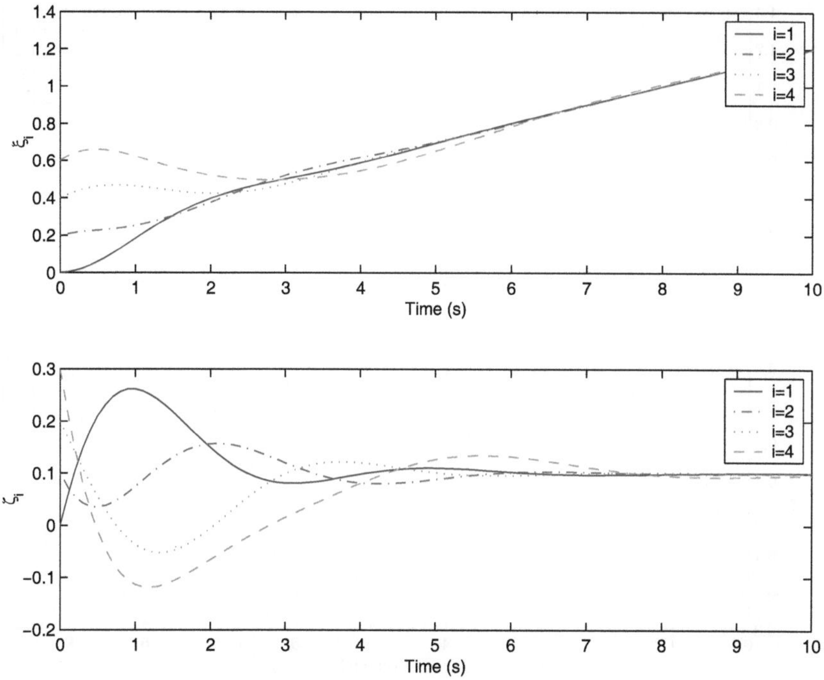

Fig. 4.10. Evolution of the information states and their derivatives under the interaction topology given by Fig. 4.9 with $\gamma = 1$

Proof: If algorithm (4.2) achieves consensus asymptotically, following Lemma 4.1, we know that Θ has exactly two zero eigenvalues. Therefore, we see that $\mathcal{L}_n$ has a simple zero eigenvalue. Corollary 2.5 implies that $\mathcal{G}_n$ has a directed spanning tree. ■

Before showing a sufficient condition for information consensus, we need the following lemma:

Lemma 4.10. *Let*

$$\rho_{\pm} = \frac{\gamma\mu - \alpha \pm \sqrt{(\gamma\mu - \alpha)^2 + 4\mu}}{2},$$

where $\rho, \mu \in \mathbb{C}$. *If* $\alpha \geq 0$, $\mathrm{Re}(\mu) < 0$, $\mathrm{Im}(\mu) > 0$, *and*

$$\gamma > \sqrt{\frac{2}{|\mu| \cos\left[\tan^{-1} \frac{\mathrm{Im}(\mu)}{-\mathrm{Re}(\mu)}\right]}}, \tag{4.9}$$

then $\mathrm{Re}(\rho_{\pm}) < 0$, *where* $\mathrm{Re}(\cdot)$ *and* $\mathrm{Im}(\cdot)$ *represent, respectively, the real and imaginary parts of a number.*

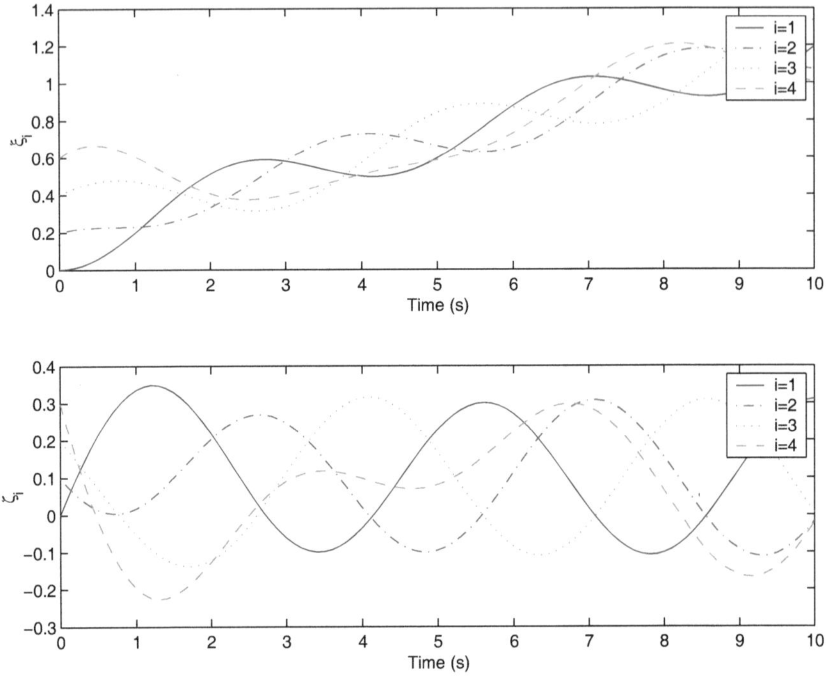

Fig. 4.11. Evolution of the information states and their derivatives under the interaction topology given by Fig. 4.9 with $\gamma = 0.4$. Consensus is not reached in this case.

Proof: [4] We use Fig. 4.12 to show the notations used in the proof. Let $a \triangleq \gamma\mu$, $b \triangleq \sqrt{(\gamma\mu)^2 + 4\mu}$, and $c \triangleq \gamma\mu - \alpha$. Also let $s_1 \triangleq (\gamma\mu)^2$, $s_2 \triangleq 4\mu$, and $s_3 \triangleq (\gamma\mu)^2 + 4\mu$. Furthermore, we let $q_1 \triangleq (\gamma\mu - \alpha)^2$, $q_2 \triangleq 4\mu$, and $q_3 \triangleq (\gamma\mu - \alpha)^2 + 4\mu$.

When $\alpha = 0$, consider the triangle formed by vectors s_1, s_2, and s_3, where η_i is the inner angle of the triangle that faces the edge s_i. According to the law of cosines, $|s_3|^2 = |s_1|^2 + |s_2|^2 - 2|s_1||s_2|\cos\eta_3$. Note that if $\gamma > \sqrt{\frac{2}{|\mu|\cos\eta_3}}$, then $|s_1|^2 > |s_3|^2$. Also note that $\eta_3 = \tan^{-1}\frac{\mathrm{Im}(\mu)}{-\mathrm{Re}(\mu)}$. Therefore, Inequality (4.9) guarantees that $|s_1| > |s_3|$, which in turn implies that $|a| > |b|$. Noting that the phase angle of b is smaller than a, we know that $\mathrm{Re}(\rho_\pm) = \mathrm{Re}(\frac{a \pm b}{2}) < 0$.

When $\alpha > 0$, note that Inequality (4.9) implies that $|s_1| > |s_3|$ and hence $\eta_1 > \eta_3$. Represent a and c in polar coordinates as (r_1, θ_1) and (r_2, θ_2), respectively, where $\theta_i \in (\frac{\pi}{2}, \pi)$, $i = 1, 2$. Then s_1 and q_1 can be represented in

[4] The proof is motivated by [203]. In particular, the proof for the case of $\alpha = 0$ follows directly from Theorem 6 in [203].

polar coordinates as $(r_1^2, 2\theta_1)$ and $(r_2^2, 2\theta_2)$, respectively. Consider another triangle composed of vectors q_1, q_2, and q_3 with inner angles given by ϕ_i, where ϕ_i faces the edge q_i. Noting that $\theta_2 > \theta_1$, we know that $2\theta_2 > 2\theta_1$. We can then show that $\phi_1 > \eta_1$ and $\phi_3 < \eta_3$ by noting that $s_2 = q_2$ and comparing the triangles composed of q_i and s_i respectively, which implies that $\phi_1 > \phi_3$. Using the law of cosines, we know that $|q_1| > |q_3|$. As a result, we know that $\text{Re}(\rho_\pm) < 0$ by following an argument similar to that for $\alpha = 0$. ■

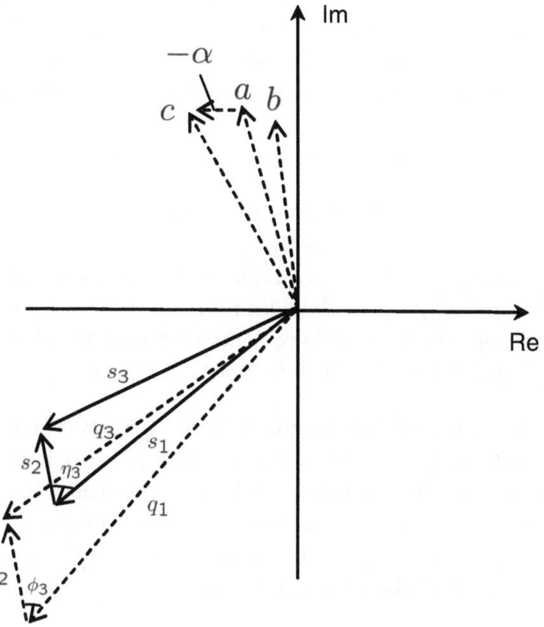

Fig. 4.12. Graphical view of notations used in the proof of Lemma 4.10

Theorem 4.11. *Let μ_i, $i = 1, \cdots, n$, denote the ith eigenvalue of $-\mathcal{L}_n$. Algorithm (4.2) achieves consensus asymptotically if directed graph $\mathcal{G}_n$ has a directed spanning tree and*

$$\gamma > \bar{\gamma}, \tag{4.10}$$

where $\bar{\gamma} \triangleq 0$ if all of the $n-1$ nonzero eigenvalues of $-\mathcal{L}_n$ are negative and

$$\bar{\gamma} = \max_{\forall \text{Re}(\mu_i)<0 \text{ and } \text{Im}(\mu_i)>0} \sqrt{\frac{2}{|\mu_i| \cos\left[\tan^{-1} \frac{\text{Im}(\mu_i)}{-\text{Re}(\mu_i)}\right]}}$$

otherwise.

Proof: If $\mathcal{G}_n$ has a directed spanning tree, Corollary 2.5 implies that $-\mathcal{L}_n$ has a simple zero eigenvalue and all other eigenvalues have negative real parts. Without loss of generality, we let $\mu_1 = 0$ and $\text{Re}(\mu_i) < 0$, $i = 2, \cdots, n$. Then we know that Θ has exactly two zero eigenvalues. Note that if $\mu_i < 0$, then (4.7) implies that $\text{Re}(\lambda_{i\pm}) < 0$ for any $\gamma > 0$, where $\lambda_{i\pm}$ are eigenvalues of Θ associated with μ_i. It is left to show that Inequality (4.10) guarantees that all eigenvalues of Θ associated with μ_i that satisfy $\text{Re}(\mu_i) < 0$ and $\text{Im}(\mu_i) \neq 0$ have negative real parts. We need only to consider μ_i that satisfies $\text{Re}(\mu_i) < 0$ and $\text{Im}(\mu_i) > 0$ because any μ_i that satisfies $\text{Re}(\mu_i) < 0$ and $\text{Im}(\mu_i) < 0$ is a complex conjugate of some μ_i that satisfies $\text{Re}(\mu_i) < 0$ and $\text{Im}(\mu_i) > 0$. This argument then follows from Lemma 4.10. As a result, Lemma 4.1 implies that consensus can be achieved asymptotically under the assumption of the theorem. ■

We also have the following lemma regarding the consensus equilibrium.

Lemma 4.12. *Suppose that Θ has exactly two zero eigenvalues and all other eigenvalues have negative real parts. If $\zeta_i(0) = 0$, $i = 1, \ldots, n$, then $\xi_i(t) \to \sum_{i=1}^n p_i \xi_i(0)$ and $\zeta_i(t) \to 0$, $i = 1, \ldots, n$, as $t \to \infty$, where $p = [p_1, \ldots, p_n]^T \geq 0$, $\mathbf{1}_n^T p = 1$, and $\mathcal{L}_n^T p = 0$. In addition, if $\zeta_i(0) = 0$, $i \in \mathcal{I}_L$, where $\mathcal{I}_L$ denotes the set of vehicles that have a directed path to all other vehicles, then $\xi_i(t) \to \sum_{i \in \mathcal{I}_L} p_i \xi_i(0)$ and $\zeta_i(t) \to 0$, $i = 1, \ldots, n$, as $t \to \infty$.*

Proof: The first part of the lemma follows directly from Lemma 4.1. For the second part of the lemma, we note that $p_i \neq 0$ if vehicle i has a directed path to all other vehicles and $p_i = 0$ otherwise (see the subsection on consensus equilibrium in Section 2.2.1). As a result, we know that $\xi_i(t) \to \sum_{i \in \mathcal{I}_L} p_i \xi_i(0) + t \sum_{i \in \mathcal{I}_L} p_i \zeta_i(0)$ and $\zeta_i(t) \to \sum_{i \in \mathcal{I}_L} p_i \zeta_i(0)$ as $t \to \infty$, and the second part of the lemma is proved. ■

Remark 4.13. Relevant to this subsection are [69] and [203], where formation stabilization and alignment problems are considered for multiple vehicles modeled by double-integrator or more complicated linear dynamics. In [69], information exchange techniques are studied to improve stability margins and the accuracy of vehicle formations. Reference [203] derives matrix theoretical conditions under which alignment can be achieved for multiple vehicles with double-integrator dynamics in a general multi-observation setting. In contrast, this subsection applies graph theoretical tools to explore explicit graphical conditions on the interaction topologies under which consensus can be achieved. As another comparison, this subsection focuses on analyzing whether a given consensus algorithm converges, whereas [69] and [203] consider whether a feedback gain exists to achieve formation stabilization or alignment. □

4.1.2 Convergence Analysis Under Switching Interaction Topologies

In the case of switching interaction topologies, the convergence analysis is more involved than that for a fixed interaction topology. For first-order consensus algorithm (2.2), we have shown in Theorem 2.33 that consensus can be achieved asymptotically under switching interaction topologies if there exist infinitely many uniformly bounded, consecutive time intervals such that the union of the directed graphs across each such interval has a directed spanning tree. However, as shown in the following example, this condition is generally not sufficient for information consensus for second-order consensus algorithm (4.2).

Example 4.14. Let

$$\mathcal{L}_{3(1)} = \begin{bmatrix} 1 & -1 & 0 \\ 0 & 0 & 0 \\ 0 & 0 & 0 \end{bmatrix}, \quad \mathcal{L}_{3(2)} = \begin{bmatrix} 1 & -1 & 0 \\ 0 & 1 & -1 \\ -2 & 0 & 2 \end{bmatrix}, \quad \text{and} \quad \mathcal{L}_{3(3)} = \begin{bmatrix} 0 & 0 & 0 \\ 0 & 1 & -1 \\ 0 & 0 & 0 \end{bmatrix}.$$

Also let $\gamma_1 = \gamma_2 = \gamma_3 = 1$. Let Θ_i be defined as

$$\Theta_i = \begin{bmatrix} 0_{3\times 3} & I_3 \\ -\mathcal{L}_{3(i)} & -\gamma_i \mathcal{L}_{3(i)} \end{bmatrix},$$

where $i = 1, 2, 3$. Note that the directed graphs of $\mathcal{L}_{3(1)}$ and $\mathcal{L}_{3(3)}$, denoted $\Gamma[\mathcal{L}_{3(1)}]$ and $\Gamma[\mathcal{L}_{3(3)}]$ respectively, do not have directed spanning trees whereas the directed graph of $\mathcal{L}_{3(2)}$, denoted $\Gamma[\mathcal{L}_{3(2)}]$, does. Also note that Θ_2 has exactly two zero eigenvalues and all other nonzero eigenvalues have negative real parts, whereas both Θ_1 and Θ_3 have exactly four zero eigenvalues and all other nonzero eigenvalues have negative real parts. At each time interval of 5 seconds, we let the interaction topology be $\Gamma[\mathcal{L}_{3(1)}]$ 90% of the time and be $\Gamma[\mathcal{L}_{3(2)}]$ the rest of the time. Note that at each time interval of 5 seconds, the union of the interaction topologies $\Gamma[\mathcal{L}_{3(1)}] \cup \Gamma[\mathcal{L}_{3(2)}]$ has a directed spanning tree. Using first-order algorithm (2.2), consensus is achieved, as shown in Fig. 4.13. However, as shown in Fig. 4.14, consensus is not achieved using second-order algorithm (4.2). However, if we increase the gain γ_2 to 10, consensus is achieved asymptotically, as shown in Fig. 4.15. Alternatively, if we reduce the length of each time interval to 1 second, consensus is achieved asymptotically, as shown in Fig. 4.16. In addition, if we let the interaction topology be $\Gamma[\mathcal{L}_{3(1)}]$ 50% of the time and be $\Gamma[\mathcal{L}_{3(2)}]$ the rest of the time, consensus is achieved asymptotically, as shown in Fig. 4.17. Next, at each time interval of 5 seconds, we let the interaction topology be $\Gamma[\mathcal{L}_{3(1)}]$ 90% of the time and be $\Gamma[\mathcal{L}_{3(3)}]$ the rest of the time. Note that at each time interval of 5 seconds, the union of the interaction topologies $\Gamma[\mathcal{L}_{3(1)}] \cup \Gamma[\mathcal{L}_{3(3)}]$ has a directed spanning tree. Also note that $\Gamma[\mathcal{L}_{3(3)}]$ is only a subgraph of $\Gamma[\mathcal{L}_{3(2)}]$. In contrast to Fig. 4.14, Fig. 4.18 shows that consensus is achieved asymptotically even if $\Gamma[\mathcal{L}_{3(3)}]$ has less information exchange than $\Gamma[\mathcal{L}_{3(2)}]$.
□

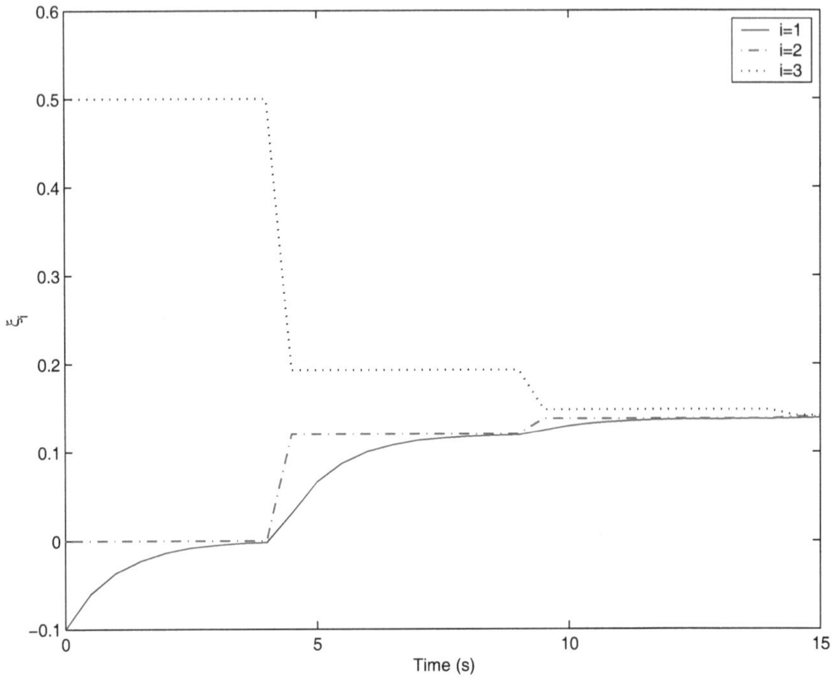

Fig. 4.13. Consensus of information under directed switching interaction topologies using first-order algorithm (2.2) $\{\Gamma[\mathcal{L}_{3(1)}] : 90\%, \Gamma(\mathcal{L}_{3(2)}) : 10\%\}$. Consensus is achieved in this case.

In the simple case where the interaction topology among vehicles is undirected and is based on their physical proximity, that is, there is information exchange between vehicle i and j if and only if the distance between them is below a certain threshold, we have the following lemma for information consensus:

Lemma 4.15. *If time-varying graph $\mathcal{G}_n(t)$ is undirected and connected at each instant, algorithm* (4.2) *achieves consensus asymptotically.*

Proof: Let $V_{ij} = \frac{1}{2}\|\xi_i - \xi_j\|^2$. The second equation in (4.1) can be rewritten as

$$\dot{\zeta}_i = -\sum_{j=1}^{n} a_{ij}\frac{\partial V_{ij}}{\partial \xi_i} - \gamma \sum_{j=1}^{n} a_{ij}(\zeta_i - \zeta_j). \tag{4.11}$$

Note that $a_{ij} = a_{ji}$ because $\mathcal{G}_n(t)$ is undirected. Also note that (4.11) can be written in matrix form as $\dot{\zeta} = -[\mathcal{L}_n(t) \otimes I_m]\xi - \gamma[\mathcal{L}_n(t) \otimes I_m]\zeta$, where $\mathcal{L}_n(t)$ is the Laplacian matrix associated with undirected graph $\mathcal{G}_n$ at time t. Noting that (4.11) has the same form as (4) in [231], we can follow a proof similar to that of Theorem VI.2 in [231] to show that $\|\zeta_i(t) - \zeta_j(t)\| \to 0$ and $\dot{\zeta}_i(t) \to 0$,

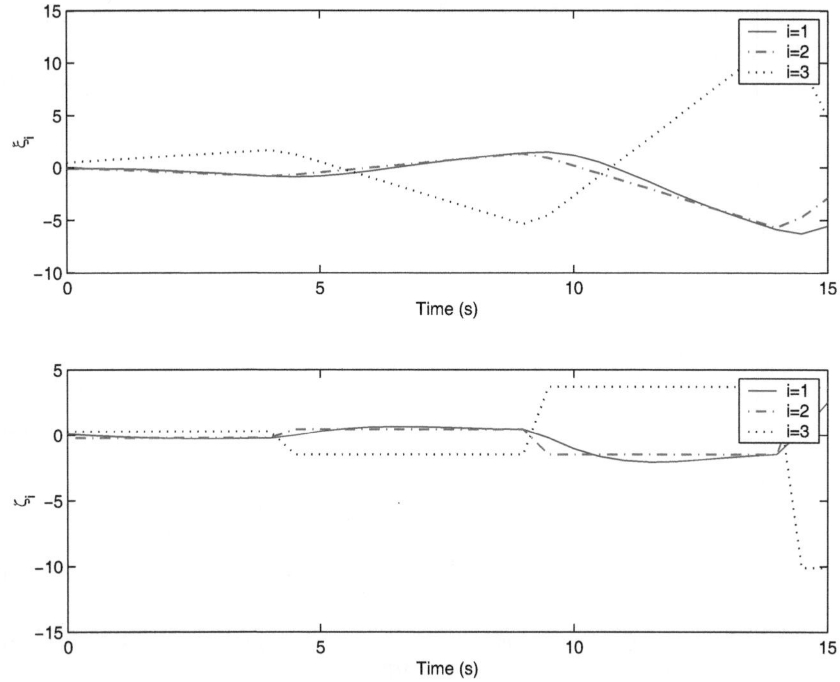

Fig. 4.14. Consensus of information under switching interaction topologies using second-order algorithm (4.2) $\{\Gamma[\mathcal{L}_{3(1)}] : 90\%, \Gamma[\mathcal{L}_{3(2)}] : 10\%\}$. Consensus is not achieved in this case.

as $t \to \infty$. As a result, we know that $[\mathcal{L}_n(t) \otimes I_m]\xi(t) \to 0$, which implies that $\|\xi_i(t) - \xi_j(t)\| \to 0$, as $t \to \infty$, because undirected graph $\mathcal{G}_n(t)$ is connected at each instant. ■

In the general case where the interaction topology among vehicles is directed and is switching randomly with time, we assume that (4.3) can be written as

$$\begin{bmatrix} \dot{\xi} \\ \dot{\zeta} \end{bmatrix} = (\Theta_\sigma \otimes I_m) \begin{bmatrix} \xi \\ \zeta \end{bmatrix},$$

where $\sigma : [0, \infty) \to \mathcal{P}$ is a piecewise constant switching signal with switching times $t_0, t_1, \cdots$ and $\mathcal{P}$ denotes a set indexing the class of all possible directed interaction topologies for the n vehicles that have directed spanning trees. Here we assume that $\Theta(t)$ is piecewise constant and satisfies $\Theta(t) = \Theta(t_i)$, $t \in [t_i, t_{i+1})$.

Let $\xi_{ij} \triangleq \xi_i - \xi_j$ and $\zeta_{ij} \triangleq \zeta_i - \zeta_j$ be the consensus error variables. Note that $\xi_{ij} = \xi_{1j} - \xi_{1i}$ and $\zeta_{ij} = \zeta_{1j} - \zeta_{1i}$. Defining the consensus error vector as $\tilde{\xi} = [\xi_{12}^T, \xi_{13}^T, \ldots, \xi_{1n}^T]^T$ and $\tilde{\zeta} = [\zeta_{12}^T, \zeta_{13}^T, \ldots, \zeta_{1n}^T]^T$, we get the following equation:

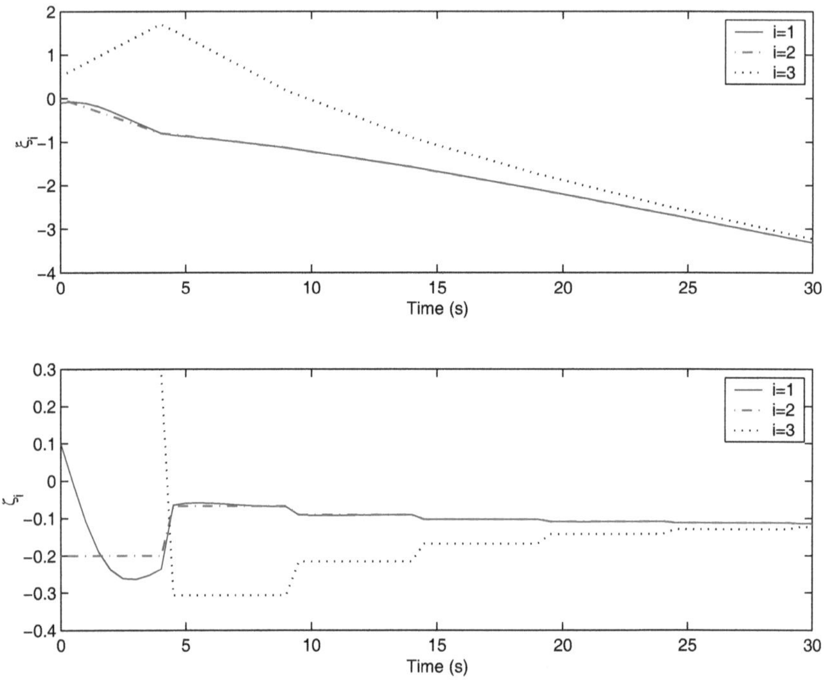

Fig. 4.15. Consensus of information under directed switching interaction topologies using second-order algorithm (4.2) with increased γ_2 $\{\Gamma[\mathcal{L}_{3(1)}]$: 90%, $\Gamma[\mathcal{L}_{3(2)}]$: 10%$\}$. Consensus is achieved in this case.

$$\begin{bmatrix} \dot{\tilde{\xi}} \\ \dot{\tilde{\zeta}} \end{bmatrix} = (\Delta_\sigma \otimes I_m) \begin{bmatrix} \tilde{\xi} \\ \tilde{\zeta} \end{bmatrix}, \tag{4.12}$$

where Δ_σ is a $2(n-1) \times 2(n-1)$ matrix that can be derived from Θ_σ. If Δ_σ is stable, we can find $a_\sigma \geq 0$ and $\chi_\sigma > 0$ such that $\left\|e^{\Delta_\sigma t}\right\| \leq e^{(a_\sigma - \chi_\sigma t)}$, $t \geq 0$.

Before stating the theorem for information consensus under directed switching interaction topologies, we need the following lemma:

Lemma 4.16. *[146, Lemma 2] Let $\{A_p : p \in \mathcal{P}\}$ be a closed, bounded set of real, $n \times n$ matrices. Suppose that for each $p \in \mathcal{P}$, A_p is stable, and let a_p and χ_p be any finite, nonnegative and positive numbers, respectively, for which $\left\|e^{A_p t}\right\| \leq e^{(a_p - \chi_p t)}$, $t \geq 0$. Suppose that τ_0 is a number satisfying $\tau_0 > \sup_{p \in \mathcal{P}}\{\frac{a_p}{\chi_p}\}$. For any admissible switching signal $\sigma : [0, \infty) \to \mathcal{P}$ with dwell time no smaller than τ_0, the transition matrix of A_σ satisfies $\|\Phi(t, \mu)\| \leq e^{(a - \chi(t-\mu))}$, $\forall t \geq \mu \geq 0$, where $a = \sup_{p \in \mathcal{P}}\{a_p\}$ and $\chi = \inf_{p \in \mathcal{P}}\{\chi_p - \frac{a_p}{\tau_0}\}$.*

Theorem 4.17. *Let $t_0, t_1, \cdots$ be the times when directed graph $\mathcal{G}_n(t)$ switches. Also let τ be the dwell time such that $t_{i+1} - t_i \geq \tau$, $\forall i = 0, 1, \cdots$. If $\mathcal{G}_n(t)$ has a*

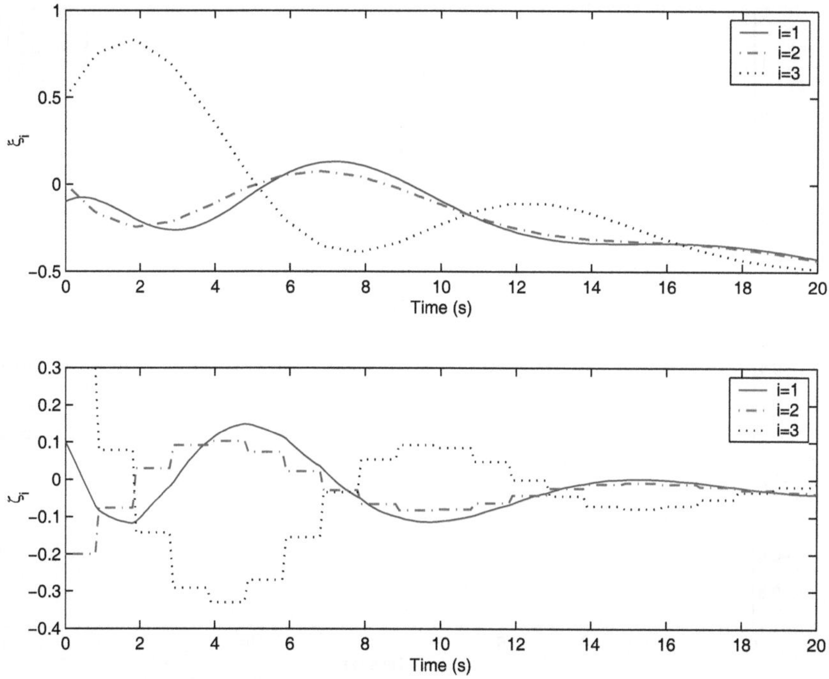

Fig. 4.16. Consensus of information under directed switching interaction topologies using second-order algorithm (4.2) with decreased interval length $\{\Gamma[\mathcal{L}_{3(1)}] : 90\%, \Gamma[\mathcal{L}_{3(2)}] : 10\%\}$. Consensus is achieved in this case.

directed spanning tree for each $t \in [t_i, t_{i+1})$, γ is constant and satisfies (4.10) *for each $\sigma \in \mathcal{P}$, and the dwell time τ satisfies $\tau > \sup_{\sigma\in\mathcal{P}}\{\frac{a_\sigma}{\chi_\sigma}\}$, then algorithm* (4.2) *achieves consensus asymptotically and is robust to perturbations under directed switching interaction topologies.*

Proof: Given a certain $\sigma_\ell \in \mathcal{P}$, note that $\mathcal{G}_n(t)$ has a directed spanning tree for $t \in [t_\ell, t_{\ell+1})$ and γ satisfies (4.10). Then we know that consensus is achieved asymptotically if $\sigma(t) \equiv \sigma_\ell$, $\forall t \geq 0$, from Theorem 4.11, that is, $\xi_i \to \xi_j$ and $\zeta_i \to \zeta_j$, $\forall i \neq j$, if $\sigma(t) \equiv \sigma_\ell$. Equivalently, we know that $\tilde{\xi} \to 0$ and $\tilde{\zeta} \to 0$ asymptotically if $\sigma(t) \equiv \sigma_\ell$, which implies that switched system (4.12) is stable for each $\sigma \in \mathcal{P}$ under the conditions of the theorem. As a result, switched system (4.12) is globally exponentially stable if dwell time τ satisfies $\tau > \sup_{\sigma\in\mathcal{P}}\{\frac{a_\sigma}{\chi_\sigma}\}$, according to Lemma 4.16. The stability of switched system (4.12) implies that consensus can be achieved asymptotically. The robustness of consensus algorithm (4.2) to perturbations comes from the fact that (4.12) is globally exponentially stable. ■

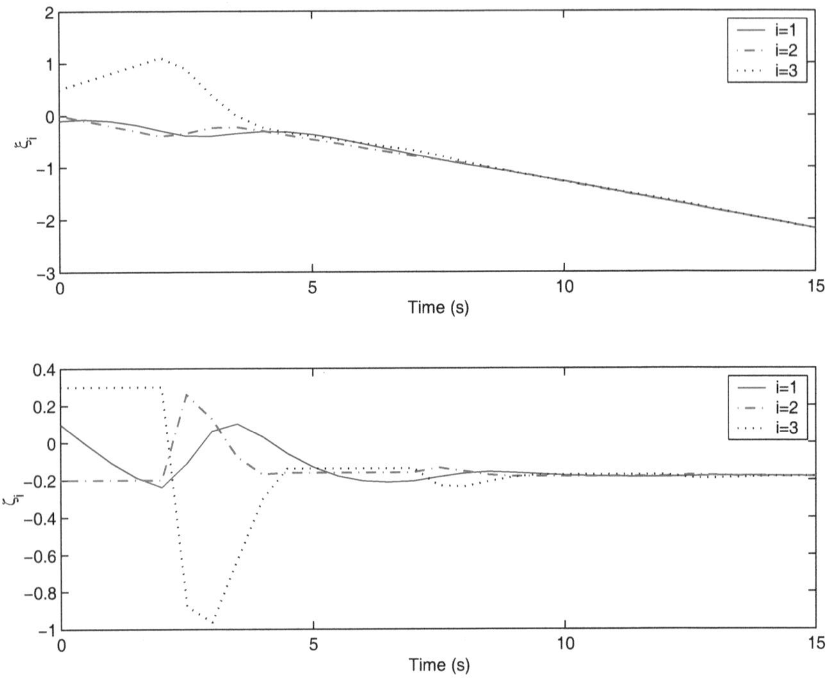

Fig. 4.17. Consensus of information under directed switching interaction topologies using second-order algorithm (4.2) $\{\Gamma[\mathcal{L}_{3(1)}] : 50\%, \Gamma[\mathcal{L}_{3(2)}] : 50\%\}$. Consensus is achieved in this case.

4.2 Consensus with Bounded Control Inputs

Note that (4.2) does not explicitly take into account a bounded control effort. We propose a consensus algorithm for (4.1) with bounded control inputs as

$$u_i = -\sum_{j=1}^{n} \{a_{ij} \tanh[K_r(\xi_i - \xi_j)] + b_{ij} \tanh[K_v(\zeta_i - \zeta_j)]\}, \quad i = 1, \ldots, n, \quad (4.13)$$

where $K_r \in \mathbb{R}^{m \times m}$ and $K_v \in \mathbb{R}^{m \times m}$ are constant positive-definite diagonal matrices, a_{ij} and b_{ij} are, respectively, the (i,j) entry of the adjacency matrix $\mathcal{A}_n \in \mathbb{R}^{n \times n}$ and $\mathcal{B}_n \in \mathbb{R}^{n \times n}$ associated with $\mathcal{G}_n^A$ and $\mathcal{G}_n^B$[5] respectively, and $\tanh(\cdot)$ is defined componentwise as in Section 3.3.2. Note that with (4.13), u_i is bounded because $\tanh(\cdot)$ is bounded. In particular, it follows that $\|u_i\|_\infty \leq \sum_{j=1}^{n}(a_{ij} + b_{ij})$, which is independent of the initial conditions of the information states and their derivatives.

Before moving on, we need the following lemma:

[5] In this case, the interaction topologies for ξ_i and ζ_i are allowed to be different.

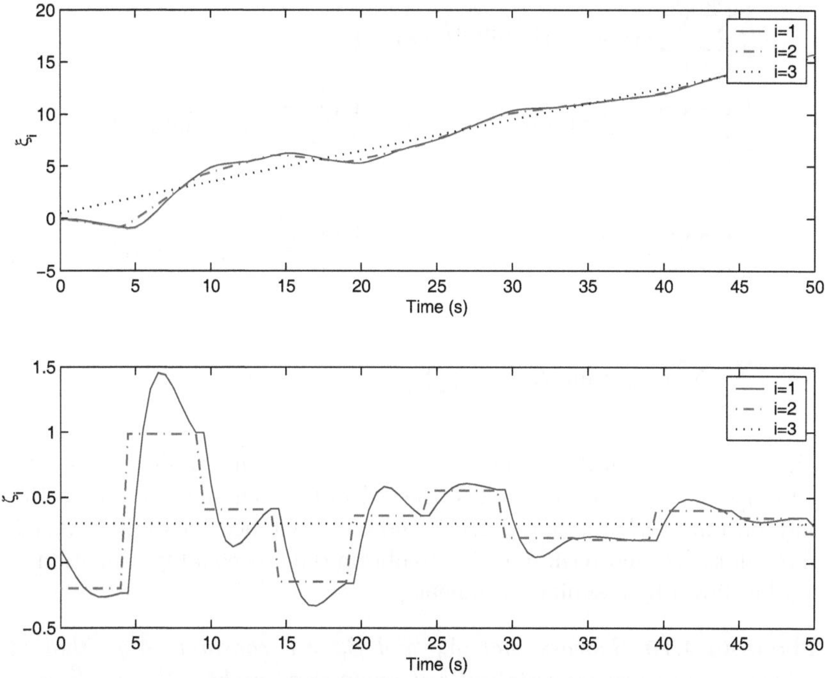

Fig. 4.18. Consensus of information under directed switching interaction topologies using second-order algorithm (4.2) $\{\Gamma[\mathcal{L}_{3(1)}] : 90\%, \Gamma[\mathcal{L}_{3(3)}] : 10\%\}$. Consensus is achieved in this case.

Lemma 4.18. *Suppose that* $\varsigma \in \mathbb{R}^m$, $\varphi \in \mathbb{R}^m$, $K \in \mathbb{R}^{m\times m}$, *and* $C = [c_{ij}] \in \mathbb{R}^{n\times n}$. *If* C *is symmetrical, then*

$$\frac{1}{2}\sum_{i=1}^{n}\sum_{j=1}^{n} c_{ij}(\varsigma_i - \varsigma_j)^T \tanh[K(\varphi_i - \varphi_j)] = \sum_{i=1}^{n}\sum_{j=1}^{n} c_{ij}\varsigma_i^T \tanh[K(\varphi_i - \varphi_j)], \tag{4.14}$$

and

$$\frac{1}{2}\sum_{i=1}^{n}\sum_{j=1}^{n} c_{ij}(\varsigma_i - \varsigma_j)^T(\varphi_i - \varphi_j) = \sum_{i=1}^{n}\sum_{j=1}^{n} c_{ij}\varsigma_i^T(\varphi_i - \varphi_j). \tag{4.15}$$

Proof: To show (4.14), note that

$$\begin{aligned}
&\frac{1}{2}\sum_{i=1}^{n}\sum_{j=1}^{n} c_{ij}(\varsigma_i - \varsigma_j)^T \tanh[K(\varphi_i - \varphi_j)] \\
=&\frac{1}{2}\sum_{i=1}^{n}\sum_{j=1}^{n} c_{ij}\varsigma_i^T \tanh[K(\varphi_i - \varphi_j)] - \frac{1}{2}\sum_{i=1}^{n}\sum_{j=1}^{n} c_{ij}\varsigma_j^T \tanh[K(\varphi_i - \varphi_j)] \qquad (4.16) \\
=&\frac{1}{2}\sum_{i=1}^{n}\sum_{j=1}^{n} c_{ij}\varsigma_i^T \tanh[K(\varphi_i - \varphi_j)] + \frac{1}{2}\sum_{j=1}^{n}\sum_{i=1}^{n} c_{ji}\varsigma_j^T \tanh[K(\varphi_j - \varphi_i)] \qquad (4.17) \\
=&\sum_{i=1}^{n}\sum_{j=1}^{n} c_{ij}\varsigma_i^T \tanh[K(\varphi_i - \varphi_j)],
\end{aligned}$$

where we have used the fact that $c_{ij} = c_{ji}$ and $\tanh[K(\varphi_j - \varphi_i)] = -\tanh[K(\varphi_i - \varphi_j)]$, have switched the order of the summation signs in the second term in (4.16) to obtain (4.17), and have switched the dummy variables i and j in the second term in (4.17) to obtain the last equality. Equation (4.15) can be shown by a similar argument. ■

Theorem 4.19. *Suppose that $\mathcal{A}_n$ and $\mathcal{B}_n$ are constant. Algorithm* (4.13) *achieves consensus asymptotically if undirected graphs $\mathcal{G}_n^A$ and $\mathcal{G}_n^B$ are connected.*

Proof: Note that with (4.13), (4.1) can be written as

$$\dot{\xi}_{ij} = \zeta_i - \zeta_j, \quad \dot{\zeta}_i = -\sum_{j=1}^{n}\{a_{ij}\tanh(K_r\xi_{ij}) + b_{ij}\tanh[K_v(\zeta_i - \zeta_j)]\}, \qquad (4.18)$$

where $\xi_{ij} \stackrel{\triangle}{=} \xi_i - \xi_j$. Consider the Lyapunov function candidate for (4.18) as

$$V = \frac{1}{2}\sum_{i=1}^{n}\sum_{j=1}^{n} a_{ij}\mathbf{1}_m^T K_r^{-1}\log[\cosh(K_r\xi_{ij})] + \frac{1}{2}\sum_{i=1}^{n}\zeta_i^T\zeta_i, \qquad (4.19)$$

where $\cosh(\cdot)$ and $\log(\cdot)$ are defined componentwise. Note that V is positive definite and radially unbounded with respect to ξ_{ij}, $\forall i \neq j$, and ζ_i if undirected graph $\mathcal{G}_n^A$ is connected. Differentiating V gives

$$\begin{aligned}\dot{V} &= \frac{1}{2}\sum_{i=1}^{n}\sum_{j=1}^{n} a_{ij}(\zeta_i - \zeta_j)^T \tanh(K_r\xi_{ij}) \\ &\quad - \sum_{i=1}^{n}\zeta_i^T\left(\sum_{j=1}^{n}\{a_{ij}\tanh(K_r\xi_{ij}) + b_{ij}\tanh[K_v(\zeta_i - \zeta_j)]\}\right) \\ &= -\sum_{i=1}^{n}\zeta_i^T\left\{\sum_{j=1}^{n} b_{ij}\tanh[K_v(\zeta_i - \zeta_j)]\right\} \\ &= -\frac{1}{2}\sum_{i=1}^{n}\sum_{j=1}^{n} b_{ij}(\zeta_i - \zeta_j)^T \tanh[K_v(\zeta_i - \zeta_j)] \le 0,\end{aligned}$$

where we have used the fact that $\frac{\mathrm{d}\log[\cosh(x)]}{dt} = \dot{x}\tanh(x)$ with $x \in \mathbb{R}$ and have used (4.18) to obtain the first equality, have used (4.14) in Lemma 4.18 to obtain the second equality by noting that $\xi_{ij} = \xi_i - \xi_j$, have used (4.14) in Lemma 4.18 again to obtain the third equality, and have used the fact that x and $\tanh(Kx)$ have the same sign componentwise, when x is a vector and K is a positive-definite diagonal matrix, to obtain the last inequality.

Let $S = \{(\xi_{ij}, \zeta_i) | \dot{V} = 0\}$. Note that $\dot{V} \equiv 0$ implies that $\zeta_i \equiv \zeta_j$, $\forall i \neq j$, when undirected graph $\mathcal{G}_n^B$ is connected, which in turn implies that $\dot{\zeta}_i \equiv \dot{\zeta}_j$, $\forall i \neq j$. Therefore, it follows that $\dot{\zeta} \in \mathrm{span}(\mathbf{1}_n \otimes \eta)$, where $\dot{\zeta} = [\dot{\zeta}_1^T, \ldots, \dot{\zeta}_n^T]^T$ and η is some $m \times 1$ real vector, when undirected graph $\mathcal{G}_n^B$ is connected. Because $\zeta_i \equiv \zeta_j$, it follows from (4.18) that

$$\dot{\zeta}_i = -\sum_{j=1}^{n} a_{ij}\tanh(K_r\xi_{ij}), \quad i = 1, \ldots, n. \tag{4.20}$$

Note that $(\mathbf{1}_n \otimes \eta)^T\dot{\zeta} \equiv 0$ from (4.20), where we have used the fact that $a_{ij} = a_{ji}$ and $\tanh(K_r\xi_{ij}) = -\tanh(K_r\xi_{ji})$. Thus it follows that $\dot{\zeta}$ is orthogonal to $\mathbf{1}_n \otimes \eta$. Therefore, we conclude that $\dot{\zeta} \equiv 0$, which in turn implies from (4.20) that $-\sum_{j=1}^{n} a_{ij}\tanh(K_r\xi_{ij}) \equiv 0$. As a result, it follows that $-\sum_{i=1}^{n}\xi_i^T[\sum_{j=1}^{n} a_{ij}\tanh(K_r\xi_{ij})] \equiv 0$, which in turn implies that $-\frac{1}{2}\sum_{i=1}^{n}\sum_{j=1}^{n} a_{ij}\xi_{ij}^T\tanh(K_r\xi_{ij}) \equiv 0$ from (4.14) in Lemma 4.18 by noting that $\xi_{ij} = \xi_i - \xi_j$. When undirected graph $\mathcal{G}_n^A$ is connected, it follows that $\xi_{ij} \equiv 0$, *i.e.*, $\xi_i \equiv \xi_j$, $\forall i \neq j$. By Theorem F.4, it follows that $\xi_i(t) \to \xi_j(t)$ and $\zeta_i(t) \to \zeta_j(t)$, $\forall i \neq j$, as $t \to \infty$. ■

Remark 4.20. Using the algorithm $u_i = -\sum_{j=1}^{n}[a_{ij}(\xi_i - \xi_j) + b_{ij}(\zeta_i - \zeta_j)]$, $i = 1, \ldots, n$, the Lyapunov function $V = \frac{1}{2}\sum_{i=1}^{n}\sum_{j=1}^{n} a_{ij}\|\xi_i - \xi_j\|^2 + \frac{1}{2}\sum_{i=1}^{n}\zeta_i^T\zeta_i$ can be used to show that the algorithm achieves consensus asymptotically under the conditions of Theorem 4.19 by following a proof similar to that of Theorem 4.19. □

Note that (4.13) guarantees that $\xi_i(t) \to \xi_j(t)$ and $\zeta_i(t) \to \zeta_j(t)$, as $t \to \infty$. When it is desirable that $\xi_i(t) \to \xi_j(t)$ and $\zeta_i(t) \to 0$, as $t \to \infty$, we propose a consensus algorithm for (4.1) with bounded control inputs as

$$u_i = -\sum_{j=1}^{n} a_{ij} \tanh[K_r(\xi_i - \xi_j)] - \tanh(K_{vi}\zeta_i), \quad i = 1, \dots, n, \tag{4.21}$$

where $K_r \in \mathbb{R}^{m\times m}$ and $K_{vi} \in \mathbb{R}^{m\times m}$, $i = 1, \dots, n$, are positive-definite diagonal matrices. Note that in this case, only ξ_i need to be exchanged among vehicles.

Corollary 4.21. *Suppose that $\mathcal{A}_n$ is constant. With (4.21), $\xi_i(t) \to \xi_j(t)$ and $\zeta_i(t) \to 0$, as $t \to \infty$, if undirected graph $\mathcal{G}_n$ is connected.*

Proof: Following the proof of Theorem 4.19, consider the Lyapunov function candidate given by (4.19). Differentiating V, gives

$$\dot{V} = -\sum_{i=1}^{n} \zeta_i^T \tanh(K_{vi}\zeta_i) \leq 0.$$

Let $S = \{(\xi_{ij}, \zeta_i) | \dot{V} = 0\}$. Note that $\dot{V} \equiv 0$ implies that $\zeta_i \equiv 0$, which in turn implies that $\dot{\zeta}_i \equiv 0$. It thus follows from (4.1) and (4.21) that $-\sum_{j=1}^{n} a_{ij} \tanh[K_r(\xi_i - \xi_j)] \equiv 0$. Therefore, an argument similar to that in the proof of Theorem 4.19 shows that $\xi_i(t) \to \xi_j(t)$ and $\zeta_i(t) \to 0$, $\forall i \neq j$, as $t \to \infty$, if undirected graph $\mathcal{G}_n$ is connected. ■

4.3 Consensus Without Relative State Derivative Measurements

Note that (4.2) requires measurements of relative information state derivatives between neighboring vehicles. We propose a consensus algorithm without measurements of relative information state derivatives based on a passivity approach as

$$\dot{\hat{x}}_i = F\hat{x}_i + \sum_{j=1}^{n} a_{ij}(\xi_i - \xi_j), \tag{4.22a}$$

$$y_i = PF\hat{x}_i + P\sum_{j=1}^{n} a_{ij}(\xi_i - \xi_j), \tag{4.22b}$$

$$u_i = -\sum_{j=1}^{n} a_{ij}(\xi_i - \xi_j) - y_i, \quad i = 1, \dots, n, \tag{4.22c}$$

where $F \in \mathbb{R}^{m\times m}$ is Hurwitz, a_{ij} is the (i,j) entry of the adjacency matrix $\mathcal{A}_n$ associated with $\mathcal{G}_n$,[6] and $P \in \mathbb{R}^{m\times m}$ is the symmetrical positive-definite solution of the Lyapunov equation $F^T P + PF = -Q$ with $Q \in \mathbb{R}^{m\times m}$ symmetrical positive-definite.

Remark 4.22. A passivity approach is used in [129,130,160,235] for controlling a single vehicle. Algorithm (4.22) extends the passivity approach in these references to achieve consensus among multiple vehicles with double-integrator dynamics. □

Theorem 4.23. *Suppose that $\mathcal{A}_n$ is constant. Algorithm (4.22) achieves consensus asymptotically if undirected graph $\mathcal{G}_n$ is connected.*

Proof: Let $\xi = [\xi_1^T,\ldots,\xi_n^T]^T$, $\zeta = [\zeta_1^T,\ldots,\zeta_n^T]^T$, $y = [y_1^T,\ldots,y_n^T]^T$, $\hat{x} = [\hat{x}_1^T,\ldots,\hat{x}_n^T]^T$, and $u = [u_1^T,\ldots,u_n^T]^T$. Algorithm (4.22) can be written as

$$\dot{\hat{x}} = (I_n \otimes F)\hat{x} + (\mathcal{L}_n \otimes I_m)\xi, \tag{4.23a}$$

$$y = (I_n \otimes P)\dot{\hat{x}}, \tag{4.23b}$$

$$u = -(\mathcal{L}_n \otimes I_m)\xi - y, \tag{4.23c}$$

where $\mathcal{L}_n \in \mathbb{R}^{n\times n}$ is the Laplacian matrix associated with undirected graph $\mathcal{G}_n$.

Note that with (4.22), (4.1) can be written as

$$\dot{\xi}_{ij} = \zeta_{ij}, \tag{4.24a}$$

$$\dot{\zeta}_{ij} = -\sum_{j=1}^{n} a_{ij}\xi_{ij} - P\dot{\hat{x}}_i + \sum_{k=1}^{n} a_{jk}\xi_{jk} + P\dot{\hat{x}}_j, \tag{4.24b}$$

$$\ddot{\hat{x}}_i = F\dot{\hat{x}}_i + \sum_{j=1}^{n} a_{ij}\zeta_{ij}, \tag{4.24c}$$

where $\xi_{ij} \stackrel{\triangle}{=} \xi_i - \xi_j$ and $\zeta_{ij} \stackrel{\triangle}{=} \zeta_i - \zeta_j$. Consider the Lyapunov function candidate for (4.24) as

$$V = \frac{1}{2}\xi^T(\mathcal{L}_n \otimes I_m)^2\xi + \frac{1}{2}\zeta^T(\mathcal{L}_n \otimes I_m)\zeta + \frac{1}{2}\dot{\hat{x}}^T(I_n \otimes P)\dot{\hat{x}}.$$

Note that from the property of the symmetrical Laplacian matrix $\mathcal{L}_n$, V is positive definite and radially unbounded with respect to ξ_{ij}, ζ_{ij}, $\forall i \neq j$, and $\dot{\hat{x}}_i$ when undirected graph $\mathcal{G}_n$ is connected. Differentiating V gives

[6] In this case, only ξ_i need to be exchanged among vehicles.

$$
\begin{aligned}
\dot{V} &= \zeta^T(\mathcal{L}_n \otimes I_m)^2\xi + \zeta^T(\mathcal{L}_n \otimes I_m)u + \frac{1}{2}\ddot{\hat{x}}^T(I_n \otimes P)\dot{\hat{x}} + \frac{1}{2}\dot{\hat{x}}^T(I_n \otimes P)\ddot{\hat{x}} \\
&= \zeta^T[(\mathcal{L}_n \otimes I_m)^2\xi + (\mathcal{L}_n \otimes I_m)u] \\
&\quad + \frac{1}{2}\dot{\hat{x}}^T(I_n \otimes F^T)(I_n \otimes P)\dot{\hat{x}} + \frac{1}{2}\zeta^T(\mathcal{L}_n^T \otimes I_m)(I_n \otimes P)\dot{\hat{x}} \\
&\quad + \frac{1}{2}\dot{\hat{x}}^T(I_n \otimes P)(I_n \otimes F)\dot{\hat{x}} + \frac{1}{2}\dot{\hat{x}}^T(I_n \otimes P)(\mathcal{L}_n \otimes I_m)\zeta \\
&= \zeta^T[(\mathcal{L}_n \otimes I_m)^2\xi + (\mathcal{L}_n \otimes I_m)u] - \frac{1}{2}\dot{\hat{x}}^T(I_n \otimes Q)\dot{\hat{x}} \\
&\quad + \zeta^T(\mathcal{L}_n \otimes I_m)(I_n \otimes P)\dot{\hat{x}} \\
&= -\frac{1}{2}\dot{\hat{x}}^T(I_n \otimes Q)\dot{\hat{x}} \leq 0,
\end{aligned}
$$

where we have used (4.23), the fact that $\mathcal{L}_n = \mathcal{L}_n^T$ when graph $\mathcal{G}_n$ is undirected, and Lemma C.8.

Let $S = \{(\xi_{ij}, \zeta_{ij}, \dot{\hat{x}}_i) | \dot{V} = 0\}$. Note that $\dot{V} \equiv 0$ implies that $\dot{\hat{x}} \equiv 0$, which in turn implies that $\ddot{\hat{x}} \equiv 0$, $(\mathcal{L}_n \otimes I_m)\zeta \equiv 0$ by differentiating (4.23a), and $y \equiv 0$ from (4.23b). Because $(\mathcal{L}_n \otimes I_m)\zeta \equiv 0$, it follows from Lemma 2.10 that $\zeta_i \equiv \zeta_j$, *i.e.*, $\zeta_{ij} \equiv 0$, $\forall i \neq j$, when undirected graph $\mathcal{G}_n$ is connected. It also follows that $\dot{\zeta}_i \equiv \dot{\zeta}_j$, which implies that $\dot{\zeta} \in \text{span}(\mathbf{1}_n \otimes \eta)$, where η is some $m \times 1$ real vector, when undirected graph $\mathcal{G}_n$ is connected. Because $y \equiv 0$, from (4.1) and (4.23c), it follows that

$$
\dot{\zeta} \equiv -(\mathcal{L}_n \otimes I_m)\xi. \tag{4.25}
$$

Note that $(\mathbf{1}_n \otimes \eta)^T\dot{\zeta} \equiv -(\mathbf{1}_n \otimes \eta)^T(\mathcal{L}_n \otimes I_m)\xi \equiv -(\mathbf{1}_n^T\mathcal{L}_n \otimes \eta^T I_m) \equiv 0$ because $\mathbf{1}_n^T\mathcal{L}_n = 0$ when graph $\mathcal{G}_n$ is undirected. Thus $\dot{\zeta}$ is orthogonal to $\mathbf{1}_n \otimes \eta$. We then conclude that $\dot{\zeta} \equiv 0$, which in turn implies from (4.25) that $(\mathcal{L}_n \otimes I_m)\xi \equiv 0$. If undirected graph $\mathcal{G}_n$ is connected, $(\mathcal{L}_n \otimes I_m)\xi \equiv 0$ implies that $\xi_i \equiv \xi_j$, *i.e.*, $\xi_{ij} \equiv 0$, $\forall i \neq j$. By Theorem F.4, it follows that $\xi_i(t) \to \xi_j(t)$ and $\zeta_i(t) \to \zeta_j(t)$, $\forall i \neq j$, as $t \to \infty$. ■

When it is desirable that $\xi_i(t) \to \xi_j(t)$ and $\zeta_i(t) \to 0$, as $t \to \infty$, we propose an algorithm as

$$
\dot{\hat{x}}_i = F\hat{x}_i + \xi_i, \tag{4.26a}
$$

$$
y_i = PF\hat{x}_i + P\xi_i, \tag{4.26b}
$$

$$
u_i = -\sum_{j=1}^{n} a_{ij}(\xi_i - \xi_j) - y_i, \quad i = 1, \ldots, n, \tag{4.26c}
$$

where F, P, and a_{ij} are defined as in (4.22).

Corollary 4.24. *Suppose that $\mathcal{A}_n$ is constant. With (4.26), $\xi_i(t) \to \xi_j(t)$ and $\zeta_i(t) \to 0$, as $t \to \infty$, if undirected graph $\mathcal{G}_n$ is connected.*

Proof: Control law (4.26) can be written as

$$\dot{\hat{x}} = (I_n \otimes F)\hat{x} + \xi, \tag{4.27a}$$

$$y = (I_n \otimes P)\dot{\hat{x}}, \tag{4.27b}$$

$$u = -(\mathcal{L}_n \otimes I_m)\xi - y. \tag{4.27c}$$

Note that with (4.26), (4.1) can be written as

$$\dot{\xi}_{ij} = \zeta_i - \zeta_j, \quad \dot{\zeta}_i = -\sum_{j=1}^{n} a_{ij}\xi_{ij} - P\dot{\hat{x}}_i, \quad \ddot{\hat{x}}_i = F\dot{\hat{x}}_i + \zeta_i. \tag{4.28}$$

Consider the Lyapunov function candidate for (4.28) as

$$V = \frac{1}{2}\xi^T(\mathcal{L}_n \otimes I_m)\xi + \frac{1}{2}\zeta^T\zeta + \frac{1}{2}\dot{\hat{x}}^T(I_n \otimes P)\dot{\hat{x}},$$

which is positive definite and radially unbounded with respect to ξ_{ij}, $\forall i \neq j$, ζ_i, and $\dot{\hat{x}}_i$. Following the proof of Theorem 4.23, the derivative of V is given as

$$\begin{aligned}\dot{V} &= \zeta^T[(\mathcal{L}_n \otimes I_m)\xi + u] - \frac{1}{2}\dot{\hat{x}}^T(I_n \times Q)\dot{\hat{x}} + \zeta^T(I_n \otimes P)\dot{\hat{x}} \\ &= -\frac{1}{2}\dot{\hat{x}}^T(I_n \times Q)\dot{\hat{x}} \leq 0.\end{aligned}$$

Let $S = \{(\xi_{ij}, \zeta_i, \dot{\hat{x}}_i) | \dot{V} = 0\}$. Note that $\dot{V} \equiv 0$ implies that $\dot{\hat{x}} \equiv 0$, which in turn implies that $\ddot{\hat{x}} \equiv 0$, $\zeta \equiv 0$ by differentiating (4.27a), and $y \equiv 0$ from (4.27b). Because $\zeta \equiv 0$, it follows that $\dot{\zeta} \equiv 0$, which implies that $-(\mathcal{L}_n \otimes I_m)\xi \equiv 0$ from (4.1) and (4.27c) by noting that $y \equiv 0$. If undirected graph $\mathcal{G}_n$ is connected, $(\mathcal{L}_n \otimes I_m)\xi \equiv 0$ implies that $\xi_i \equiv \xi_j$, *i.e.*, $\xi_{ij} \equiv 0$, $\forall i \neq j$. By Theorem F.4, it follows that $\xi_i(t) \to \xi_j(t)$ and $\zeta_i(t) \to 0$, $\forall i \neq j$, as $t \to \infty$. ■

4.4 Notes

The results in this chapter are based mainly on [181–183, 185, 186]. For other theoretical results related to consensus algorithms for double-integrator dynamics, see [69, 89, 115, 203].

Acknowledgment is given to

©2006 John Wiley & Sons. Reprinted, with permission, from Wei Ren and Ella M. Atkins, "Distributed multi-vehicle coordinated control *via* local information exchange," *International Journal of Robust and Nonlinear Control*, vol. 17, no. 10–11, pp. 1002–1033, July 2007.

©2007 IEEE. Reprinted, with permission, from Wei Ren, "Second-order consensus algorithm with extensions to switching topologies and reference models," *Proceedings of the American Control Conference*, pp. 1431–1436, New York City, 2007.

5

Extensions to a Reference Model

This chapter extends the consensus algorithms for double-integrator dynamics in Chapter 4 to a reference model. We first consider consensus algorithms that ensure that the derivatives of the information states follow a reference. Both strategies with and without coupling between neighbors' information state derivatives are addressed. We then consider consensus algorithms that ensure that the information states and their derivatives evolve according to a reference model. Three strategies, full access to the reference model, leader-following strategy, and partial access to the reference model, are addressed.

5.1 Problem Statement

Suppose that the team consists of the same n vehicles as in Chapter 4, together with an additional (virtual) vehicle labeled $n+1$, which acts as the unique (virtual) leader of the team. As in Chapter 3, we call vehicle $n+1$ the *team leader* and vehicles $1,\ldots,n$ the *followers*. Vehicle $n+1$ has the information state $\xi_{n+1} \triangleq \xi^r \in \mathbb{R}^m$ and its derivative $\zeta_{n+1} \triangleq \zeta^r \in \mathbb{R}^m$, where ξ^r and ζ^r represent the reference states for ξ_i and ζ_i, $i=1,\ldots,n$, respectively. Note that ξ_i and ζ_i satisfy double-integrator dynamics (4.1). The reference states ξ^r and ζ^r satisfy the reference model

$$\dot{\xi}^r = \zeta^r, \quad \dot{\zeta}^r = f(t, \xi^r, \zeta^r), \tag{5.1}$$

where $f(\cdot,\cdot,\cdot)$ is piecewise continuous in t and locally Lipschitz in ξ^r and ζ^r.

We use directed graphs $\mathcal{G}_{n+1}^A \triangleq (\mathcal{V}_{n+1}, \mathcal{E}_{n+1}^A)$, where $\mathcal{V}_{n+1} = \{1,\ldots,n+1\}$ is the node set and $\mathcal{E}_{n+1}^A \subseteq \mathcal{V}_{n+1} \times \mathcal{V}_{n+1}$ is the edge set, and $\mathcal{G}_{n+1}^B \triangleq (\mathcal{V}_{n+1}, \mathcal{E}_{n+1}^B)$, where $\mathcal{V}_{n+1}$ is the same node set as above and $\mathcal{E}_{n+1}^B \subseteq \mathcal{V}_{n+1} \times \mathcal{V}_{n+1}$ is the edge set, to model the interaction topologies for ξ_i and ζ_i, respectively, among the $n+1$ vehicles. Let $\mathcal{A}_{n+1} = [a_{ij}] \in \mathbb{R}^{(n+1)\times(n+1)}$ and $\mathcal{B}_{n+1} = [b_{ij}] \in \mathbb{R}^{(n+1)\times(n+1)}$ be the adjacency matrix associated with

$\mathcal{G}_{n+1}^A$ and $\mathcal{G}_{n+1}^B$, respectively. In particular, $a_{ij} > 0$ (respectively, $b_{ij} > 0$) if $(j,i) \in \mathcal{E}_{n+1}^A$ [respectively, $(j,i) \in \mathcal{E}_{n+1}^B$] and $a_{ij} = 0$ (respectively, $b_{ij} = 0$) otherwise for all $i = 1, \ldots, n$ and $j = 1, \ldots, n+1$, and $a_{(n+1)j} = 0$ [respectively, $b_{(n+1)j} = 0$] for all $j = 1, \ldots, n+1$. Also let $\mathcal{L}_{n+1}^A = [\ell_{ij}] \in \mathbb{R}^{(n+1)\times(n+1)}$ (*i.e.*, $\ell_{ij} = -a_{ij}$, $i \neq j$, $\ell_{ii} = \sum_{j=1, j\neq i}^{n+1} a_{ij}$) and $\mathcal{L}_{n+1}^B = [\ell_{ij}] \in \mathbb{R}^{(n+1)\times(n+1)}$ (*i.e.*, $\ell_{ij} = -b_{ij}$, $i \neq j$, $\ell_{ii} = \sum_{j=1, j\neq i}^{n+1} b_{ij}$) be the nonsymmetrical Laplacian matrix associated with $\mathcal{G}_{n+1}^A$ and $\mathcal{G}_{n+1}^B$, respectively. When there is only one interaction topology associated with the $n+1$ vehicles, we simply use $\mathcal{G}_{n+1} \triangleq (\mathcal{V}_{n+1}, \mathcal{E}_{n+1})$ to model the interaction topology among the $n+1$ vehicles, as in Chapter 3. Similarly, $\mathcal{A}_{n+1}$ and $\mathcal{L}_{n+1}$ are, respectively, the adjacency matrix and the nonsymmetrical Laplacian matrix associated with $\mathcal{G}_{n+1}$.

Remark 5.1. When we need to focus only on vehicles 1 to n (*i.e.*, the followers), we still use $\mathcal{G}_n^A \triangleq (\mathcal{V}_n, \mathcal{E}_n^A)$ and $\mathcal{G}_n^B \triangleq (\mathcal{V}_n, \mathcal{E}_n^B)$ to model the interaction topologies for ξ_i and ζ_i, respectively, among the n followers, as in Chapter 4. In addition, $\mathcal{A}_n \in \mathbb{R}^{n\times n}$ (respectively, $\mathcal{B}_n \in \mathbb{R}^{n\times n}$) and $\mathcal{L}_n^A \in \mathbb{R}^{n\times n}$ (respectively, $\mathcal{L}_n^B \in \mathbb{R}^{n\times n}$) are still, respectively, the adjacency matrix and the nonsymmetrical Laplacian matrix associated with $\mathcal{G}_n^A$ (respectively, $\mathcal{G}_n^B$), as in Chapter 4. When there is only one interaction topology associated with the n followers, we simply use $\mathcal{G}_n \triangleq (\mathcal{V}_n, \mathcal{E}_n)$ to model the interaction topology among the n followers, as in Chapter 2. Similarly, $\mathcal{A}_n$ and $\mathcal{L}_n$ are, respectively, the adjacency matrix and the nonsymmetrical Laplacian matrix associated with $\mathcal{G}_n$. □

In this chapter, there is only one interaction topology associated with the $n+1$ vehicles. In addition, we assume that the interaction topology is time invariant and all weights in the consensus algorithms are constant.

5.2 Consensus with a Reference for Information State Derivatives

In this section, the objective is to guarantee that $\xi_i(t) \to \xi_j(t)$ and $\zeta_i(t) \to \zeta^r(t)$, as $t \to \infty$. We consider two strategies with and without coupling between neighbors' information state derivatives.

5.2.1 Consensus with Coupling Between Neighbors' Information State Derivatives

When there is coupling between neighbors' information state derivatives, we propose an algorithm as

$$u_i = \dot{\zeta}^r - \alpha(\zeta_i - \zeta^r) - \sum_{j=1}^{n} a_{ij}[(\xi_i - \xi_j) + \gamma(\zeta_i - \zeta_j)], \tag{5.2}$$

where α and γ are positive scalars and a_{ij} is the (i,j) entry of the adjacency matrix $\mathcal{A}_n \in \mathbb{R}^{n\times n}$ associated with $\mathcal{G}_n$. Note that with (5.2), ζ^r and $\dot{\zeta}^r$ are available to all followers, which corresponds to the case where $(n+1,i) \in \mathcal{E}_{n+1}$, $i = 1,\dots,n$.

Lemma 5.2. *Let* $p = [p_1,\dots,p_n]^T \geq 0$ *satisfy* $\mathbf{1}_n^T p = 1$ *and* $\mathcal{L}_n^T p = 0$, *where* $\mathcal{L}_n \in \mathbb{R}^{n\times n}$ *is the nonsymmetrical Laplacian matrix associated with* $\mathcal{G}_n$. *Also let*

$$\Sigma = \begin{bmatrix} 0_{n\times n} & I_n \\ -\mathcal{L}_n & -\alpha I_n - \gamma \mathcal{L}_n \end{bmatrix}, \tag{5.3}$$

where α *and* γ *are positive scalars. Then*

$$\lim_{t\to\infty} e^{\Sigma t} \to \begin{bmatrix} \mathbf{1}_n p^T & \frac{1}{\alpha}\mathbf{1}_n p^T \\ 0_{n\times n} & 0_{n\times n} \end{bmatrix}.$$

if and only if Σ *has a simple zero eigenvalue and all other eigenvalues have negative real parts.*

Proof: (Sufficiency.) The eigenvalues of Σ can be found by solving the equation $\det(\lambda I_{2n} - \Sigma) = 0$. Note that

$$\begin{aligned} \det(\lambda I_{2n} - \Sigma) &= \det(\lambda^2 I_n + \gamma\lambda\mathcal{L}_n + \alpha\lambda I_n + \mathcal{L}_n) \\ &= \det\left[(\lambda^2 + \alpha\lambda) I_n + (1+\gamma\lambda)\mathcal{L}_n\right]. \end{aligned} \tag{5.4}$$

Also note that

$$\det(\lambda I_n + \mathcal{L}_n) = \prod_{i=1}^{n} (\lambda - \mu_i), \tag{5.5}$$

where μ_i is the ith eigenvalue of $-\mathcal{L}_n$.

By comparing (5.4) and (5.5), we see that the roots of (5.4) can be obtained by solving $\lambda^2 + \alpha\lambda = \mu_i(1+\gamma\lambda)$. Therefore, it is straightforward to see that the eigenvalues of Σ are given by

$$\rho_{i\pm} = \frac{\gamma\mu_i - \alpha \pm \sqrt{(\gamma\mu_i - \alpha)^2 + 4\mu_i}}{2}, \tag{5.6}$$

where $\rho_{i\pm}$ are called eigenvalues of Σ that are associated with μ_i.

Note from (5.6) that Σ has k zero eigenvalues if and only if $-\mathcal{L}_n$ has k zero eigenvalues. Without loss of generality, we let $\mu_1 = 0$, which implies that $\rho_{1+} = 0$ and $\rho_{1-} = -\alpha$. The proof then follows a line similar to that of Lemma 4.1 by noting that $\nu_1 = [p^T, \frac{1}{\alpha}p^T]^T$, $w_1 = [\mathbf{1}_n^T, \mathbf{0}_n^T]^T$, and

$$J = \begin{bmatrix} 0 & 0_{1\times(2n-1)} \\ 0_{(2n-1)\times 1} & J' \end{bmatrix},$$

where J' is the Jordan upper diagonal block matrix corresponding to the $2n-1$ eigenvalues that have negative real parts.

(Necessity.) If $\lim_{t\to\infty} e^{\Sigma t} \to \begin{bmatrix} \mathbf{1}_n p^T & \frac{1}{\alpha}\mathbf{1}_n p^T \\ 0_{n\times n} & 0_{n\times n} \end{bmatrix}$, we know that $\lim_{t\to\infty} Pe^{Jt}P^{-1}$ has rank one, which in turn implies that $\lim_{t\to\infty} e^{Jt}$ has rank one. However, if the sufficient condition does not hold, we know that $\lim_{t\to\infty} e^{Jt}$ has a rank larger than one by following an argument similar to the necessity proof of Lemma 4.1. This results in a contradiction. ■

Corollary 5.3. *Let* p *be defined as in Lemma 5.2. With consensus algorithm (5.2),* $\xi_i(t) \to \sum_{i=1}^n p_i\xi_i(0) + \int_0^t \zeta^r(\tau)d\tau + \frac{1}{\alpha}\sum_{i=1}^n p_i\zeta_i(0) - \frac{1}{\alpha}\zeta^r(0)$ *and* $\zeta_i(t) \to \zeta^r(t)$, $i = 1,\dots,n$, *asymptotically, as* $t\to\infty$, *if and only if* Σ *defined in (5.3) has a simple zero eigenvalue and all other eigenvalues have negative real parts.*

Proof: (Sufficiency.) Let $\tilde{\xi}_i \triangleq \xi_i - \xi^r$, where $\xi^r \triangleq \int_0^t \zeta^r(\tau)d\tau$, and $\tilde{\zeta}_i = \zeta_i - \zeta^r$. With (5.2), (4.1) can be written as

$$\dot{\xi}_i - \dot{\xi}^r = \zeta_i - \zeta^r$$
$$\dot{\zeta}_i = -\alpha(\zeta_i - \zeta^r) - \sum_{j=1}^n a_{ij}\{(\xi_i - \xi^r) - (\xi_j - \xi^r) + \gamma[(\zeta_i - \zeta^r) - (\zeta_j - \zeta^r)]\},$$

which implies that

$$\dot{\tilde{\xi}}_i = \tilde{\zeta}_i$$
$$\dot{\tilde{\zeta}}_i = -\alpha\tilde{\zeta}_i - \sum_{j=1}^n a_{ij}[(\tilde{\xi}_i - \tilde{\xi}_j) + \gamma(\tilde{\zeta}_i - \tilde{\zeta}_j)]. \tag{5.7}$$

Equation (5.7) can be written in matrix form as

$$\begin{bmatrix} \dot{\tilde{\xi}} \\ \dot{\tilde{\zeta}} \end{bmatrix} = (\Sigma \otimes I_m)\begin{bmatrix} \tilde{\xi} \\ \tilde{\zeta} \end{bmatrix},$$

where $\tilde{\xi} \triangleq [\tilde{\xi}_1^T,\dots,\tilde{\xi}_n^T]^T$ and $\tilde{\zeta} \triangleq [\tilde{\zeta}_1^T,\dots,\tilde{\zeta}_n^T]^T$.

If Σ has a simple zero eigenvalue and all other eigenvalues have negative real parts, we know from Lemma 5.2 that

$$\lim_{t\to\infty}\begin{bmatrix} \tilde{\xi}(t) \\ \tilde{\zeta}(t) \end{bmatrix} = \lim_{t\to\infty}(e^{\Sigma t}\otimes I_m)\begin{bmatrix} \tilde{\xi}(0) \\ \tilde{\zeta}(0) \end{bmatrix} = \left(\begin{bmatrix} \mathbf{1}_n p^T & \frac{1}{\alpha}\mathbf{1}_n p^T \\ 0_{n\times n} & 0_{n\times n} \end{bmatrix}\otimes I_m\right)\begin{bmatrix} \tilde{\xi}(0) \\ \tilde{\zeta}(0) \end{bmatrix},$$

which implies that $\tilde{\xi}(t) \to (\mathbf{1}_n p^T \otimes I_m)\tilde{\xi}(0) + \frac{1}{\alpha}(\mathbf{1}_n p^T \otimes I_m)\tilde{\zeta}(0)$ and $\tilde{\zeta}(t) \to 0$. This in turn gives the sufficient part.

(Necessity.) If $\xi_i(t) \to \sum_{i=1}^n p_i\xi_i(0) + \int_0^t \zeta^r(\tau)d\tau + \frac{1}{\alpha}\sum_{i=1}^n p_i\zeta_i(0) - \frac{1}{\alpha}\zeta^r(0)$ and $\zeta_i(t) \to \zeta^r(t)$, $i = 1,\dots,n$, as $t\to\infty$, we know that $\tilde{\xi}(t) \to (\mathbf{1}_n p^T \otimes I_m)\tilde{\xi}(0) + \frac{1}{\alpha}(\mathbf{1}_n p^T \otimes I_m)\tilde{\zeta}(0)$ and $\tilde{\zeta}(t) \to 0$, which implies that

$$\lim_{t\to\infty} e^{\Sigma t} \to \begin{bmatrix} \mathbf{1}_n p^T & \frac{1}{\alpha}\mathbf{1}_n p^T \\ 0_{n\times n} & 0_{n\times n} \end{bmatrix}.$$

The necessary part then comes from Lemma 5.2. ∎

We have the following theorem for consensus algorithm (5.2).

Theorem 5.4. *Let p be defined as in Lemma 5.2. Also let μ_i be the ith eigenvalue of $-\mathcal{L}_n$. Consensus algorithm (5.2) guarantees that $\xi_i(t) \to \sum_{i=1}^n p_i\xi_i(0) + \int_0^t \zeta^r(\tau)d\tau + \frac{1}{\alpha}\sum_{i=1}^n p_i\zeta_i(0) - \frac{1}{\alpha}\zeta^r(0)$ and $\zeta_i(t) \to \zeta^r(t)$, $i = 1,\ldots,n$, as $t \to \infty$, if directed graph $\mathcal{G}_n$ has a directed spanning tree and Inequality (4.10) is satisfied.*

Proof: Let $\rho_{i\pm}$ be given by (5.6). If $\mathcal{G}_n$ has a directed spanning tree, Corollary 2.5 implies that $-\mathcal{L}_n$ has a simple zero eigenvalue and all other eigenvalues have negative real parts. Without loss of generality, we let $\mu_1 = 0$ and $\text{Re}(\mu_i) < 0$, $i = 2,\cdots,n$. Then we can find that $\rho_{1+} = 0$ and $\rho_{1-} = -\alpha$. Note that if $\mu_i < 0$, then (5.6) implies that $\text{Re}(\rho_{i\pm}) < 0$. It is left to show that Inequality (4.10) guarantees that all $\rho_{i\pm}$ associated with μ_i that satisfy $\text{Re}(\mu_i) < 0$ and $\text{Im}(\mu_i) \neq 0$ have negative real parts. We need only to consider μ_i that satisfies $\text{Re}(\mu_i) < 0$ and $\text{Im}(\mu_i) > 0$ because any μ_i that satisfies $\text{Re}(\mu_i) < 0$ and $\text{Im}(\mu_i) < 0$ is a complex conjugate of some μ_i that satisfies $\text{Re}(\mu_i) < 0$ and $\text{Im}(\mu_i) > 0$. This argument then follows from Lemma 4.10. Therefore, Corollary 5.3 implies that consensus can be achieved asymptotically under the assumption of the theorem. ∎

Remark 5.5. Under some circumstances, it might be desirable that $\xi_i(t) \to \xi_j(t)$ and $\zeta_i(t) \to 0$, as $t \to \infty$. For example, in formation stabilization applications, we want each vehicle to agree on its *a priori* unknown fixed formation center, which has a constant position and zero velocity. In this case, we simply let $\zeta^r \equiv 0$. □

5.2.2 Consensus Without Coupling Between Neighbors' Information State Derivatives

When there is coupling between neighbors' information state derivatives, we propose a consensus algorithm as

$$u_i = \dot{\zeta}^r - \alpha(\zeta_i - \zeta^r) - \sum_{j=1}^n a_{ij}(\xi_i - \xi_j), \quad i = 1,\ldots,n, \tag{5.8}$$

where α is a positive scalar and a_{ij} is the (i,j) entry of the adjacency matrix $\mathcal{A}_n \in \mathbb{R}^{n\times n}$ associated with $\mathcal{G}_n$. Note that with (5.8), ζ^r and $\dot{\zeta}^r$ are available to all followers, which corresponds to the case where $(n+1,i) \in \mathcal{E}_{n+1}$, $i = 1,\ldots,n$.

We have the following theorem for algorithm (5.8):

Theorem 5.6. *Let μ_i be the ith eigenvalue of $-\mathcal{L}_n$. With (5.8), $\xi_i(t) \to \xi_j(t)$ and $\zeta_i(t) \to \zeta^r(t)$, $i = 1, \ldots, n$, as $t \to \infty$, if directed graph $\mathcal{G}_n$ has a directed spanning tree and*

$$\alpha > \bar{\alpha}, \tag{5.9}$$

where $\bar{\alpha} \triangleq 0$ if all $n-1$ nonzero eigenvalues of $-\mathcal{L}_n$ are negative and

$$\bar{\alpha} \triangleq \max_{\forall \mathrm{Re}(\mu_i)<0 \text{ and } \mathrm{Im}(\mu_i)>0} |\mu_i| \sqrt{\frac{2}{-\mathrm{Re}(\mu_i)}}$$

otherwise.

Proof: Let $\tilde{\xi}_i \triangleq \xi_i - \int_0^t \zeta^r(\tau)d\tau$, $\tilde{\zeta}_i \triangleq \zeta_i - \zeta^r$, $\tilde{\xi} \triangleq [\tilde{\xi}_1^T, \ldots, \tilde{\xi}_n^T]^T$, and $\tilde{\zeta} \triangleq [\tilde{\zeta}_1^T, \ldots, \tilde{\zeta}_n^T]^T$. With (5.8), (4.1) can be written in matrix form as $\begin{bmatrix} \dot{\tilde{\xi}} \\ \dot{\tilde{\zeta}} \end{bmatrix} = (\Phi \otimes I_m) \begin{bmatrix} \tilde{\xi} \\ \tilde{\zeta} \end{bmatrix}$, where $\Phi \triangleq \begin{bmatrix} 0_{n\times n} & I_n \\ -\mathcal{L}_n & -\alpha I_n \end{bmatrix}$.

Noting that $\mathcal{L}_n \mathbf{1}_n = 0$ from Corollary 2.5, it follows that $[\mathbf{1}_n^T, \mathbf{0}_n^T]^T$ is an eigenvector for Φ associated with the zero eigenvalue, which implies that $\mathrm{span}\left(\begin{bmatrix} \mathbf{1}_n \\ \mathbf{0}_n \end{bmatrix}\right)$ is contained in the kernel of Φ. If Φ has a simple zero eigenvalue and all other eigenvalues have negative real parts, then $\begin{bmatrix} \tilde{\xi}(t) \\ \tilde{\zeta}(t) \end{bmatrix} \to \mathrm{span}\left(\begin{bmatrix} \mathbf{1}_n \\ \mathbf{0}_n \end{bmatrix} \otimes \eta\right)$, as $t \to \infty$, where η is some $m \times 1$ vector, which is equivalent to $\xi_i(t) \to \xi_j(t)$ and $\zeta_i \to \zeta^r(t)$, as $t \to \infty$.

Next, we show that if $\mathcal{G}_n$ has a directed spanning tree and Inequality (5.9) is satisfied, then Φ has a simple zero eigenvalue and all other eigenvalues have negative real parts.

Let λ be an eigenvalue of Φ and $s = [p^T, q^T]^T$ be its associated eigenvector, where p and q are $n \times 1$ column vectors. Note that

$$\Phi s = \lambda s \Longleftrightarrow \begin{bmatrix} 0_{n\times n} & I_n \\ -\mathcal{L}_n & -\alpha I_n \end{bmatrix} \begin{bmatrix} p \\ q \end{bmatrix} = \lambda \begin{bmatrix} p \\ q \end{bmatrix}$$
$$\Longleftrightarrow q = \lambda p \text{ and } -\mathcal{L}_n p - \alpha q = \lambda q \Longleftrightarrow -\mathcal{L}_n p = (\lambda^2 + \alpha\lambda)p,$$

which implies that $\lambda^2 + \alpha\lambda$ is an eigenvalue of $-\mathcal{L}_n$ with an associated eigenvector p. Letting $\mu \triangleq \lambda^2 + \alpha\lambda$ gives $\lambda^2 + \alpha\lambda - \mu = 0$, which implies that given each μ, there are two roots for λ, denoted by $\lambda_\pm = \frac{-\alpha \pm \sqrt{\alpha^2 + 4\mu}}{2}$. As a result, each eigenvalue of $-\mathcal{L}_n$, μ_i, $i = 1, \ldots, n$, corresponds to two eigenvalues of Φ, denoted by $\lambda_{i\pm}$.

If $\mathcal{G}_n$ has a directed spanning tree, then Corollary 2.5 implies that $-\mathcal{L}_n$ has a simple zero eigenvalue and all other eigenvalues have negative real parts. Without loss of generality, let $\mu_1 = 0$ and $\mathrm{Re}(\mu_i) < 0$, $i = 2, \ldots, n$.

Then it follows that $\lambda_{1+} = 0$ and $\lambda_{1-} = -\alpha$. Note that if $\mu_i < 0$, then $\mathrm{Re}(\lambda_{i\pm}) = \mathrm{Re}(\frac{-\alpha \pm \sqrt{\alpha^2+4\mu_i}}{2}) < 0$ for any $\alpha > 0$. It is left to show that Inequality (5.9) guarantees that all eigenvalues of Φ associated with μ_i that satisfy $\mathrm{Re}(\mu_i) < 0$ and $\mathrm{Im}(\mu_i) \neq 0$ have negative real parts. As in Lemma 4.10, we use Fig. 5.1 to show the notations used in the proof. We need only to consider μ_i that satisfies $\mathrm{Re}(\mu_i) < 0$ and $\mathrm{Im}(\mu_i) > 0$ because any μ_i that satisfies $\mathrm{Re}(\mu_i) < 0$ and $\mathrm{Im}(\mu_i) < 0$ is a complex conjugate of some μ_i that satisfies $\mathrm{Re}(\mu_i) < 0$ and $\mathrm{Im}(\mu_i) > 0$. Consider the triangle formed by vectors α^2, $4\mu_i$, and $\alpha^2 + 4\mu_i$. According to the law of cosines, $|\alpha^2 + 4\mu_i|^2 = (\alpha^2)^2 + (4|\mu_i|)^2 - 8\alpha^2 |\mu_i| \cos(\phi_i)$, where $\cos(\phi_i) = \frac{-\mathrm{Re}(\mu_i)}{|\mu_i|}$. Note that if $\alpha > |\mu_i| \sqrt{\frac{2}{-\mathrm{Re}\mu_i}}$, then $|\alpha^2 + 4\mu_i|^2 < \alpha^4$, which implies that $\left|\sqrt{\alpha^2 + 4\mu_i}\right| < \alpha$. Therefore, it follows that $\left|\mathrm{Re}(\sqrt{\alpha^2 + 4\mu_i})\right| < \alpha$, which in turn implies that $\mathrm{Re}(\lambda_{i\pm}) = \mathrm{Re}(\frac{-\alpha \pm \sqrt{\alpha^2+4\mu_i}}{2}) < 0$.

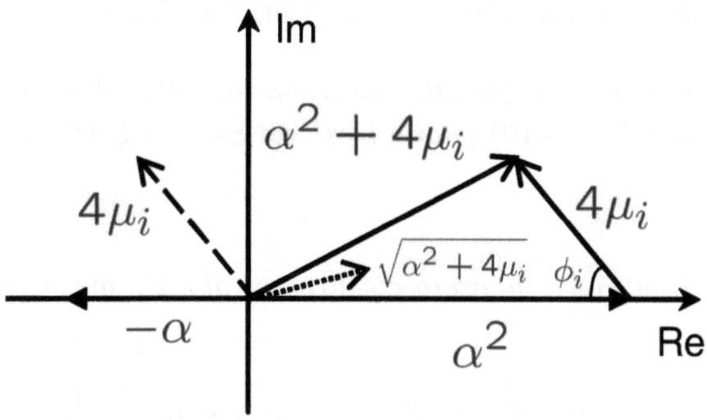

Fig. 5.1. Graphical view of notations used in the proof of Theorem 5.6

Combining the above arguments, it follows that if $\mathcal{G}_n$ has a directed spanning tree and Inequality (5.9) is valid, then $\xi_i(t) \to \xi_j(t)$ and $\zeta_i(t) \to \zeta^r(t)$, as $t \to \infty$. ■

Corollary 5.7. *Suppose that $\mathcal{G}_n$ is undirected. With* (5.8), *$\xi_i(t) \to \xi_j(t)$ and $\zeta_i(t) \to \zeta^r(t)$, as $t \to \infty$, if $\mathcal{G}_n$ is connected.*

5.3 Consensus with References for Information States and Their Derivatives

In this section, the objective is to guarantee that $\xi_i(t) \to \xi^r(t)$ and $\zeta_i(t) \to \zeta^r(t)$, as $t \to \infty$, where $\xi^r(t)$ and $\zeta^r(t)$ satisfy reference model (5.1). We first consider two special cases where either the reference model is available to all followers on the team or the interaction topology for the n followers is itself a directed spanning tree. We then consider the general case where the interaction topology for the $n+1$ vehicles has a directed spanning tree.

5.3.1 Full Access to the Reference Model

In this strategy, we incorporate in reference model (5.1), to each vehicle's consensus algorithm. The consensus algorithm for each vehicle is designed as

$$\begin{aligned} u_i = {} & \dot{\zeta}^r - \alpha[(\xi_i - \xi^r) + \gamma(\zeta_i - \zeta^r)] \\ & - \sum_{j=1}^{n} a_{ij}[(\xi_i - \xi_j) + \gamma(\zeta_i - \zeta_j)], \end{aligned} \tag{5.10}$$

where α and γ are positive scalars and a_{ij} is the (i,j) entry of the adjacency matrix $\mathcal{A}_n \in \mathbb{R}^{n \times n}$ associated with $\mathcal{G}_n$. Note that with (5.10), ξ^r, ζ^r, and $\dot{\zeta}^r$ are available to all followers, which corresponds to the case where $(n+1, i) \in \mathcal{E}_{n+1}$, $i = 1, \ldots, n$.

We have the following theorem for algorithm (5.10):

Theorem 5.8. *Let μ_i be the ith eigenvalue of $-\mathcal{L}_n$. Also let $\nu_i \stackrel{\triangle}{=} -\alpha + \mu_i$. Consensus algorithm (5.10) guarantees that $\xi_i(t) \to \xi^r(t)$ and $\zeta_i(t) \to \zeta^r(t)$, $i = 1, \ldots, n$, if*

$$\gamma > \bar{\gamma}, \tag{5.11}$$

where $\bar{\gamma} \stackrel{\triangle}{=} 0$, if all $n-1$ nonzero eigenvalues of $-\mathcal{L}_n$ are negative and

$$\bar{\gamma} = \max_{\forall \mathrm{Re}(\nu_i)<0 \text{ and } \mathrm{Im}(\nu_i)>0} \sqrt{\frac{2}{|\nu_i| \cos(\tan^{-1} \frac{\mathrm{Im}(\nu_i)}{-\mathrm{Re}(\nu_i)})}}$$

otherwise.

Proof: With algorithm (5.10), (4.1) can be written in matrix form as

$$\begin{bmatrix} \dot{\tilde{\xi}} \\ \dot{\tilde{\zeta}} \end{bmatrix} = (\Upsilon \otimes I_m) \begin{bmatrix} \tilde{\xi} \\ \tilde{\zeta} \end{bmatrix},$$

where $\tilde{\xi}_i \stackrel{\triangle}{=} \xi_i - \xi^r$, $\tilde{\zeta}_i \stackrel{\triangle}{=} \zeta_i - \zeta^r$, $\tilde{\xi} \stackrel{\triangle}{=} [\tilde{\xi}_1^T, \cdots, \tilde{\xi}_n^T]^T$, $\tilde{\zeta} \stackrel{\triangle}{=} [\tilde{\zeta}_1^T, \cdots, \tilde{\zeta}_n^T]^T$, and $\Upsilon \stackrel{\triangle}{=} \begin{bmatrix} 0_{n\times n} & I_n \\ -\alpha I_n - \mathcal{L}_n & -\gamma(\alpha I_n + \mathcal{L}_n) \end{bmatrix}$. Note that the eigenvalues of $-(\alpha I_n + \mathcal{L}_n)$ are given by ν_i and $\mathrm{Re}(\nu_i) < 0$, $i = 1, \ldots, n$, for any $\mathcal{G}_n$. Following (4.4) and (4.7) with $\alpha I_n + \mathcal{L}_n$ and ν_i playing the roles of $\mathcal{L}_n$ and μ_i, respectively, the eigenvalues of Υ are given by $\rho_{i\pm} = \frac{\gamma\nu_i \pm \sqrt{\gamma^2\nu_i^2 + 4\nu_i}}{2}$. If Inequality (5.11) is true, we know that all eigenvalues of Υ have negative real parts by following the proof of Theorem 4.11. Therefore, we see that $\tilde{\xi}(t) \to 0$ and $\tilde{\zeta}(t) \to 0$, as $t \to \infty$, which in turn proves the theorem. ■

Remark 5.9. Note that unlike Theorems 5.4 and 5.6, $\mathcal{G}_n$ does not affect the convergence result in Theorem 5.8, as long as the scaling factor γ is sufficiently large. As a result, even in the worst case that there is no information exchange among the followers (*i.e.*, $\mathcal{A}_n = 0_{n\times n}$), we can still guarantee that $\xi_i(t) \to \xi^r(t)$ and $\zeta_i(t) \to \zeta^r(t)$, $i = 1, \ldots, n$, as $t \to \infty$, as long as Inequality (5.11) is valid. However, better transient performance is achieved [*i.e.*, $\xi_i(t) \to \xi_j(t)$ and $\zeta_i(t) \to \zeta_j(t)$ during the transition] when $\mathcal{G}_n$ has a directed spanning tree due to coupling among the vehicles. □

Example 5.10. Theorem 5.8 can be illustrated graphically. Consider the interaction topology shown in Fig. 5.2 where reference model (5.1) is available to all followers. Here we use node ξ^r to denote the team leader. Note that there exists an edge from node ξ^r to all followers on the team. □

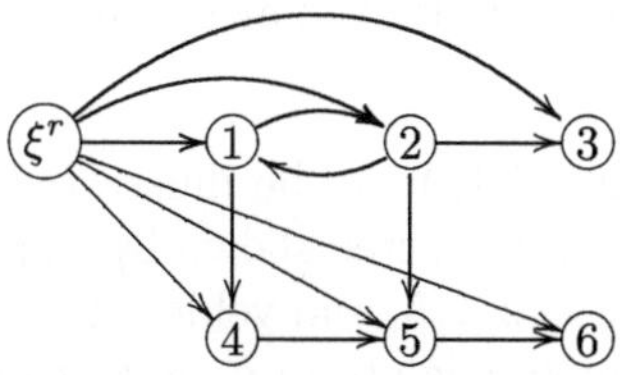

Fig. 5.2. Interaction topology where reference model (5.1) is available to all followers

5.3.2 Leader-following Strategy

In the leader-following strategy, $\mathcal{G}_n$ is itself a directed spanning tree with vehicle k the root, and reference model (5.1) is only available to vehicle k, that is, vehicle k is the child of vehicle $n+1$. Accordingly, in the leader-following strategy, $\mathcal{G}_{n+1}$ is also a directed spanning tree with vehicle $n+1$ (*i.e.*, the unique team leader) the root, that is, each vehicle except vehicle $n+1$ has exactly one parent.

The consensus algorithm for all followers (*i.e.*, vehicles 1 to n) is designed as

$$u_i = \dot{\zeta}^r - K_{ri}(\xi_i - \xi^r) - K_{vi}(\zeta_i - \zeta^r), \quad i = k \tag{5.12a}$$

$$u_i = \dot{\zeta}_{i_\ell} - K_{ri}(\xi_i - \xi_{i_\ell}) - K_{vi}(\zeta_i - \zeta_{i_\ell}), \quad i \neq k, \tag{5.12b}$$

where K_{ri} and K_{vi} are $m \times m$ symmetrical positive-definite matrices and vehicle i_ℓ is the parent of vehicle i, $i \neq k$.[1] With algorithm (5.12), (4.1) can be written as

$$\dot{\tilde{\zeta}}_i = -K_{ri}\tilde{\xi}_i - K_{vi}\tilde{\zeta}_i, \tag{5.13}$$

[1] That is, an edge (i_ℓ, i) exists in $\mathcal{G}_n$. However, vehicle i_ℓ is not the unique team leader.

where $\tilde{\xi}_k = \xi_k - \xi^r$, $\tilde{\xi}_i = \xi_i - \xi_{i_\ell}$, $i \neq k$, $\tilde{\zeta}_k = \zeta_k - \zeta^r$, and $\tilde{\zeta}_i = \zeta_i - \zeta_{i_\ell}$, $i \neq k$. Note that (5.13) implies that $\tilde{\xi}_i(t) \to 0$ and $\tilde{\zeta}_i(t) \to 0$, as $t \to \infty$, because K_{ri} and K_{vi} are symmetrical positive-definite matrices, which in turn implies that $\xi_k(t) \to \xi^r(t)$, $\zeta_k(t) \to \zeta^r(t)$, $\xi_i(t) \to \xi_{i_\ell}(t)$, and $\zeta_i(t) \to \zeta_{i_\ell}(t)$, $i \neq k$, as $t \to \infty$. Therefore, it follows that $\xi_i(t) \to \xi^r(t)$ and $\zeta_i(t) \to \zeta^r(t)$, $i = 1, \ldots, n$, as $t \to \infty$, by renumbering the vehicles consecutively by depth in the directed spanning tree.

Remark 5.11. Note that in the leader-following strategy, where the interaction topology is itself a directed spanning tree, information flows only from parents to children. When a child is perturbed by a disturbance, the parent is unaware of this disturbance, and its motion remains unaffected. It might be intuitive to introduce information flow from children to parents to introduce feedback to improve group robustness. However, it is not clear how information from children can be incorporated into the algorithm for parents without affecting stability. □

Example 5.12. Algorithm (5.12) can be illustrated by the following example. Consider the directed spanning tree shown by Fig. 5.3, where vehicle j is the parent of vehicle $j+1$, $j = 2, 4, 5$, and vehicle 1 is the parent of vehicles 2 and 4. Note that there exists an edge from node ξ^r to vehicle 1. □

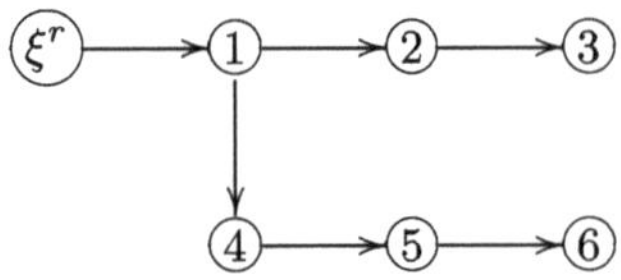

Fig. 5.3. A directed spanning tree where reference model (5.1) is available only to one vehicle

5.3.3 General Case

In the general case where $\mathcal{G}_n$ is arbitrary and reference model (5.1) may be available to one or multiple followers, we propose the consensus algorithm

$$u_i = \frac{1}{\kappa_i} \sum_{j=1}^{n} a_{ij} [\dot{\zeta}_j - K_{ri}(\xi_i - \xi_j) - K_{vi}(\zeta_i - \zeta_j)] + \frac{1}{\kappa_i} a_{i(n+1)} [\dot{\zeta}^r - K_{ri}(\xi_i - \xi^r) - K_{vi}(\zeta_i - \zeta^r)], \quad (5.14)$$

where a_{ij} is the (i,j) entry of the adjacency matrix $\mathcal{A}_{n+1} \in \mathbb{R}^{n \times n}$ associated with $\mathcal{G}_{n+1}$, $\kappa_i \triangleq \sum_{j=1}^{n+1} a_{ij}$ and K_{ri} and K_{vi} are $m \times m$ symmetrical positive-definite matrices. Note that in (5.14), each vehicle needs its neighbors' information states and their first-, and second-order derivatives.[2] In practical implementation, the second-order derivatives of the neighbors' information states can be calculated by numerical differentiation.

We have the following theorem for consensus with a reference model:

Theorem 5.13. *With consensus algorithm* (5.14), *there exists a unique solution for* u_i *and* $\xi_i(t) \to \xi^r(t)$ *and* $\zeta_i(t) \to \zeta^r(t)$, *as* $t \to \infty$, *if and only if directed graph* $\mathcal{G}_{n+1}$ *has a directed spanning tree (with vehicle* $n+1$ *as the root).*[3]

Proof: Note that $\kappa_i = \sum_{j=1}^{n+1} a_{ij} \neq 0$, $i = 1, \ldots, n$, under the condition of the theorem by following the proof of Theorem 3.8. As in the proof of Theorem 3.8, define $M = [m_{ij}] \in \mathbb{R}^{n\times n}$ as $m_{ij} = -a_{ij}$, $i \neq j$, and $m_{ii} = \sum_{j=1, j\neq i}^{n+1} a_{ij}$. Also define $b \in \mathbb{R}^n$ as $b \triangleq [-a_{1(n+1)}, \ldots, -a_{n(n+1)}]^T$. Let $\xi \triangleq [\xi_1^T, \cdots, \xi_n^T]^T$, $e \triangleq (M \otimes I_m)\xi + (b \otimes I_m)\xi^r$, $u \triangleq [u_1^T, \ldots, u_n^T]^T$, $K_r \triangleq \text{diag}(K_{r1}, \ldots, K_{rn})$, and $K_v \triangleq \text{diag}(K_{v1}, \ldots, K_{vn})$. Noting that $\dot{\zeta}_j = u_j$, we rewrite (5.14) as

$$(M \otimes I_m)u + (b \otimes I_m)\dot{\zeta}^r = -K_r e - K_v \dot{e}. \tag{5.15}$$

If M has full rank, then it is straightforward to show that there is a unique solution for u. Under the condition of the theorem, M has full rank by following the proof of Theorem 3.8. This proves the first statement of the theorem.

Note that (5.15) can be written as $\ddot{e} = -K_r e - K_v \dot{e}$. It follows that $e(t) \to 0$ and $\dot{e}(t) \to 0$, as $t \to \infty$, because K_r and K_v are symmetrical positive-definite matrices. Then by following the proof of Theorem 3.15, we know that $\xi_i(t) \to \xi^r(t)$ and $\zeta_i(t) \to \zeta^r(t)$, as $t \to \infty$, if and only if $\mathcal{G}_{n+1}$ has a directed spanning tree. ■

Remark 5.14. Note that no constraints are imposed on $f(t, \xi^r, \zeta^r)$ in (5.1) in the proof of Theorem 5.13, as long as $f(\cdot, \cdot, \cdot)$ is piecewise continuous in t and locally Lipschitz in ξ^r and ζ^r. □

Example 5.15. Theorem 5.13 is illustrated by the following example. Consider the interaction topology given by Fig. 5.4, where reference model (5.1) is available only to vehicles 1 and 5. Note that although neither vehicle 1 nor vehicle 5 has a directed path to all other followers on the team, there exists a directed path from node ξ^r to all followers on the team. Also note that unlike the leader-following strategy where the interaction topology is constrained to be a directed spanning tree, algorithm (5.14) allows information to flow arbitrarily among the followers.

---

[2] The second-order derivatives of the information states are equivalent to the information control inputs by noting from (4.1) that $\ddot{\xi}_j \equiv \dot{\zeta}_j \equiv u_j$.

[3] Equivalently, from Lemma 2.11, vehicle $n+1$ is the only vehicle that has a directed path to all other vehicles on the team by noting that all entries of the last row of $\mathcal{A}_{n+1}$ are zero.

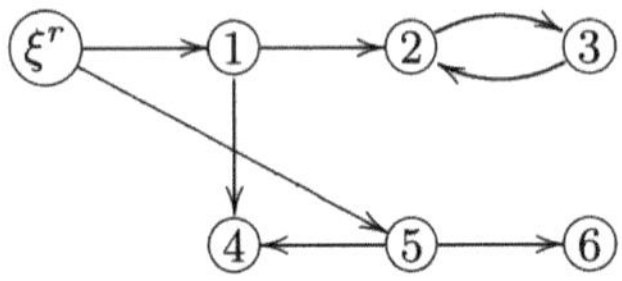

Fig. 5.4. A general interaction topology where reference model (5.1) is available only to a subgroup of followers and the original topology without node ξ^r does not have a directed spanning tree

To further illustrate, let $f(t, \xi^r, \zeta^r) = -\frac{\sin(\xi^r)}{1+e^{-t}}$ in (5.1), where $\xi^r(0) = \frac{\pi}{2}$ and $\zeta^r(0) = 0$. Figure 5.5 demonstrates that using (5.14) all followers' information states and their derivatives reach consensus on the solution of reference model (5.1). □

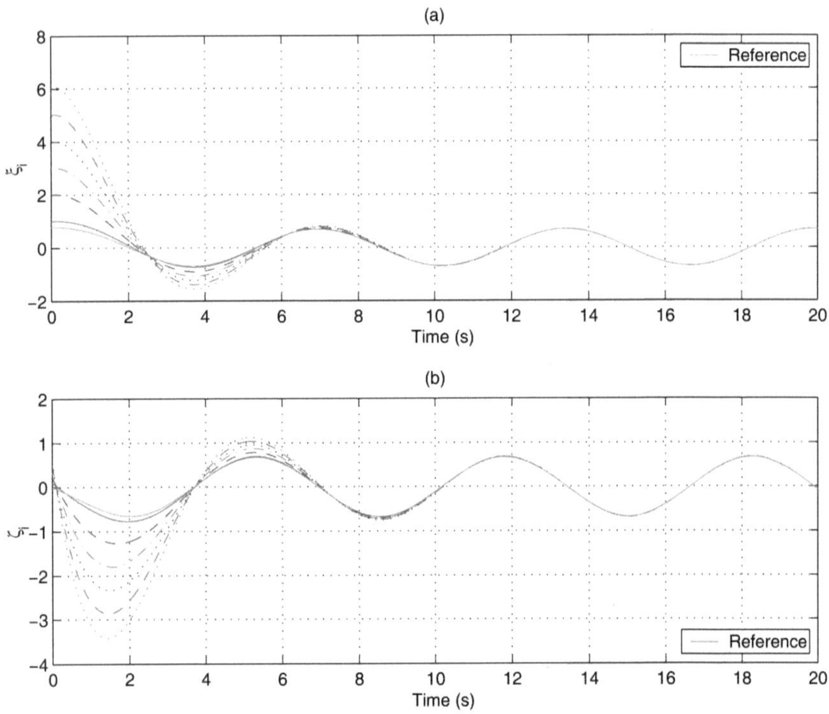

Fig. 5.5. Information states and their derivatives using algorithm (5.14) when reference model (5.1) is available only to a subgroup of followers

Note that (5.14) does not explicitly take into account a bounded control effort. Next, we propose a consensus algorithm with bounded control inputs as

$$\begin{aligned}u_i = &\frac{1}{\kappa_i}(\sum_{j=1}^n a_{ij}\dot{\zeta}_j + a_{i(n+1)}\dot{\zeta}^r)\\ &- \frac{1}{\kappa_i}K_{ri}\tanh[\sum_{j=1}^n a_{ij}(\xi_i - \xi_j) + a_{i(n+1)}(\xi_i - \xi^r)]\\ &- \frac{1}{\kappa_i}K_{vi}\tanh[\sum_{j=1}^n a_{ij}(\zeta_i - \zeta_j) + a_{i(n+1)}(\zeta_i - \zeta^r)], \quad i = 1, \ldots, n, \quad (5.16)\end{aligned}$$

where a_{ij} and κ_i are defined as in (5.14), K_{ri} and K_{vi} are $m \times m$ positive-definite diagonal matrices, and $\tanh(\cdot)$ is defined componentwise. Note that as in (5.14), each vehicle needs the information states and their first-, and second-order derivatives (*i.e.*, the information control inputs u_j) from its neighbors.

Theorem 5.16. *Suppose that $f(t, \xi^r, \zeta^r)$ in (5.1) is bounded. With (5.16), there exists a unique bounded solution for u_i and $\xi_i(t) \to \xi^r(t)$ and $\zeta_i(t) \to \zeta^r(t)$, as $t \to \infty$, if and only if directed graph $\mathcal{G}_{n+1}$ has a directed spanning tree.*

Proof: Let M and b be defined as in the proof of Theorem 5.13. Define $d \triangleq [d_1^T, \ldots, d_n^T]^T \in \mathbb{R}^{mn}$, where

$$\begin{aligned}d_i = &-K_{ri}\tanh[\sum_{j=1}^n a_{ij}(\xi_i - \xi_j) + a_{i(n+1)}(\xi_i - \xi^r)]\\ &- K_{vi}\tanh[\sum_{j=1}^n a_{ij}(\zeta_i - \zeta_j) + a_{i(n+1)}(\zeta_i - \zeta^r)].\end{aligned}$$

Noting that $\dot{\zeta}_j = u_j$, we rewrite (5.16) as

$$(M \otimes I_m)u + (b \otimes I_m)\dot{\zeta}^r = d, \tag{5.17}$$

where $u \triangleq [u_1^T, \ldots, u_n^T]^T$. Note that b, $\dot{\zeta}^r$, M, and d are all bounded. The uniqueness and boundedness of u_i follows from the proof of Theorem 3.8.

Note that (5.17) can be written as

$$\ddot{e}_i = -K_{ri}\tanh(e_i) - K_{vi}\tanh(\dot{e}_i), \tag{5.18}$$

where $e_i \triangleq \sum_{j=1}^n a_{ij}(\xi_i - \xi_j) + a_{i(n+1)}(\xi_i - \xi^r)$. Consider the Lyapunov function candidate

$$V = \sum_{i=1}^n \{\mathbf{1}_m^T K_{ri}\log[\cosh(e_i)] + \frac{1}{2}\dot{e}_i^T \dot{e}_i\},$$

which is positive definite and radially unbounded with respect to e_i and $\dot{e}_i$. Differentiating V gives

$$\begin{aligned}\dot{V} &= \sum_{i=1}^{n}(\dot{e}_i^T K_{ri}\tanh(e_i) + \dot{e}_i^T[-K_{ri}\tanh(e_i) - K_{vi}\tanh(\dot{e}_i)]) \\ &= -\sum_{i=1}^{n}\dot{e}_i^T K_{vi}\tanh(\dot{e}_i) \leq 0.\end{aligned}$$

Let $S = \{(e_i, \dot{e}_i)|\dot{V} = 0\}$. Note that $\dot{V} \equiv 0$ implies that $\dot{e}_i \equiv 0$, which in turn implies that $\ddot{e}_i \equiv 0$. Because $\dot{e}_i \equiv 0$ and $\ddot{e}_i \equiv 0$, it follows that $e_i \equiv 0$ from (5.18). By Theorem F.4, it follows that $e_i(t) \to 0$ and $\dot{e}_i(t) \to 0$ asymptotically, as $t \to \infty$. Letting $e \triangleq [e_1^T, \ldots, e_n^T]^T$ and $\xi \triangleq [\xi_1^T, \ldots, \xi_n^T]^T$ gives $e = (M \otimes I_m)r + (b \otimes I_m)\xi^r$. Then by following the proof of Theorem 3.15, we know that $\xi_i(t) \to \xi^r(t)$ and $\zeta_i(t) \to \zeta^r(t)$, as $t \to \infty$, if and only if $\mathcal{G}_{n+1}$ has a directed spanning tree. ■

Remark 5.17. Note that with the full access strategy (5.10), the reference model must be available to all followers. In contrast, consensus algorithms (5.14) and (5.16) do not impose this constraint and allow the reference model to be available to one or more followers. Also note that with the leader-following strategy (5.12), each vehicle except the unique team leader has exactly one parent (*i.e.*, no information loops are allowed). In contrast, consensus algorithms (5.14) and (5.16) allow information to flow from any follower to any other follower while guaranteeing that the stability remains unchanged as long as the minimum connectivity requirements in Theorems 5.13 and 5.16 are satisfied. As a result, information feedback can be introduced through the general information exchange and coupling between neighboring vehicles, which increases redundancy and robustness to failures of interaction links. The full access strategy (5.10) can be considered a special case of (5.14), where the reference model is available to all followers. The leader-following strategy (5.12) can also be considered a special case of (5.14), where the reference model is available only to one follower and each follower has exactly one parent. □

5.4 Notes

The results in this chapter are based mainly on [179, 181–183]. Section 5.2.1 follows [170, 174]. A variant of algorithm (5.8) is studied in [261] over an undirected interaction topology. A variant of algorithm (5.12) is studied in [244]. See [155, 232, 238] for flocking algorithms for double-integrator dynamics.

Acknowledgment is given to

©2007 IEEE. Reprinted, with permission, from Wei Ren, "Second-order consensus algorithm with extensions to switching topologies and reference models," *Proceedings of the American Control Conference*, pp. 1431–1436, New York City, 2007.

Part IV

Consensus Algorithms for Rigid Body Attitude Dynamics

6

Consensus Algorithms for Rigid Body Attitude Dynamics

This chapter considers consensus algorithms for rigid body attitude dynamics. We propose algorithms for attitude consensus among multiple rigid bodies for three different cases. In the first case, multiple rigid bodies align their attitudes with zero final angular velocities. In the second case, multiple rigid bodies converge to a constant reference attitude while aligning their attitudes during the transition with zero final angular velocities without the requirement for absolute and relative angular velocity measurements. In the third case, multiple rigid bodies converge to the same angular velocity while aligning their attitudes during the transition. Simulation results for attitude consensus among six rigid bodies demonstrate the effectiveness of the algorithms.

6.1 Problem Statement

Attitude control of a rigid body has been studied extensively in the literature (see [129, 235, 249, 251, 254] and references therein). Motivated by the benefits gained by having multiple inexpensive, simple rigid bodies working together, cooperative attitude control of multiple rigid bodies has received recent attention. For example, in space-based interferometry applications, a formation of networked spacecraft could be used to synthesize a space-based interferometer with baselines reaching tens to hundreds of kilometers as an alternative to traditional monolithic spacecraft, where it is essential that the spacecraft maintain relative or the same attitude during formation maneuvers.

The main purpose of this chapter is to extend the consensus algorithms for single- or double-integrator dynamics in Chapters 2–5 to rigid body attitude dynamics. We use the term *attitude consensus* or *attitude alignment* to refer to the case where multiple rigid bodies maintain the same attitude.

It is worthwhile to mention that although we use PD-like control laws for attitude alignment, the algorithms developed for the double-integrator dynamics are not directly applicable to rigid body attitude dynamics due to the inherent nonlinearity in attitude kinematics. The extension from double-integrator dynamics to rigid body attitude dynamics is nontrivial. It is also worthwhile to mention that in contrast to work in attitude control of a single rigid body, the novelty of this chapter lies in the analysis of the way interrigid body information exchange plays a key role in attitude alignment from a consensus point of view.

We use Euler parameters (*i.e.*, unit quaternions) to represent rigid body attitudes in this chapter. Let $q_i = [\widehat{q_i}^T, \overline{q_i}]^T \in \mathbb{R}^4$ be the unit quaternion, where $\widehat{q_i} \in \mathbb{R}^3$ and $\overline{q_i} \in \mathbb{R}$ are, respectively, the vector and scalar parts of q_i, $\omega_i \in \mathbb{R}^3$ be the angular velocity, and $J_i \in \mathbb{R}^{3\times 3}$ and $\tau_i \in \mathbb{R}^3$ be, respectively, the inertial tensor and control torque of the ith rigid body. Rigid body attitude dynamics using Euler parameters are given by (D.2). Attitude consensus or alignment is *achieved* or *reached* among n networked rigid bodies if for all $q_i(0)$ and $\omega_i(0)$ and all $i, j = 1, \ldots, n$, $\|q_i(t) - q_j(t)\| \to 0$ and $\|\omega_i(t) - \omega_j(t)\| \to 0$, as $t \to \infty$.

In this chapter, we consider an undirected, time-invariant interaction topology and assume that all weights in the control laws are constant. We assume that the vectors in each control law have been appropriately transformed and represented in the same coordinate frame.

As in Chapter 4, the interaction topologies for q_i and ω_i among the n rigid bodies are allowed to be different under some circumstances. We use undirected graphs $\mathcal{G}_n^A \triangleq (\mathcal{V}_n, \mathcal{E}_n^A)$ and $\mathcal{G}_n^B \triangleq (\mathcal{V}_n, \mathcal{E}_n^B)$ to model the interaction topologies for q_i and ω_i, respectively, among the n rigid bodies. Let $\mathcal{A}_n = [a_{ij}] \in \mathbb{R}^{n\times n}$ and $\mathcal{B}_n = [b_{ij}] \in \mathbb{R}^{n\times n}$ be the adjacency matrix associated with $\mathcal{G}_n^A$ and $\mathcal{G}_n^B$, respectively. Also let $\mathcal{L}_n^A = [\ell_{ij}] \in \mathbb{R}^{n\times n}$ (*i.e.*, $\ell_{ij} = -a_{ij}$, $i \neq j$, $\ell_{ii} = \sum_{j=1, j\neq i}^n a_{ij}$) and $\mathcal{L}_n^B = [\ell_{ij}] \in \mathbb{R}^{n\times n}$ (*i.e.*, $\ell_{ij} = -b_{ij}$, $i \neq j$, $\ell_{ii} = \sum_{j=1, j\neq i}^n b_{ij}$) be the Laplacian matrix associated with $\mathcal{G}_n^A$ and $\mathcal{G}_n^B$, respectively. When there is only one interaction topology associated with the n rigid bodies, we simply use $\mathcal{G}_n \triangleq (\mathcal{V}_n, \mathcal{E}_n)$ to model the interaction topology among the n rigid bodies, as in Chapter 2. Similarly, $\mathcal{A}_n$ and $\mathcal{L}_n$ are, respectively, the adjacency matrix and Laplacian matrix associated with $\mathcal{G}_n$.

6.2 Attitude Consensus with Zero Final Angular Velocities

In this section, we consider the case where multiple rigid bodies align their attitudes during the transition and their angular velocities approach zero. The proposed control torque for the ith rigid body is

$$\tau_i = -k_G \widehat{q^{r*} q_i} - D_{Gi}\omega_i - \sum_{j=1}^n [a_{ij}\widehat{q_j^* q_i} + b_{ij}(\omega_i - \omega_j)], \quad i = 1, \ldots, n, \tag{6.1}$$

where k_G is a nonnegative scalar, $D_{Gi} \in \mathbb{R}^{3\times 3}$ is symmetrical positive definite, $q^r \in \mathbb{R}^4$ denotes the constant reference attitude for each rigid body, and a_{ij}

and b_{ij} are, respectively, the (i, j) entry of the adjacency matrix $\mathcal{A}_n \in \mathbb{R}^{n\times n}$ and $\mathcal{B}_n \in \mathbb{R}^{n\times n}$ associated with, respectively, $\mathcal{G}_n^A$ and $\mathcal{G}_n^B$

Remark 6.1. Note that control law (6.1) is model independent (*i.e.*, no J_i). Also note that although certain torque feedback can be chosen to linearize (D.2b), the quaternion kinematics (D.2a) are inherently nonlinear. This feature makes the rigid body attitude alignment problem more complicated than consensus algorithms for systems modeled by single or double-integrator dynamics. □

We have the following theorem for attitude alignment among multiple networked rigid bodies with control torque (6.1):

Theorem 6.2. *Suppose that the control torque is given by* (6.1) *and* $\mathcal{G}_n^A$ *and* $\mathcal{G}_n^B$ *are undirected.*[1] *If* $k_G > 2\sum_{j=1}^n a_{ij}$*, then* $q_i(t) \to q^r$ *and* $\omega_i(t) \to 0$*,* $i = 1, \ldots, n$*, as* $t \to \infty$*. If* $k_G = 0$ *and* $\mathcal{G}_n^A$ *is a tree,*[2] *then* $q_i(t) \to q_j(t)$ *and* $\omega_i(t) \to 0$*,* $i = 1, \ldots, n$*, as* $t \to \infty$*.*

Proof:
Subcase A: $k_G > 2\sum_{j=1}^n a_{ij}$**.**
Consider the Lyapunov function candidate

$$V = k_G \sum_{i=1}^{n} \|q^{r*}q_i - \mathbf{q}_I\|^2 + \frac{1}{2}\sum_{i=1}^{n}\sum_{j=1}^{n} a_{ij}\,\|q^{r*}q_i - q^{r*}q_j\|^2 + \frac{1}{2}\sum_{i=1}^{n}(\omega_i^T J_i \omega_i),$$

where $\mathbf{q}_I \stackrel{\triangle}{=} [0, 0, 0, 1]^T$. Note that V is positive definite with respect to $q^{r*}q_i - \mathbf{q}_I$ and ω_i, $i = 1, \ldots, n$. The set $\{q^{r*}q_i - \mathbf{q}_I, \omega_i | V \le c\}$, where $c > 0$, is compact with respect to $q^{r*}q_i - \mathbf{q}_I$ and ω_i, $i = 1, \ldots, n$.

Applying Lemma D.1, the derivative of V is

$$\begin{aligned}\dot{V} =& k_G \sum_{i=1}^{n} \omega_i^T \widehat{q^{r*}q_i} + \frac{1}{2}\sum_{i=1}^{n}\sum_{j=1}^{n} a_{ij}(\omega_i - \omega_j)^T \widehat{q_j^* q_i} \\ &+ \sum_{i=1}^{n} \omega_i^T(\tau_i - \omega_i \times J_i\omega_i).\end{aligned} \tag{6.2}$$

Note that $\omega_i^T(\omega_i \times J_i\omega_i) = 0$ and that

[1] That is, $\mathcal{A}_n$ and $\mathcal{B}_n$ are symmetrical.

[2] The condition that $\mathcal{G}_n^A$ is a tree implies that $|\mathcal{E}_n^A| = n - 1$, where $|\cdot|$ denotes the cardinality of a set.

$$\begin{aligned}
&\frac{1}{2}\sum_{i=1}^{n}\sum_{j=1}^{n} a_{ij}(\omega_i-\omega_j)^T\widehat{q_j^* q_i}\\
=&\frac{1}{2}\sum_{i=1}^{n}\omega_i^T(\sum_{j=1}^{n} a_{ij}\widehat{q_j^* q_i})-\frac{1}{2}\sum_{i=1}^{n}\sum_{j=1}^{n} a_{ij}\omega_j^T\widehat{q_j^* q_i}\\
=&\frac{1}{2}\sum_{i=1}^{n}\omega_i^T(\sum_{j=1}^{n} a_{ij}\widehat{q_j^* q_i})-\frac{1}{2}\sum_{i=1}^{n}\sum_{j=1}^{n} a_{ji}\omega_j^T\widehat{q_j^* q_i}\\
=&\frac{1}{2}\sum_{i=1}^{n}\omega_i^T(\sum_{j=1}^{n} a_{ij}\widehat{q_j^* q_i})+\frac{1}{2}\sum_{j=1}^{n}\sum_{i=1}^{n} a_{ji}\omega_j^T\widehat{q_i^* q_j}\\
=&\frac{1}{2}\sum_{i=1}^{n}\omega_i^T(\sum_{j=1}^{n} a_{ij}\widehat{q_j^* q_i})+\frac{1}{2}\sum_{j=1}^{n}\omega_j^T(\sum_{i=1}^{n} a_{ji}\widehat{q_i^* q_j})\\
=&\sum_{i=1}^{n}\omega_i^T(\sum_{j=1}^{n} a_{ij}\widehat{q_j^* q_i}),
\end{aligned} \tag{6.3}$$

where we have used the fact that $a_{ij}=a_{ji}$ to obtain the second equality and we have switched the order of the summation signs and have used the fact that $\widehat{q_j^* q_i}=-\widehat{q_i^* q_j}$ to obtain the third equality.

As a result, (6.2) becomes

$$\dot{V}=\sum_{i=1}^{n}\omega_i^T(k_G\widehat{q^{r*} q_i}+\sum_{j=1}^{n} a_{ij}\widehat{q_j^* q_i}+\tau_i).$$

With control law (6.1), the derivative of V becomes

$$\dot{V}=-\sum_{i=1}^{n}(\omega_i^T D_{Gi}\omega_i)-\frac{1}{2}\sum_{i=1}^{n}\sum_{j=1}^{n} b_{ij}\left\|\omega_i-\omega_j\right\|^2\leq 0, \tag{6.4}$$

where we have used the fact that

$$\sum_{i=1}^{n}\omega_i^T\sum_{j=1}^{n} b_{ij}(\omega_i-\omega_j)=\frac{1}{2}\sum_{i=1}^{n}\sum_{j=1}^{n} b_{ij}\left\|\omega_i-\omega_j\right\|^2,$$

following (4.15) in Lemma 4.18 by noting that $b_{ij}=b_{ji}$.

Let $\Omega=\{q^{r*}q_i-\mathbf{q}_I,\omega_i|\dot{V}=0\}$. Note that $\dot{V}\equiv 0$ implies that $\omega_i\equiv 0$, $i=1,\ldots,n$. Because $\omega_i\equiv 0$, we know that

$$k_G\widehat{q^{r*} q_i}+\sum_{j=1}^{n} a_{ij}\widehat{q_j^* q_i}=0, \quad i=1,\ldots,n \tag{6.5}$$

from (D.2b) and (6.1).

Noting that $q_j^* q_i=q_j^*(q^r q^{r*})q_i=(q_j^* q^r)(q^{r*} q_i)$, we rewrite (6.5) as

$$\widehat{p_i^* q^{r*} q_i} = 0, \tag{6.6}$$

where

$$p_i = k_G \mathbf{q}_I + \sum_{j=1}^{n} a_{ij} q^{r*} q_j. \tag{6.7}$$

Also note that (6.6) is equivalent to

$$-\overline{q^{r*} q_i} \widehat{p_i} + \overline{p_i} \widehat{q^{r*} q_i} + \widehat{q^{r*} q_i} \times \widehat{p_i} = 0. \tag{6.8}$$

Multiplying (6.8) by $(\widehat{q^{r*} q_i} \times \widehat{p_i})^T$, gives

$$\left\| \widehat{q^{r*} q_i} \times \widehat{p_i} \right\|^2 = 0. \tag{6.9}$$

Combining (6.8) and (6.9) gives

$$-\overline{q^{r*} q_i} \widehat{p_i} + \overline{p_i} \widehat{q^{r*} q_i} = 0. \tag{6.10}$$

Using (6.7), we rewrite (6.10) as

$$-\overline{q^{r*} q_i} \sum_{j=1}^{n} a_{ij} \widehat{q^{r*} q_j} + (k_G + \sum_{j=1}^{n} a_{ij} \overline{q^{r*} q_j}) \widehat{q^{r*} q_i} = 0, \quad i = 1, \ldots, n. \tag{6.11}$$

Note that (6.11) can be written in matrix form as

$$[P(t) \otimes I_3] \, \widehat{q_s} = 0,$$

where $\widehat{q_s} \in \mathbb{R}^{3n}$ is a column vector stack composed of $\widehat{q^{r*} q_\ell}$, $\ell = 1, \ldots, n$, and $P(t) = [p_{ij}(t)] \in \mathbb{R}^{n \times n}$ is given by $p_{ii}(t) = k_G + \sum_{j=1}^{n} a_{ij} \overline{q^{r*} q_j}$ and $p_{ij}(t) = -a_{ij} \overline{q^{r*} q_i}$.

Noting that $|\overline{q^{r*} q_j}| \leq 1$, $j = 1, \ldots, n$, and $k_G > 2 \sum_{j=1}^{n} a_{ij}$, we see that $P(t)$ is strictly diagonally dominant and therefore has full rank, which in turn implies that $\widehat{q_s} \equiv 0$. Thus, we see that $\widehat{q^{r*} q_i} \equiv 0$, $i = 1, \ldots, n$, which implies that $q^{r*} q_i - \mathbf{q}_I \equiv 0$ if $\dot{V} \equiv 0$.

Therefore, by Theorem F.3, it follows that $q^{r*} q_i(t) - \mathbf{q}_I \to 0$ and $\omega_i(t) \to 0$, as $t \to \infty$. Equivalently, we know that $q_i(t) \to q^r$ and $\omega_i(t) \to 0$, $i = 1, \ldots, n$, as $t \to \infty$.

Subcase B: $k_G = 0$ and $\mathcal{G}_n^A$ is a tree.

Consider the Lyapunov function candidate

$$V = \frac{1}{2} \sum_{i=1}^{n} \sum_{j=1}^{n} a_{ij} \left\| q_j^* q_i - \mathbf{q}_I \right\|^2 + \frac{1}{2} \sum_{i=1}^{n} (\omega_i^T J_i \omega_i),$$

which is positive definite with respect to $q_j^* q_i - \mathbf{q}_I$ and ω_i, $i = 1, \ldots, n$. The set $\{(q_j^* q_i - \mathbf{q}_I, \omega_i) | V \leq c\}$, where $c > 0$, is compact with respect to $q_j^* q_i - \mathbf{q}_I$

and ω_i, $i = 1, \ldots, n$. Then by following a procedure similar to that in Subcase A, we get (6.4).

Let $\Omega = \{(q_j^* q_i - \mathbf{q}_I, \omega_i) | \dot{V} = 0\}$ and note that $\dot{V} \equiv 0$ implies that $\omega_i \equiv 0$, $i = 1, \ldots, n$. Because $\omega_i \equiv 0$, we know that

$$\sum_{j=1}^{n} a_{ij} \widehat{q_j^* q_i} = 0, \quad i = 1, \ldots, n, \tag{6.12}$$

from (D.2b) and (6.1).

Let $\widehat{q_j^* q_i}$ be a variable associated with an edge $(i,j) \in \mathcal{E}_n^A$. Note that (i,j) and (j,i) denote the same element in the edge set $\mathcal{E}_n^A$ when $\mathcal{G}_n^A$ is undirected. Also note that $\widehat{q_i^* q_j} = -\widehat{q_j^* q_i}$. In the following, we assume that $i < j$ without loss of generality. Noting that $\mathcal{G}_n^A$ is a tree, we know that $|\mathcal{E}_n^A| = n-1$, which implies that there are $n-1$ variables associated with $\mathcal{E}_n^A$. Let $\widehat{q_u} \in \mathbb{R}^{3(n-1)}$ be a column vector stack composed of all $\widehat{q_j^* q_i}$, where $(i,j) \in \mathcal{E}_n^A$. By noting that $\widehat{q_i^* q_j} = -\widehat{q_j^* q_i}$, (6.12) can be rewritten as

$$(Q \otimes I_3)\, \widehat{q_u} = 0, \tag{6.13}$$

where $Q \in \mathbb{R}^{n \times n-1}$.

Consider a system given by $Q\tilde{x} = 0$, where $\tilde{x}$ is a column vector stack composed of $x_{ij} = x_i - x_j$, where $(i,j) \in \mathcal{E}_n^A$, and $x_k \in \mathbb{R}$, $k = 1, \ldots, n$. Note that $Q\tilde{x} = 0$ can be written as $\mathcal{L}_n^A x = 0$, where $x = [x_1, \ldots, x_n]^T$ and $\mathcal{L}_n^A$ is the $n \times n$ (symmetrical) Laplacian matrix associated with undirected graph $\mathcal{G}_n^A$. Noting that $\mathcal{G}_n^A$ is a tree, we know that $x_1 = \cdots = x_n$, which in turn implies that $\tilde{x} = 0$. As a result, we know that Q can be transformed to a row echelon form to show that $Q\tilde{x} = 0$ implies that $x_{ij} = 0$, where $(i,j) \in \mathcal{E}_n^A$. The same transformation procedure can be used on (6.13) to show that $(Q \otimes I_3)\widehat{q_u} = 0$ implies that $\widehat{q_j^* q_i} = 0$, where $(i,j) \in \mathcal{E}_n^A$. Thus, we see that $q_j^* q_i - \mathbf{q}_I \equiv 0$, $\forall i \neq j$, if $\dot{V} \equiv 0$.

Therefore, by Theorem F.3, it follows that $q_j^*(t) q_i(t) - \mathbf{q}_I \to 0$ and $\omega_i(t) \to 0$, as $t \to \infty$. Equivalently, we know that $q_i(t) \to q_j(t)$, $\forall i \neq j$, and $\omega_i(t) \to 0$, $i = 1, \ldots, n$, as $t \to \infty$. ∎

6.3 Attitude Consensus Without Absolute and Relative Angular Velocity Measurements

Note that (6.1) requires both absolute and relative angular velocity measurements. We propose a control torque without absolute and relative angular velocity measurements based on a passivity approach as

$$\dot{\hat{x}}_i = F\hat{x}_i + \sum_{j=1}^{n} a_{ij}(q_i - q_j) + \kappa q_i, \tag{6.14a}$$

$$y_i = PF\hat{x}_i + P\sum_{j=1}^{n} a_{ij}(q_i - q_j) + \kappa P q_i, \tag{6.14b}$$

$$\tau_i = -k_G \widehat{q^{r*}q_i} - \sum_{j=1}^{n} a_{ij}\widehat{q_j^* q_i} - \widehat{q_i^* y_i}, \quad i = 1, \ldots, n, \tag{6.14c}$$

where $F \in \mathbb{R}^{4\times 4}$ is Hurwitz, κ is a positive scalar, a_{ij} is the (i,j) entry of the adjacency matrix $\mathcal{A}_n \in \mathbb{R}^{n\times n}$ associated with $\mathcal{G}_n$, k_G is a positive scalar, $q^r \in \mathbb{R}^4$ denotes the constant reference attitude for each rigid body, and $P \in \mathbb{R}^{4\times 4}$ is the symmetrical positive-definite solution of the Lyapunov equation $F^T P + PF = -Q$ with $Q \in \mathbb{R}^{4\times 4}$ symmetrical positive definite. Note that only q_i need to be exchanged among rigid bodies. Also note that the term $\sum_{j=1}^{n} a_{ij}(q_i - q_j)$ in (6.14a) introduces relative damping between neighboring rigid bodies, whereas the term $\sum_{j=1}^{n} a_{ij}\widehat{q_j^* q_i}$ in (6.14c) introduces mutual consensus between neighbors.

Remark 6.3. A passivity approach is used in [129, 130, 160, 235] for controlling a single vehicle. Algorithm (4.22) extends the passivity approach in these references to achieve attitude consensus among multiple rigid bodies. □

Theorem 6.4. *Suppose that $\mathcal{G}_n$ is undirected. With (6.14), $q_i(t) \to q^r$ and $\omega_i(t) \to 0$, as $t \to \infty$, if $k_G > 2\sum_{j=1}^{n} a_{ij}$.*

Proof: Consider the Lyapunov function candidate

$$V = k_G \sum_{i=1}^{n} \left\| q^{r*}q_i - \mathbf{q}_I \right\|^2 + \frac{1}{2}\sum_{i=1}^{n}\sum_{j=1}^{n} a_{ij} \left\| q^{r*}q_i - q^{r*}q_j \right\|^2 + \frac{1}{2}\sum_{i=1}^{n}(\omega_i^T J_i \omega_i) \\ + \dot{\hat{x}}^T (M \otimes I_4)^{-1}(I_n \otimes P)\dot{\hat{x}},$$

where $\hat{x} \triangleq [\hat{x}_1^T, \ldots, \hat{x}_n^T]^T$ and $M \triangleq \mathcal{L}_n + \kappa I_n$ with $\mathcal{L}_n \in \mathbb{R}^{n\times n}$ the Laplacian matrix associated with undirected graph $\mathcal{G}_n$. Note that $\mathcal{L}_n$ is symmetrical positive semidefinite because $\mathcal{G}_n$ is undirected from which it follows that both M and M^{-1} are symmetrical positive definite. From Lemma C.8, $(M \otimes I_4)^{-1}(I_n \otimes P) = (M^{-1} \otimes I_4)(I_n \otimes P) = M^{-1}I_n \otimes I_4 P = I_n M^{-1} \otimes P I_4 = (I_n \otimes P)(M^{-1} \otimes I_4) = (I_n \otimes P)(M \otimes I_4)^{-1}$, that is, $(M \otimes I_4)^{-1}$ and $I_n \otimes P$ commute. Similarly, it is straightforward to show that $(M \otimes I_4)^{-1}$ and $I_n \otimes F^T$ also commute. Note that both $M^{-1}I_n \otimes I_4 P$ and $(M \otimes I_4)^{-1}(I_n \otimes P)$ are symmetrical positive definite. Therefore, V is positive definite with respect to $q^{r*}q_i - \mathbf{q}_I$, ω_i, and $\dot{\hat{x}}_i$. The set $\{(q^{r*}q_i - \mathbf{q}_I, \omega_i, \dot{\hat{x}}_i) | V \le c\}$, where $c > 0$, is compact with respect to $q^{r*}q_i - \mathbf{q}_I$, ω_i, and $\dot{\hat{x}}_i$.

By applying Lemma D.1, we obtain the derivative of V as

$$\begin{aligned}\dot{V} = k_G \sum_{i=1}^{n} \omega_i^T \widehat{q^{r*} q_i} + \frac{1}{2}\sum_{i=1}^{n}\sum_{j=1}^{n} a_{ij}(\omega_i - \omega_j)^T \widehat{q_j^* q_i} + \sum_{i=1}^{n} \omega_i^T [\tau_i - \omega_i \times (J_i \omega_i)] \\ + \dot{\hat{x}}^T (I_n \otimes F^T)(M \otimes I_4)^{-1}(I_n \otimes P)\dot{\hat{x}} \\ + \dot{q}^T (M \otimes I_4)^T (M \otimes I_4)^{-1}(I_n \otimes P)\dot{\hat{x}} \\ + \dot{\hat{x}}^T (M \otimes I_4)^{-1}(I_n \otimes P)(I_n \otimes F)\dot{\hat{x}} \\ + \dot{\hat{x}}^T (M \otimes I_4)^{-1}(I_n \otimes P)(M \otimes I_4)\dot{q},\end{aligned}$$

where we have used the relation

$$\ddot{\hat{x}} = (I_n \otimes F)\dot{\hat{x}} + (M \otimes I_4)\dot{q} \tag{6.15}$$

with $\dot{q} = [\dot{q}_1^T, \ldots, \dot{q}_n^T]^T$. From (6.3), note that when $\mathcal{G}_n$ is undirected, then

$$\frac{1}{2}\sum_{i=1}^{n}\sum_{j=1}^{n} a_{ij}(\omega_i - \omega_j)^T \widehat{q_j^* q_i} = \sum_{i=1}^{n} \omega_i^T (\sum_{j=1}^{n} a_{ij} \widehat{q_j^* q_i}).$$

Further note that $(M \otimes I_4)^{-1}$ and $I_n \otimes F^T$ commute, $(M \otimes I_4)^{-1}$ and $I_n \otimes P$ commute, $M \otimes I_4 = (M \otimes I_4)^T$, $y = (I_n \otimes P)\dot{\hat{x}}$ with $y = [y_1^T, \ldots, y_n^T]^T$, and $F^T P + PF = -Q$. The derivative of V, therefore, becomes

$$\dot{V} = -\sum_{i=1}^{n} \omega_i^T \widehat{q_i^* y_i} - \dot{\hat{x}}^T (M \otimes I_4)^{-1}(I_n \otimes Q)\dot{\hat{x}} + 2\dot{q}^T y.$$

Note that $2\dot{q}^T y = 2\sum_{i=1}^{n} \dot{q}_i^T y_i = 2\sum_{i=1}^{n} \overline{\dot{q}_i^* y_i} = \sum_{i=1}^{n} \overline{\breve{\omega}_i{}^* q_i^* y_i} = \sum_{i=1}^{n} \omega_i^T \widehat{q_i^* y_i}$, where we have used (D.1), (D.3), and the fact that $q_i^* = [-\widehat{q}_i{}^T, \overline{q}_i]^T$ and $\breve{\omega}_i{}^* = [-\omega_i^T, 0]^T$. Therefore, the derivative of V is given by

$$\dot{V} = -\dot{\hat{x}}^T (M \otimes I_4)^{-1}(I_n \otimes Q)\dot{\hat{x}},$$

which is negative semidefinite because $(M \otimes I_4)^{-1}(I_n \otimes Q) = M^{-1} I_n \otimes Q I_4$ is symmetrical positive definite.

Let $S = \{(q^{r*} q_i - \mathbf{q}_I, \omega_i, \dot{\hat{x}}_i) | \dot{V} = 0\}$. Note that $\dot{V} \equiv 0$ implies that $\dot{\hat{x}} \equiv 0$, which in turn implies that $(M \otimes I_4)\dot{q} \equiv 0$ according to (6.15) and $y_i \equiv 0$ by noting that $y_i = P\dot{\hat{x}}_i$ according to (6.14b). Because $M \otimes I_4$ is symmetrical positive definite, it follows that $\dot{q}_i \equiv 0$. Because $\breve{\omega}_i = 2q_i^* \dot{q}_i$ from (D.3), it follows that $\omega_i \equiv 0$. From (D.2b) and (6.14c), it then follows that

$$k_G \widehat{q^{r*} q_i} + \sum_{j=1}^{n} a_{ij} \widehat{q_j^* q_i} \equiv 0, \quad i = 1, \ldots, n.$$

Then by following the proof for Subcase A in Theorem 6.2, we know that $q^{r*} q_i \equiv \mathbf{q}_I$ if $k_G > 2\sum_{j=1}^{n} a_{ij}$. By Theorem F.3, it follows that $q_i(t) \to q^r$ and $\omega_i(t) \to 0$, as $t \to \infty$. ■

6.4 Attitude Consensus with Nonzero Final Angular Velocities

In this section, we consider the case where multiple rigid bodies align their attitudes but with possibly nonzero final angular velocities. The proposed control torque on the ith rigid body is

$$\tau_i = \omega_i \times J_i\omega_i - J_i \sum_{j=1}^{n} [a_{ij}\widehat{q_j^* q_i} + b_{ij}(\omega_i - \omega_j)], \quad i = 1, \ldots, n, \tag{6.16}$$

where a_{ij} and b_{ij} are defined as in (6.1). Note that (6.16) is model dependent in the sense that J_i must be known.

Theorem 6.5. *Suppose that $\mathcal{G}_n^A$ and $\mathcal{G}_n^B$ are undirected. With the control torque given by (6.16), if $\mathcal{G}_n^A$ is a tree and $\mathcal{G}_n^B$ is connected, then $q_i(t) \to q_j(t)$, and $\omega_i(t) \to \omega_j(t)$, $\forall i \neq j$, as $t \to \infty$.*

Proof: Consider the Lyapunov function candidate

$$V = \frac{1}{2}\sum_{i=1}^{n}\sum_{j=1}^{n} a_{ij} \left\|q_j^* q_i - \mathbf{q}_I\right\|^2 + \frac{1}{2}\sum_{i=1}^{n} \omega_i^T \omega_i,$$

which is positive definite with respect to $q_j^* q_i - \mathbf{q}_I$ and ω_i. The set $\{(q_j^* q_i - \mathbf{q}_I, \omega_i) | V \leq c\}$, where $c > 0$, is compact with respect to $q_j^* q_i - \mathbf{q}_I$ and ω_i, $i = 1, \ldots, n$.

Following a procedure similar to that of Theorem 6.2, we obtain

$$\dot{V} = -\frac{1}{2}\sum_{i=1}^{n}\sum_{j=1}^{n} b_{ij} \left\|\omega_i - \omega_j\right\|^2 \leq 0.$$

Let $\Omega = \{(q_j^* q_i - \mathbf{q}_I, \omega_i) | \dot{V} = 0\}$. Note and observe that $\dot{V} \equiv 0$ implies that $\omega_i \equiv \omega_j$, $\forall i \neq j$, because undirected graph $\mathcal{G}_n^B$ is connected. Therefore, we see that $\dot{\omega}_i \equiv \dot{\omega}_j$. As a result, we know that $\dot{\omega} \in \text{span}\{\mathbf{1}_n \otimes \eta\}$, where $\dot{\omega} = [\dot{\omega}_1^T, \cdots, \dot{\omega}_n^T]^T$ and η is some 3×1 vector.

Because $\omega_i \equiv \omega_j$, $\forall i \neq j$, we know that

$$\dot{\omega}_i = -\sum_{j=1}^{n} a_{ij}\widehat{q_j^* q_i}, \quad i = 1, \ldots, n, \tag{6.17}$$

from (D.2b) and (6.16). We also know that

$$\sum_{i=1}^{n} \eta^T \dot{\omega}_i = -\sum_{i=1}^{n} \eta^T (\sum_{j=1}^{n} a_{ij}\widehat{q_j^* q_i}) = 0,$$

where we have used the fact that $a_{ij} = a_{ji}$ and $\widehat{q_j^* q_i} = -\widehat{q_i^* q_j}$. As a result, we see that $\dot{\omega}$ is orthogonal to span$\{\mathbf{1}_n \otimes \eta\}$. Therefore, we conclude that $\dot{\omega} \equiv 0$. From (6.17), we get

$$\sum_{j=1}^{n} a_{ij} \widehat{q_j^* q_i} \equiv 0, \quad i = 1, \ldots, n.$$

Under the assumption that undirected graph $\mathcal{G}_n^A$ is a tree, by following the proof for Subcase B in Theorem 6.2, we see that $q_j^* q_i - \mathbf{q}_I \equiv 0$, $\forall i \neq j$, if $\dot{V} \equiv 0$. By Theorem F.3, it follows that $q_j^*(t) q_i(t) - \mathbf{q}_I \to 0$ and $\omega_i(t) \to \omega_j(t)$, $\forall i \neq j$, as $t \to \infty$. Equivalently, $q_i(t) \to q_j(t)$ and $\omega_i(t) \to \omega_j(t)$, $\forall i \neq j$, as $t \to \infty$. ■

Remark 6.6. Theorems 6.2, 6.4, and 6.5 do not rely on the assumption that the scalar parts of the unit quaternions must be nonnegative for all time. □

6.5 Simulation Results

In this section, we simulate a scenario where six rigid bodies align their attitudes through local information exchange. We apply, respectively, (6.1), (6.14), and (6.16) for attitude alignment. For simplicity, we assume that undirected graphs $\mathcal{G}_n^A$ and $\mathcal{G}_n^B$ are identical (*i.e.*, $\mathcal{E}_n^A = \mathcal{E}_n^B$). The undirected graph for six rigid bodies with (6.1), where $k_G > 0$, and (6.14) is shown by Fig. 6.1. The undirected graph for the six rigid bodies with (6.1), where $k_G = 0$, and (6.16) is shown by Fig. 6.2.

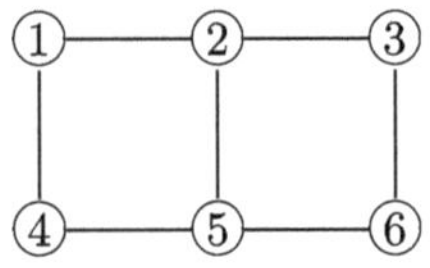

Fig. 6.1. Undirected graph for six rigid bodies with (6.1), where $k_G > 0$, and (6.14). The graph is a tree.

The rigid body specifications are shown in Table 6.1. The control parameters used for each control law are shown in Table 6.2. In the following, we let $q^r = [0, 0, 0, 1]^T$ and choose $q_i(0) \in \mathbb{R}^4$ and $\omega_i(0) \in \mathbb{R}^3$ randomly. In the following, we use a superscript (j) to denote the jth component of a quaternion or a vector.

Figures 6.3, 6.4, and 6.5 show, respectively, the attitudes, angular velocities, and control torques of rigid bodies 1, 3, and 5 with (6.1) in Subcase A.

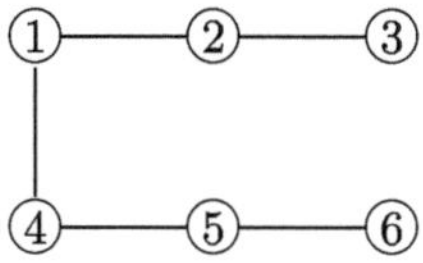

Fig. 6.2. Undirected graph for the six rigid bodies with (6.1), where $k_G = 0$, and (6.16)

Table 6.1. Rigid body specifications

J_1	[1 0.1 0.1; 0.1 0.1 0.1; 0.1 0.1 0.9] kg m^2
J_2	[1.5 0.2 0.3; 0.2 0.9 0.4; 0.3 0.4 2.0] kg m^2
J_3	[0.8 0.1 0.2; 0.1 0.7 0.3; 0.2 0.3 1.1] kg m^2
J_4	[1.2 0.3 0.7; 0.3 0.9 0.2; 0.7 0.2 1.4] kg m^2
J_5	[0.9 0.15 0.3; 0.15 1.2 0.4; 0.3 0.4 1.2] kg m^2
J_6	[1.1 0.35 0.45; 0.35 1.0 0.5; 0.45 0.5 1.3] kg m^2

Table 6.2. Control parameters for each control law

(6.1) Subcase A	Fig. 6.1
	$k_G = 1$, $D_{Gi} = 2I_3$, $a_{ij} = 5$, and $b_{ij} = 10$ if $(j,i) \in \mathcal{E}_n^A = \mathcal{E}_n^B$
(6.1) Subcase B	Fig. 6.2
	$k_G = 0$, $D_{Gi} = 2I_3$, $a_{ij} = 5$, and $b_{ij} = 10$ if $(j,i) \in \mathcal{E}_n^A = \mathcal{E}_n^B$
(6.14)	Fig. 6.1
	$F = -I_4$, $\kappa = 1$, $P = 2I_4$, $k_G = 1$, $a_{ij} = 2$ if $(j,i) \in \mathcal{E}_n^A$
(6.16)	Fig. 6.2
	$a_{ij} = 5$, and $b_{ij} = 10$ if $(j,i) \in \mathcal{E}_n^A = \mathcal{E}_n^B$

Note that all rigid bodies converge to the constant reference attitude while aligning their attitudes during the transition. Also note that their angular velocities converge to zero.

Figures 6.6, 6.7, and 6.8 show, respectively, the attitudes, angular velocities, and control toques of rigid bodies 1, 3, and 5 with (6.1) in Subcase B. Note that all rigid bodies converge to the same attitude, and their angular velocities converge to zero.

Figures 6.9, 6.10, and 6.11 show, respectively, the attitudes, angular velocities, and control torques of rigid bodies 1, 3, and 5 with (6.14). Note that all rigid bodies converge to the constant reference attitude while aligning their attitudes during the transition even without relative angular velocity measurements. Note that using the passivity approach there exist wiggles in Figs. 6.9, 6.10, and 6.11 .

Figures 6.12, 6.13, and 6.14 show, respectively, the attitudes, angular velocities, and control torques of rigid bodies 1, 3, and 5 with (6.16). Note that

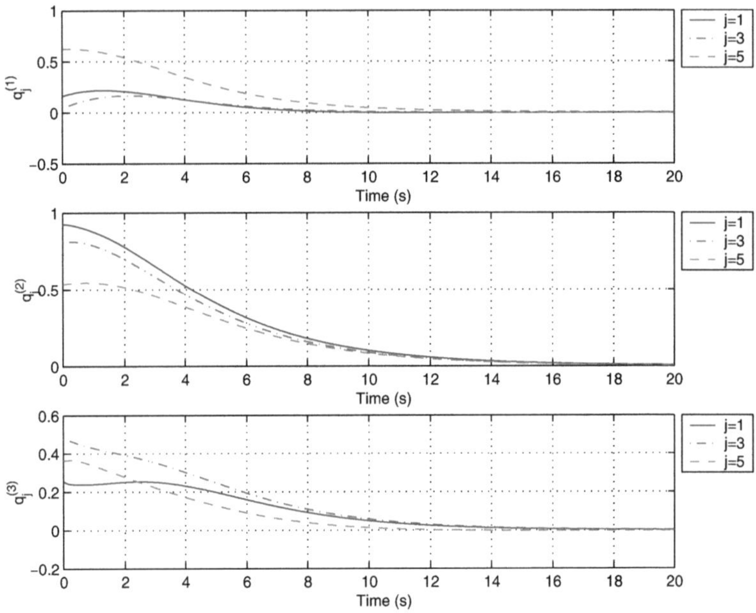

Fig. 6.3. Rigid body attitudes with (6.1) in Subcase A

all rigid bodies converge to the same attitude and the same nonzero angular velocity.

6.6 Notes

The results in this chapter are based mainly on [172, 176, 178]. See [60] for consensus algorithms for rigid body attitude dynamics where attitudes are represented by modified Rodriguez parameters (MRPs).

Acknowledgment is given to

©2006 John Wiley & Sons. Reprinted, with permission, from Wei Ren, "Distributed attitude alignment in spacecraft formation flying," *International Journal of Adaptive Control and Signal Processing*, vol. 21, no. 2–3, pp. 95–113, March–April 2007.

©2006 IEEE. Reprinted, with permission, from Wei Ren, "Distributed attitude consensus among multiple networked spacecraft," *Proceedings of the American Control Conference*, pp. 1760–1765, Minneapolis, MN, 2006.

©2007 IEEE. Reprinted, with permission, from Wei Ren, "Distributed cooperative attitude synchronization and tracking for multiple rigid bodies," *IEEE Transactions on Control Systems Technology*, 2007 (under review).

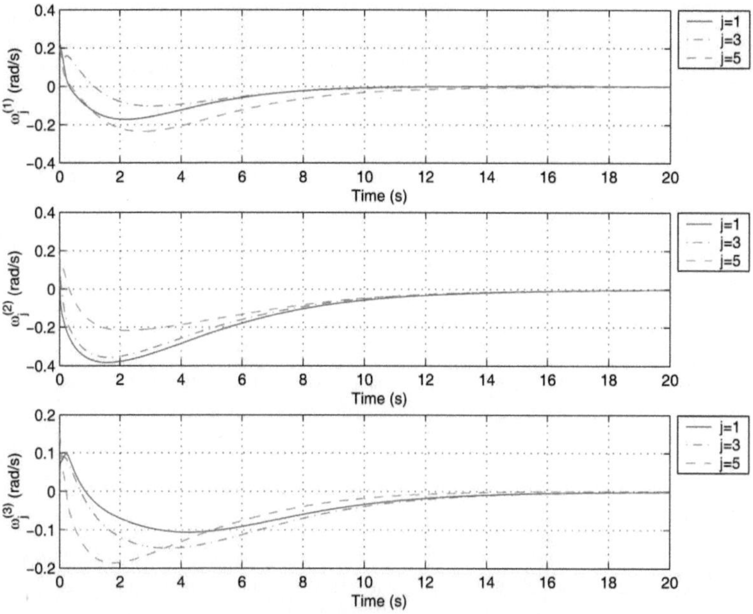

Fig. 6.4. Rigid body angular velocities with (6.1) in Subcase A

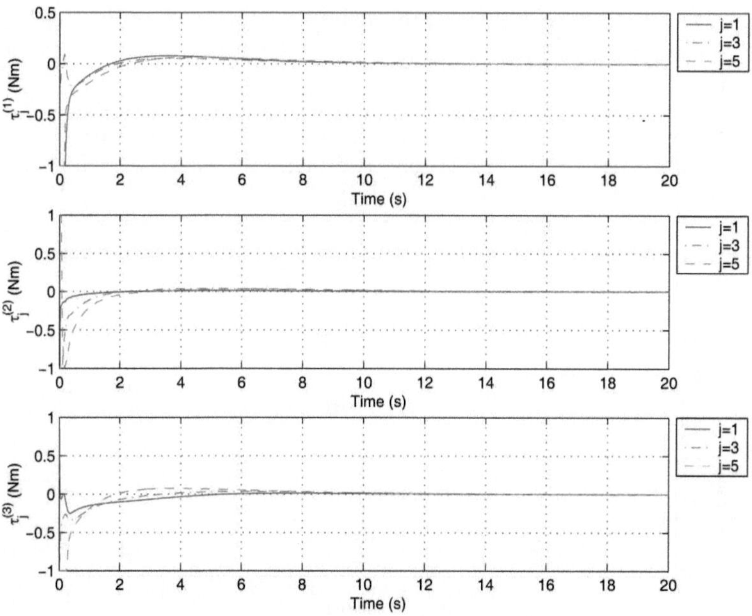

Fig. 6.5. Rigid body control torques with (6.1) in Subcase A

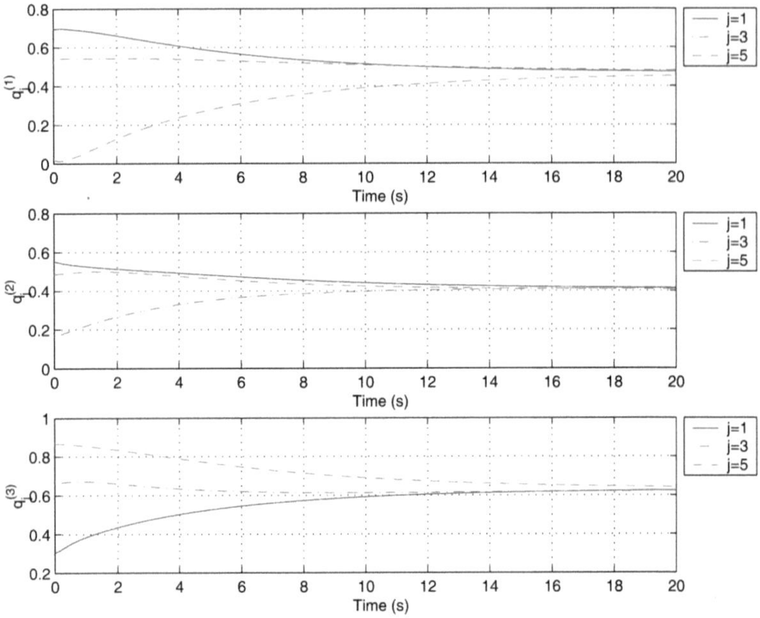

Fig. 6.6. Rigid body attitudes with (6.1) in Subcase B

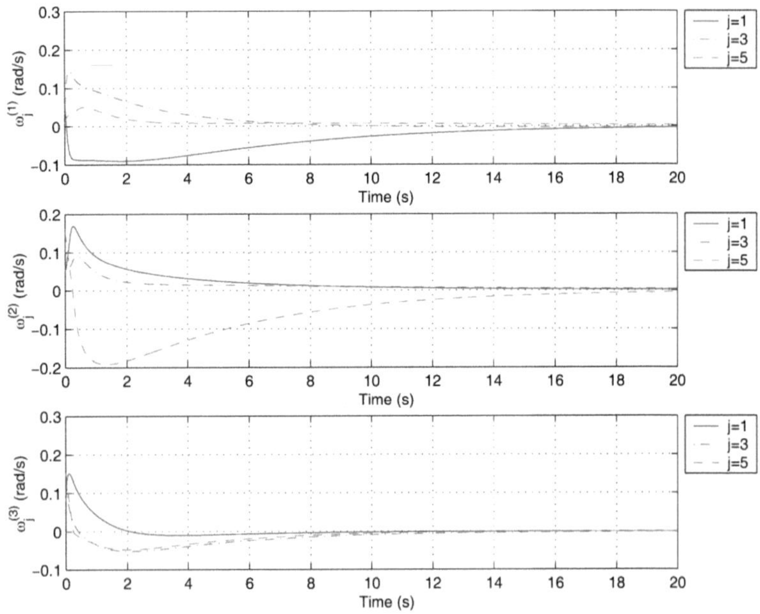

Fig. 6.7. Rigid body angular velocities with (6.1) in Subcase B

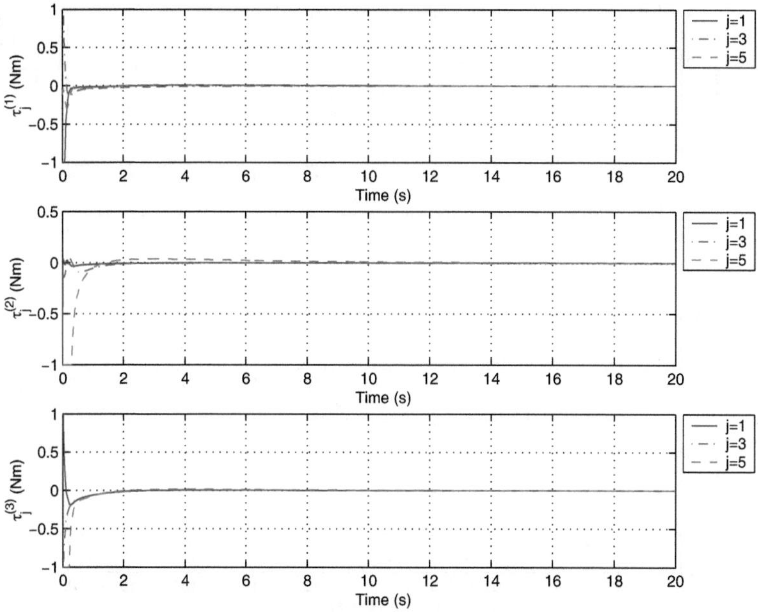

Fig. 6.8. Rigid body control torques with (6.1) in Subcase B

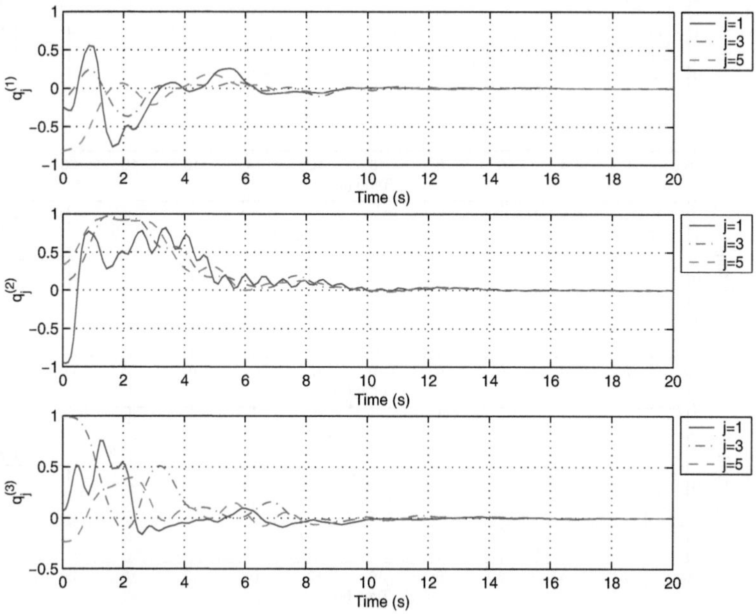

Fig. 6.9. Rigid body attitudes with (6.14)

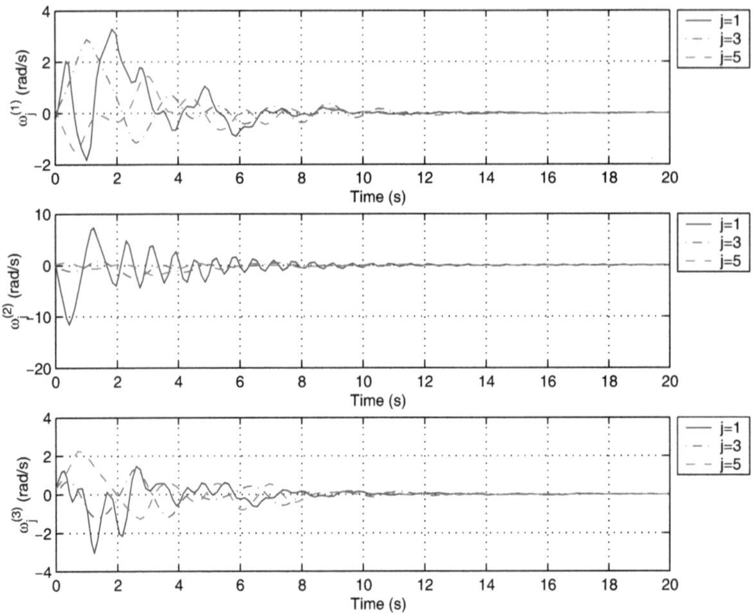

Fig. 6.10. Rigid body angular velocities with (6.14)

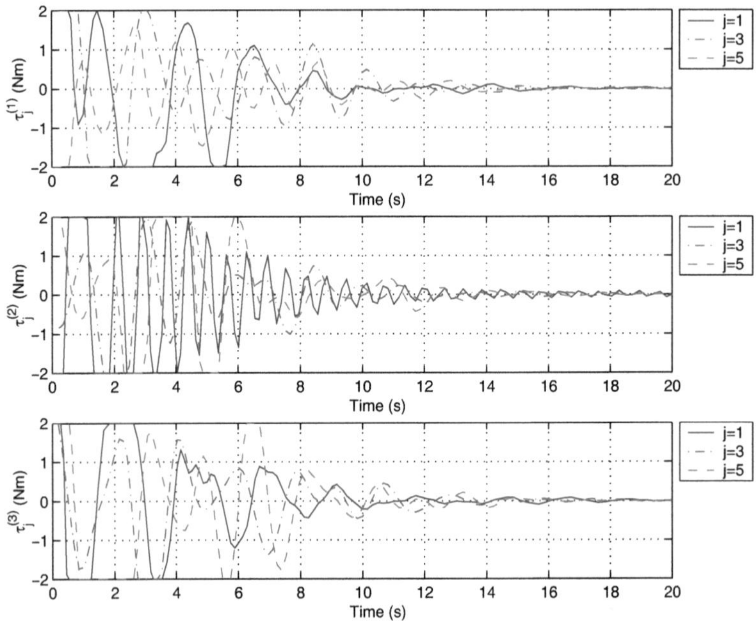

Fig. 6.11. Rigid body control torques with (6.14)

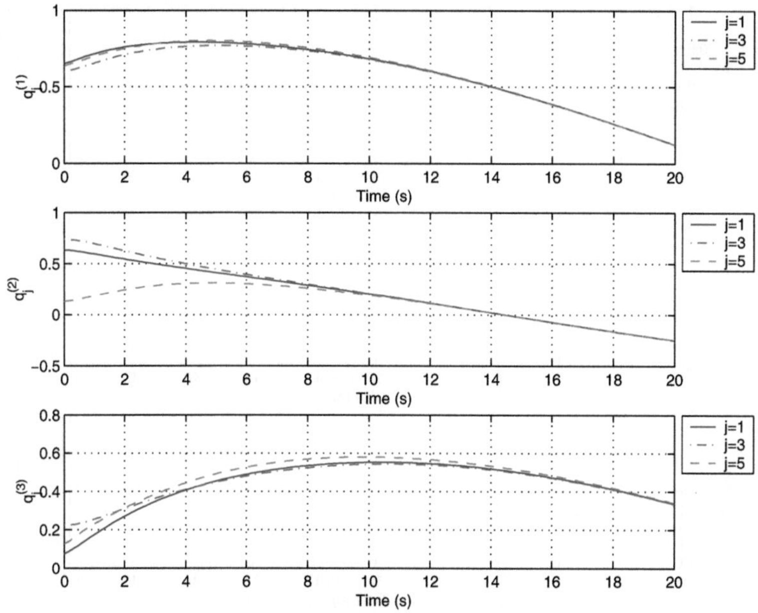

Fig. 6.12. Rigid body attitudes with (6.16)

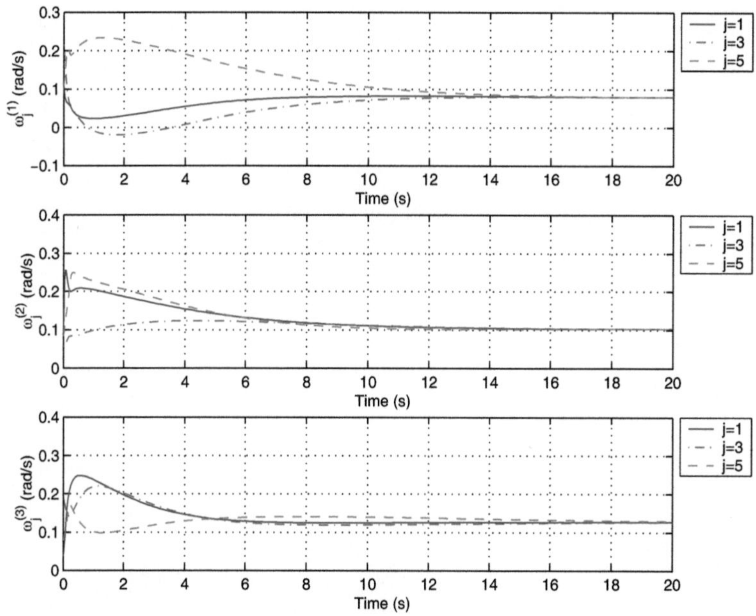

Fig. 6.13. Rigid body angular velocities with (6.16)

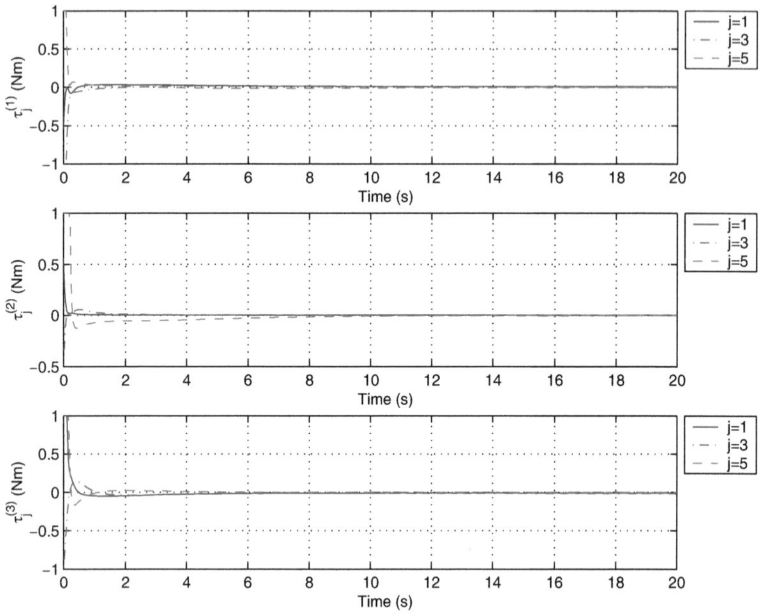

Fig. 6.14. Rigid body control torques with (6.16)

7

Relative Attitude Maintenance and Reference Attitude Tracking

This chapter extends the consensus algorithms for rigid body attitude dynamics in Chapter 6 in two ways. First, we propose algorithms to guarantee that multiple rigid bodies can maintain given relative attitudes and angular velocities during the transition. Both fixed and time-varying relative attitude maintenance are considered. Second, we propose algorithms to guarantee that multiple rigid bodies can track a given, time-varying, reference attitude when the reference attitude is available only to a subgroup of team members. Both cases where the attitudes are represented by Euler parameters and modified Rodriguez parameters are considered. Simulation results on reference attitude tracking are presented to validate our algorithms.

7.1 Relative Attitude Maintenance

In this section, a team of rigid bodies is required to maintain given relative attitudes and angular velocities during the transition. We consider both fixed and time-varying relative attitude maintenance. We assume an undirected, time-invariant interaction topology and constant weights in the control laws. As in Chapter 6, we use undirected graphs $\mathcal{G}_n^A \triangleq (\mathcal{V}_n, \mathcal{E}_n^A)$ and $\mathcal{G}_n^B \triangleq (\mathcal{V}_n, \mathcal{E}_n^B)$ to model the interaction topologies for q_i and ω_i, respectively, among the n rigid bodies. Let $\mathcal{A}_n = [a_{ij}] \in \mathbb{R}^{n\times n}$ and $\mathcal{B}_n = [b_{ij}] \in \mathbb{R}^{n\times n}$ be the adjacency matrix associated with $\mathcal{G}_n^A$ and $\mathcal{G}_n^B$, respectively.

7.1.1 Fixed Relative Attitudes with Zero Final Angular Velocities

When multiple rigid bodies are required to maintain fixed relative attitudes with zero final angular velocities, we propose the control torque on the ith rigid body as

$$\tau_i = -k_G\widehat{q^{r*}q_iq_{\delta_i}} - D_{Gi}\omega_i - \sum_{j=1}^{n}[a_{ij}q_{\delta_j}^*\widehat{q_j^*q_iq_{\delta_i}} + b_{ij}(\omega_i - \omega_j)], \quad i = 1,\ldots,n, \tag{7.1}$$

where k_G is a nonnegative scalar, $D_{Gi} \in \mathbb{R}^{3\times3}$ is symmetrical positive definite, $q^r \in \mathbb{R}^4$ denotes the constant reference attitude for each rigid body, a_{ij} is the (i,j) entry of the adjacency matrix $\mathcal{A}_n \in \mathbb{R}^{n\times n}$ associated with $\mathcal{G}_n^A$, b_{ij} is the (i,j) entry of the adjacency matrix $\mathcal{B}_n \in \mathbb{R}^{n\times n}$ associated with $\mathcal{G}_n^B$, and $q_{\delta_i} \in \mathbb{R}^4$, $i = 1,\ldots,n$, are constant quaternions defining the desired relative attitudes between q_i and q^r.

Remark 7.1. Note that the product $q_{\delta_j}q_{\delta_i}^*$ defines the desired relative attitudes between the ith rigid body and the jth rigid body. As a result, relative attitudes among the rigid bodies can be achieved by appropriately choosing q_{δ_i}, $i = 1,\ldots,n$. □

Theorem 7.2. *Suppose that the control torque is given by (7.1), and $\mathcal{G}_n^A$ and $\mathcal{G}_n^B$ are undirected. If $k_G > 2\sum_{j=1}^n a_{ij}$, then $q_i(t) \to q^r q_{\delta_i}^*$ and $\omega_i(t) \to 0$, $i = 1,\ldots,n$, as $t \to \infty$. If $k_G = 0$ and $\mathcal{G}_n^A$ is a tree, then $q_j^*(t)q_i(t) \to q_{\delta_j}q_{\delta_i}^*$ and $\omega_i(t) \to 0$, as $t \to \infty$.*

Proof: Replacing each q_i by $q_iq_{\delta_i}$ in the proof of Theorem 6.2, we know that $q_i(t)q_{\delta_i} \to q^r$, that is, $q_i(t) \to q^r q_{\delta_i}^*$, $i = 1,\ldots,n$, as $t \to \infty$, if $k_G > 2\sum_{j=1}^n a_{ij}$, and $q_i(t)q_{\delta_i} \to q_j(t)q_{\delta_j}$, that is, $q_j^*(t)q_i(t) \to q_{\delta_j}q_{\delta_i}^*$, as $t \to \infty$, if $k_G = 0$ and $\mathcal{G}_n^A$ is a tree. ■

7.1.2 Time-varying Relative Attitudes and Angular Velocities

When multiple rigid bodies are required to maintain time-varying relative attitudes and angular velocities, we propose the control torque on the ith rigid body as

$$\tau_i = -J_i\dot{\omega}_{\delta_i} + \omega_i \times (J_i\omega_i) - J_i\sum_{j=1}^{n}\{a_{ij}(q_{\delta_j}^*\widehat{q_j^*q_iq_{\delta_i}}) + b_{ij}[(\omega_i - \omega_j) + (\omega_{\delta_i} - \omega_{\delta_j})]\}, \quad i = 1,\ldots,n, \tag{7.2}$$

where a_{ij} and b_{ij} are defined as in (7.1) and the unit quaternion and angular velocity pair $(q_{\delta_i}, \omega_{\delta_i})$ satisfies the quaternion kinematics (D.2a) and defines the desired time-varying relative attitudes and angular velocities among the rigid bodies. Note that (6.16) corresponds to the case where $q_{\delta_i} = \mathbf{q}_I$ and $\omega_{\delta_i} = 0$.

Theorem 7.3. *Suppose that the control torque is given by (7.2) and $\mathcal{G}_n^A$ and $\mathcal{G}_n^B$ are undirected. If $\mathcal{G}_n^A$ is a tree and $\mathcal{G}_n^B$ is connected, then $q_j^*(t)q_i(t) \to q_{\delta_j}(t)q_{\delta_i}^*(t)$, and $\omega_i(t) \to \omega_j(t) + \omega_{\delta_j}(t) - \omega_{\delta_i}(t)$, $\forall i \neq j$, as $t \to \infty$.*

Proof: Consider the Lyapunov function candidate

$$V = \frac{1}{2}\sum_{i=1}^{n}\sum_{j=1}^{n} a_{ij}\left\|\tilde{q}_j^*\tilde{q}_i - \mathbf{q}_I\right\|^2 + \frac{1}{2}\sum_{i=1}^{n}\tilde{\omega}_i^T\tilde{\omega}_i,$$

where $\tilde{q}_i = q_i q_{\delta_i}$ and $\tilde{\omega}_i = \omega_i + \omega_{\delta_i}$. Note that V is positive definite with respect to $\tilde{q}_j^*\tilde{q}_i - \mathbf{q}_I$ and $\tilde{\omega}_i$, $i = 1, \ldots, n$. The set $\{(\tilde{q}_j^*\tilde{q}_i - \mathbf{q}_I, \tilde{\omega}_i) | V \leq c\}$, where $c > 0$, is compact with respect to $\tilde{q}_j^*\tilde{q}_i - \mathbf{q}_I$ and $\tilde{\omega}_i$, $i = 1, \ldots, n$. Following a procedure similar to that of Theorem 6.2, we obtain

$$\dot{V} = -\frac{1}{2}\sum_{i=1}^{n}\sum_{j=1}^{n} b_{ij}\left\|\tilde{\omega}_i - \tilde{\omega}_j\right\|^2 \leq 0.$$

Let $S = \{(\tilde{q}_j^*\tilde{q}_i - \mathbf{q}_I, \tilde{\omega}_i) | \dot{V} = 0\}$. Note that $\dot{V} \equiv 0$ implies that $\tilde{\omega}_i \equiv \tilde{\omega}_j$, $\forall i \neq j$, because undirected graph $\mathcal{G}_n^B$ is connected. Following the proof of Theorem 6.5 with q_i and ω_i replaced by $\tilde{q}_i$ and $\tilde{\omega}_i$, it follows that $\tilde{q}_i(t) \to \tilde{q}_j(t)$ and $\tilde{\omega}_i(t) \to \tilde{\omega}_j(t)$, $\forall i \neq j$, as $t \to \infty$. Therefore, $q_j^*(t)q_i(t) \to q_{\delta_j}(t)q_{\delta_i}^*(t)$ and $\omega_i(t) \to \omega_j(t) + \omega_{\delta_j}(t) - \omega_{\delta_i}(t)$, $\forall i \neq j$, as $t \to \infty$. ■

7.2 Reference Attitude Tracking

In this section, a team of rigid bodies is required to track a time-varying reference attitude. We consider reference attitude tracking with attitudes represented by, respectively, Euler parameters and modified Rodriguez parameters. We assume a directed, time-invariant interaction topology and constant weights in the control laws.

7.2.1 Reference Attitude Tracking with Attitudes Represented by Euler Parameters

Suppose that the team consists of the same n rigid bodies as in Chapter 6, together with an additional (virtual) rigid body labeled $n+1$, which acts as the unique (virtual) leader of the team. As in Chapter 3, we call rigid body $n+1$ the *team leader* and rigid bodies $1, \ldots, n$ the *followers*. Rigid body $n+1$ has attitude $q_{n+1} \stackrel{\triangle}{=} q^r \in \mathbb{R}^4$ and angular velocity $\omega_{n+1} \stackrel{\triangle}{=} \omega^r \in \mathbb{R}^3$, where q^r and ω^r represent the time-varying reference attitude and angular velocity. The time-varying reference attitude and angular velocity satisfy the reference attitude dynamics given by

$$\begin{aligned} \dot{\widehat{q}}^r &= -\frac{1}{2}\omega^r \times \widehat{q}^r + \frac{1}{2}\overline{q}^r\omega^r, \qquad \dot{\overline{q}}^r = -\frac{1}{2}\omega^r \cdot \widehat{q}^r, \\ J^r\dot{\omega}^r &= -\omega^r \times (J^r\omega^r) + \tau^r, \end{aligned} \tag{7.3}$$

where $\widehat{q}^r$ and $\overline{q}^r$ are, respectively, the vector and scalar parts of q^r and J^r and τ^r denote, respectively, the reference inertial tensor and control torque for the team. The objective of this section is to guarantee that $q_i(t) \to q^r(t)$ and $\omega_i(t) \to \omega^r(t)$, $i = 1, \ldots, n$, as $t \to \infty$.

As in Chapter 5, we use directed graphs $\mathcal{G}_{n+1}^A \triangleq (\mathcal{V}_{n+1}, \mathcal{E}_{n+1}^A)$ and $\mathcal{G}_{n+1}^B \triangleq (\mathcal{V}_{n+1}, \mathcal{E}_{n+1}^B)$ to model the interaction topologies for q_i and ω_i, respectively, among the $n+1$ rigid bodies. Let $\mathcal{A}_{n+1} = [a_{ij}] \in \mathbb{R}^{(n+1)\times(n+1)}$ and $\mathcal{B}_{n+1} = [b_{ij}] \in \mathbb{R}^{(n+1)\times(n+1)}$ be the adjacency matrix associated with $\mathcal{G}_{n+1}^A$ and $\mathcal{G}_{n+1}^B$, respectively. When there is only one interaction topology associated with the $n+1$ rigid bodies, we simply use $\mathcal{G}_{n+1} \triangleq (\mathcal{V}_{n+1}, \mathcal{E}_{n+1})$ to model the interaction topology among the $n+1$ rigid bodies, as in Chapter 3. Similarly, $\mathcal{A}_{n+1}$ is the adjacency matrix associated with $\mathcal{G}_{n+1}$.

Remark 7.4. When we need to focus only on rigid bodies 1 to n (*i.e.*, the followers), we still use $\mathcal{G}_n^A \triangleq (\mathcal{V}_n, \mathcal{E}_n^A)$ and $\mathcal{G}_n^B \triangleq (\mathcal{V}_n, \mathcal{E}_n^B)$ to model the interaction topologies for q_i and ω_i, respectively, among the n followers, as in Chapter 6. In addition, $\mathcal{A}_n \in \mathbb{R}^{n\times n}$ and $\mathcal{B}_n \in \mathbb{R}^{n\times n}$ are still the adjacency matrix associated with $\mathcal{G}_n^A$ and $\mathcal{G}_n^B$, respectively, as in Chapter 6. When there is only one interaction topology associated with the n followers, we simply use $\mathcal{G}_n \triangleq (\mathcal{V}_n, \mathcal{E}_n)$ to model the interaction topology among the n followers, as in Chapter 2. Similarly, $\mathcal{A}_n$ is the adjacency matrix associated with $\mathcal{G}_n$. □

Time-varying Reference Attitude: Full Access

When reference attitude dynamics (7.3) is available to all followers, we propose the control torque on the ith rigid body as

$$\begin{aligned}\tau_i =& \omega_i \times (J_i\omega_i) + J_i\dot{\omega}^r - k_G\widehat{q^{r*}q_i} - D_{Gi}(\omega_i - \omega^r) \\ &- \sum_{j=1}^{n}[a_{ij}\widehat{q_j^* q_i} + b_{ij}(\omega_i - \omega_j)], \quad i = 1, \ldots, n,\end{aligned} \tag{7.4}$$

where k_G is a nonnegative scalar, $D_{Gi} \in \mathbb{R}^{3\times 3}$ is symmetrical positive definite, a_{ij} is the (i,j) entry of the adjacency matrix $\mathcal{A}_n \in \mathbb{R}^{n\times n}$ associated with $\mathcal{G}_n^A$, and b_{ij} is the (i,j) entry of the adjacency matrix $\mathcal{B}_n \in \mathbb{R}^{n\times n}$ associated with $\mathcal{G}_n^B$. Note that (7.4) corresponds to the case where $(n+1, i) \in \mathcal{E}_{n+1}^A$ and $(n+1, i) \in \mathcal{E}_{n+1}^B$, $i = 1, \ldots, n$.

Theorem 7.5. *Suppose that the control torque is given by (7.4) and that $\mathcal{G}_n^A$ and $\mathcal{G}_n^B$ are undirected. Also suppose that $q^r(t)$ and $\omega^r(t)$ satisfy (7.3). If $k_G > 2\sum_{j=1}^n a_{ij}$, then $q_i(t) \to q^r(t)$ and $\omega_i(t) \to \omega^r(t)$, as $t \to \infty$. If $k_G = 0$ and $\mathcal{G}_n^A$ is a tree, then $q_i(t) \to q_j(t)$ and $\omega_i(t) \to \omega^r(t)$, as $t \to \infty$.*

Proof: If $k_G > 2\sum_{j=1}^n a_{ij}$, we let $\tilde{q}_i = q^{r*}q_i$ and $\tilde{\omega}_i = \omega_i - \omega^r$. Note that according to Lemma D.1, $\tilde{q}_i$ and $\tilde{\omega}_i$ also satisfy the quaternion kinematics (D.2a).

Following the proof of Theorem 7.2 with q_i, ω_i, q^r, and q_{δ_i} replaced, respectively, by $\tilde{q}_i$, $\tilde{\omega}_i$, $\mathbf{q}_I$, and $\mathbf{q}_I$, we see that $\tilde{q}_i(t) \to \mathbf{q}_I$ and $\tilde{\omega}_i(t) \to 0$, that is, $q_i(t) \to q^r(t)$ and $\omega_i(t) \to \omega^r(t)$, as $t \to \infty$. If $k_G = 0$ and $\mathcal{G}_n^A$ is a tree, we see that $\tilde{q}_i(t) \to \tilde{q}_j(t)$ and $\tilde{\omega}_i(t) \to 0$, that is, $q_i(t) \to q_j(t)$ and $\omega_i(t) \to \omega^r(t)$, as $t \to \infty$. ∎

Time-varying Reference Attitude: Partial Access

Note that control law (7.4) assumes that reference attitude dynamics (7.3) is available to each rigid body on the team, which might be restrictive in some applications. In the general case where (7.3) is available only to some team members, we propose the control torque on the ith rigid body as

$$\tau_i = \omega_i \times (J_i\omega_i) + \frac{1}{|\mathcal{N}_i|+1} J_i(\dot{\omega}^r + \sum_{j\in\mathcal{N}_i} \dot{\omega}_j)$$
$$- \frac{1}{|\mathcal{N}_i|+1}\{k_{qi}\widehat{p_{\pi_i}} + K_{\omega i}[(\omega_i - \omega^r) + \sum_{j\in\mathcal{N}_i}(\omega_i - \omega_j)]\}, \quad i \in \mathcal{V}_L, \tag{7.5a}$$

$$\tau_i = \omega_i \times (J_i\omega_i) + \frac{1}{|\mathcal{N}_i|} J_i \sum_{j\in\mathcal{N}_i} \dot{\omega}_j$$
$$- \frac{1}{|\mathcal{N}_i|}[k_{qi}\widehat{q_{\pi_i}} + K_{\omega i}\sum_{j\in\mathcal{N}_i}(\omega_i - \omega_j)], \quad i \notin \mathcal{V}_L, \tag{7.5b}$$

where $\mathcal{N}_i \subset \mathcal{V}_n$ denotes the set of neighbors of rigid body i, $|\mathcal{N}_i|$ denotes the cardinality of $\mathcal{N}_i$, $\mathcal{V}_L \subseteq \mathcal{V}_n$ denotes the set of rigid bodies to which the reference attitude dynamics (7.3) is available, k_{qi} is a positive scalar, $K_{\omega i} \in \mathbb{R}^{3\times 3}$ is symmetrical positive definite, $p_{\pi_i} = [\prod_{j\in\mathcal{N}_i}(q_j^* q_i)]q^{r*}q_i$, and $q_{\pi_i} = \prod_{j\in\mathcal{N}_i}(q_j^* q_i)$. Note that $j \in \mathcal{N}_i$ does not imply that $i \in \mathcal{N}_j$ in the case of directed information exchange. Also note that each rigid body's control torque depends on its neighbors' attitudes, angular velocities, and angular velocity derivatives. In practical implementation, the derivatives of the neighbors' angular velocities can be calculated by numerical differentiation.

Remark 7.6. Note that (7.5) corresponds to directed graph $\mathcal{G}_{n+1}$ whose adjacency matrix $\mathcal{A}_{n+1} = [a_{ij}] \in \mathbb{R}^{(n+1)\times(n+1)}$ is defined as $a_{ij} = 1, i, j = 1, \ldots, n$, if $j \in \mathcal{N}_i$ and $a_{ij} = 0$ otherwise, $a_{i(n+1)} = 1$, $i = 1, \ldots, n$, if $i \in \mathcal{V}_L$ and $a_{i(n+1)} = 0$ otherwise, and $a_{(n+1)j} = 0$, $j = 1, \ldots, n+1$. □

Theorem 7.7. *Let the adjacency matrix $\mathcal{A}_{n+1}$ associated with directed graph $\mathcal{G}_{n+1}$ be defined as in Remark 7.6. With control torque* (7.5)*, if $\mathcal{G}_{n+1}$ has a directed spanning tree,*[1] *then $\widehat{p_{\pi_i}}(t) \to 0$, $i \in \mathcal{V}_L$, $\widehat{q_{\pi_i}}(t) \to 0$, $i \notin \mathcal{V}_L$, and $\omega_i(t) \to \omega^r(t)$, $i = 1, \ldots, n$, as $t \to \infty$*

[1] With $\mathcal{A}_{n+1} = [a_{ij}] \in \mathbb{R}^{(n+1)\times(n+1)}$ defined in Remark 7.6, where $a_{(n+1)j} = 0$, $j = 1, \ldots, n+1$, the condition that $\mathcal{G}_{n+1}$ has a directed spanning tree implies that rigid body $n+1$ is the root of the directed spanning tree. This condition

Proof: Note that control torque (7.5b) has a denominator $|\mathcal{N}_i|$. As in the proof of Theorem 3.8, we know that $|\mathcal{N}_i| \neq 0$, $i = 1, \ldots, n$, if $\mathcal{G}_{n+1}$ has a directed spanning tree. Also note that with (7.5), $\dot{\omega}_i$ exists on both sides of (D.2b). We next show that under the assumption of the theorem, there is a unique solution for $\dot{\omega}_i$.

Note that $q_{n+1} \equiv q^r$ and $\omega_{n+1} \equiv \omega^r$. Also let $\mathcal{J}_i = \mathcal{N}_i$ if $i \notin \mathcal{V}_L$ and $\mathcal{J}_i = \mathcal{N}_i \cup \{n+1\}$ if $i \in \mathcal{V}_L$. Then (7.5) can be rewritten as

$$\tau_i = \omega_i \times (J_i \omega_i) + \frac{1}{|\mathcal{J}_i|} J_i \sum_{j \in \mathcal{J}_i} \dot{\omega}_j - \frac{1}{|\mathcal{J}_i|} [k_{qi} \widehat{s_{\pi_i}} + K_{\omega i} \sum_{j \in \mathcal{J}_i} (\omega_i - \omega_j)], \quad i = 1, \ldots, n, \tag{7.6}$$

where $s_{\pi_i} \triangleq \prod_{j \in \mathcal{J}_i} (q_j^* q_i)$. Combining (D.2b) and (7.6), gives

$$J_i \dot{\omega}_{\sigma_i} = -k_{qi} \widehat{s_{\pi_i}} - K_{\omega i} \omega_{\sigma_i}, \quad i = 1, \ldots, n, \tag{7.7}$$

where $\omega_{\sigma_i} \triangleq \sum_{j \in \mathcal{J}_i} (\omega_i - \omega_j)$. Note that (7.7) can be written in matrix form as

$$J[(M \otimes I_3)\dot{\omega} + (b \otimes I_3)\dot{\omega}^r] = d,$$

where $J \triangleq \mathrm{diag}(J_1, \ldots, J_n)$, $\dot{\omega} \triangleq [\dot{\omega}_1^T, \ldots, \dot{\omega}_n^T]^T$, $M = [m_{ij}] \in \mathbb{R}^{n \times n}$ with $m_{ij} = -a_{ij}$ and $m_{ii} = \sum_{j=1, j \neq i}^{n+1} a_{ij}$, $b = [-a_{1(n+1)}, \ldots, -a_{n(n+1)}]^T \in \mathbb{R}^n$, and $d = [d_1^T, \ldots, d_n^T]^T$ with $d_i = -k_{qi} \widehat{s_{\pi_i}} - K_{\omega i} \omega_{\sigma_i}$. Note that J is invertible. Also note that M is invertible under the assumption of the theorem according to the proof of Theorem 3.8. It follows that there exists a unique solution for $\dot{\omega}_i$, $i = 1, \ldots, n$.

Lemma D.1 shows that if the unit quaternion and angular velocity pairs (q_k, ω_k) and (q_ℓ, ω_ℓ) both satisfy quaternion kinematics (D.2a), then the unit quaternion and angular velocity pair $(q_\ell^* q_k, \omega_k - \omega_\ell)$ also satisfy (D.2a). It is straightforward to extend this argument by induction to show that the unit quaternion and angular velocity pair $(s_{\pi_i}, \omega_{\sigma_i})$ also satisfies (D.2a). Thus given (7.7), Lemma D.2 implies that $\widehat{s_{\pi_i}}(t) \to 0$ and $\omega_{\sigma_i}(t) \to 0$, $i = 1, \ldots, n$, as $t \to \infty$. Note that $\widehat{s_{\pi_i}}(t) \to 0$ implies that $\widehat{p_{\pi_i}}(t) \to 0$, $i \in \mathcal{V}_L$, and $\widehat{q_{\pi_i}}(t) \to 0$, $i \notin \mathcal{V}_L$, as $t \to \infty$. Note that $\omega_{\sigma_i}(t) \to 0$, $i = 1, \ldots, n$, as $t \to \infty$, implies that $(M \otimes I_3)\omega(t) + (b \otimes I_3)\omega^r(t) \to 0$, as $t \to \infty$, where $\omega \triangleq [\omega_1^T, \ldots, \omega_n^T]^T$. Then, by following the proof of Theorem 3.15, we know that $\omega_i(t) \to \omega^r(t)$, $i = 1, \ldots, n$, as $t \to \infty$, if $\mathcal{G}_{n+1}$ has a directed spanning tree. ∎

Remark 7.8. Note that Theorem 7.7 shows that all followers' angular velocities approach the reference angular velocity, but it is unclear whether all followers'

is also equivalent to the condition that rigid body $n+1$ is the only rigid body that has a directed path to all of the other rigid bodies on the team, that is, the reference states q^r, ω^r, and $\dot{\omega}^r$ can flow directly or indirectly to any follower on the team.

attitudes approach the reference attitude under the assumption of the theorem. However, when directed graph $\mathcal{G}_{n+1}$ has a special structure, convergence to the reference attitude can be concluded. In $\mathcal{G}_{n+1}$, if node k has exactly one parent, node ℓ, then $\widehat{s_{\pi_k}}(t) = \widehat{q_\ell^* q_k}(t) \to 0$ implies that $q_k(t) \to q_\ell(t)$, as $t \to \infty$, that is, rigid body k approaches the reference attitude q^r if $\ell = n+1$, or rigid bodies k and ℓ approach the same attitude if $\ell \neq n+1$. As a result, edge (ℓ, k) can be deleted from $\mathcal{G}_{n+1}$, and nodes k and ℓ can be combined as one single node whose incoming and outgoing edges are the union of the incoming and outgoing edges of nodes k and ℓ. By repeating this procedure, we can simplify $\mathcal{G}_{n+1}$. If $\mathcal{G}_{n+1}$ can be simplified to a directed graph with only one node, then $\widehat{p_{\pi_i}}(t) \to 0$, $i \in \mathcal{V}_L$, and $\widehat{q_{\pi_i}}(t) \to 0$, $i \notin \mathcal{V}_L$, as $t \to \infty$, directly imply that $q_i(t) \to q^r(t)$, $i = 1, \ldots, n$, as $t \to \infty$. The leader-following approach for attitude alignment (*e.g.*, [244, 245]) can be considered a special case of control law (7.5), where $\mathcal{G}_{n+1}$ is itself a directed spanning tree. □

Example 7.9. Theorem 7.7 can be illustrated by the following example. Figure 7.1 shows four different interaction topologies among five rigid bodies, where node q^r denotes the team leader and node i, $i = 1, \ldots, 4$, denote the ith follower. In Fig. 7.1, an edge from node q^r to node j denotes that reference attitude dynamics (7.3) is available to rigid body j. In particular, the leader-following approach (*e.g.*, [244, 245]) requires graphs like Figs. 7.1a and 7.1b, where the interaction topologies are themselves directed spanning trees, implying that each rigid body except the team leader has exactly one parent. In contrast, control law (7.5) is used for the graphs shown in Figs. 7.1c and 7.1d, which are more general than Figs. 7.1a and 7.1b in the sense that information can flow among all followers to introduce feedback between neighbors and reference attitude dynamics (7.3) may be available to one or more followers on the team. Note that node q^r has a directed path to all followers in Figs. 7.1a–d. Also note that using the approach described in Remark 7.8, Figs. 7.1a–d can be simplified to a directed graph with only one node, which directly implies that using control law (7.5), $q_i(t) \to q^r(t)$, as $t \to \infty$. □

7.2.2 Reference Attitude Tracking with Attitudes Represented by Modified Rodriguez Parameters

As described in Remark 7.8, Theorem 7.7 shows that with (7.5), all followers can track their reference angular velocity if $\mathcal{G}_{n+1}$ has a directed spanning tree. However, to conclude that they can track their reference attitude, $\mathcal{G}_{n+1}$ must satisfy the condition that it can be simplified to a graph with only one node. In this section, we use modified Rodriguez parameters (MRPs) (see Appendix D for definitions) to represent rigid body attitudes and derive a control law that guarantees reference attitude tracking under a general directed interaction topology.

Let $\sigma_i \in \mathbb{R}^3$ be the attitude represented using MRPs and $\omega_i \in \mathbb{R}^3$ be the angular velocity of the ith rigid body. Attitude dynamics using MRPs are given by (D.4). Note that (D.4) can be written as (D.5).

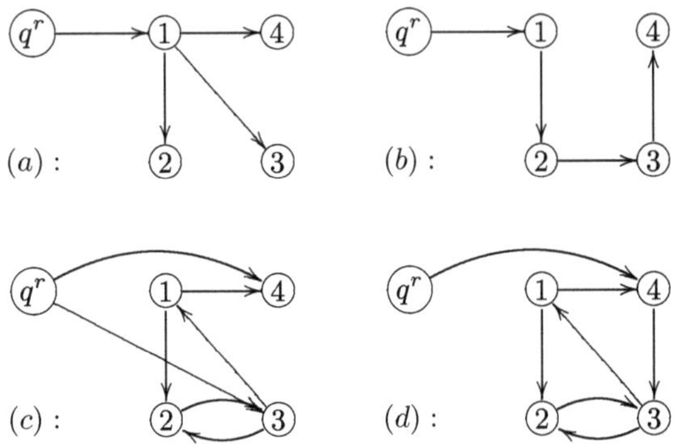

Fig. 7.1. Interaction topologies among five rigid bodies where Figs. 7.1a and 7.1b correspond to the leader-following approach and Figs. 7.1c and 7.1d correspond to control law (7.5)

Suppose that rigid body $n+1$ is the (virtual) team leader with $\sigma_{n+1} \stackrel{\triangle}{=} \sigma^r \in \mathbb{R}^3$ and $\omega_{n+1} \stackrel{\triangle}{=} \omega^r \in \mathbb{R}^3$, where σ^r and ω^r are, respectively, the reference attitude and angular velocity of the team. Suppose that σ^r, $\dot{\sigma}^r$, and $\ddot{\sigma}^r$ are bounded. We assume that there is only one interaction topology associated with the $n+1$ rigid bodies. We simply use $\mathcal{G}_{n+1} \stackrel{\triangle}{=} (\mathcal{V}_{n+1}, \mathcal{E}_{n+1})$ to model the interaction topology among the $n+1$ rigid bodies as in Chapter 3. Similarly, $\mathcal{A}_{n+1}$ is the adjacency matrix associated with $\mathcal{G}_{n+1}$.

We propose a control torque for (D.5) as

$$\begin{aligned} \tau_i = F^T(\sigma_i)[H_i^*(\sigma_i)(\ddot{\sigma}_i^d - \Lambda_i \dot{\tilde{\sigma}}_i) + C_i^*(\sigma_i, \dot{\sigma}_i)(\dot{\sigma}_i^d - \Lambda_i \tilde{\sigma}_i) \\ - K_i(\dot{\tilde{\sigma}}_i + \Lambda_i \tilde{\sigma}_i)], \quad i = 1, \ldots, n, \end{aligned} \tag{7.8}$$

where

$$\sigma_i^d \stackrel{\triangle}{=} \frac{\sum_{j=1}^n a_{ij}\sigma_j + a_{i(n+1)}\sigma^r}{\sum_{j=1}^n a_{ij} + a_{i(n+1)}}, \tag{7.9}$$

a_{ij} is the (i,j) entry of the adjacency matrix $\mathcal{A}_{n+1} \in \mathbb{R}^{(n+1)\times(n+1)}$ associated with $\mathcal{G}_{n+1}$, $\tilde{\sigma}_i \stackrel{\triangle}{=} \sigma_i - \sigma_i^d$, and Λ_i and K_i are symmetrical positive-definite matrices.

Remark 7.10. Note that in contrast to traditional trajectory tracking control laws for a single robotic manipulator (*e.g.*, [218]), σ_i^d defined by (7.9) is not an external desired state but depends on the attitudes of each rigid body's neighbors. Also note that each rigid body's control torque depends on its neighbors' attitudes and their first- and second-order derivatives. In practical implementation, the second-order derivatives of the neighbors' attitudes can be calculated by numerical differentiation. □

Theorem 7.11. *With (7.8), if directed graph $\mathcal{G}_{n+1}$ has a directed spanning tree, then $\sigma_i(t) \to \sigma^r(t)$ and $\dot{\sigma}_i(t) \to \dot{\sigma}^r(t)$, $i = 1, \ldots, n$, as $t \to \infty$.*

Proof: Note that σ_i^d defined by (7.9) has a denominator $\sum_{j=1}^n a_{ij} + a_{i(n+1)}$. As in the proof of Theorem 3.8, we know that $\sum_{j=1}^n a_{ij} + a_{i(n+1)} \neq 0$, $i = 1, \ldots, n$, if $\mathcal{G}_{n+1}$ has a directed spanning tree.

Also note that with (7.8), $\ddot{\sigma}_i$ exists on both sides of (D.5). Next, we show that under the assumption of the theorem, there is a unique solution for $\ddot{\sigma}_i$. With (7.8), (D.5) can be written as

$$H_i^*(\sigma_i)(\ddot{\tilde{\sigma}}_i + \Lambda_i\dot{\tilde{\sigma}}_i) + C_i^*(\sigma_i, \dot{\sigma}_i)(\dot{\tilde{\sigma}}_i + \Lambda_i\tilde{\sigma}_i) + K_i(\dot{\tilde{\sigma}}_i + \Lambda_i\tilde{\sigma}_i) = 0, \quad i = 1, \ldots, n. \tag{7.10}$$

Multiplying $\sum_{i=1}^n a_{ij} + a_{i(n+1)}$ on both sides of (7.10) gives

$$H_i^*(\sigma_i)(\ddot{\hat{\sigma}}_i + \Lambda_i\dot{\hat{\sigma}}_i) + C_i^*(\sigma_i, \dot{\sigma}_i)(\dot{\hat{\sigma}}_i + \Lambda_i\hat{\sigma}_i) + K_i(\dot{\hat{\sigma}}_i + \Lambda_i\hat{\sigma}_i) = 0, \quad i = 1, \ldots, n, \tag{7.11}$$

where $\hat{\sigma}_i \stackrel{\triangle}{=} [\sum_{i=1}^n a_{ij} + a_{i(n+1)}]\tilde{\sigma}_i = \sum_{i=1}^n a_{ij}(\sigma_i - \sigma_j) + a_{i(n+1)}(\sigma_i - \sigma^r)$. Letting $\hat{\sigma} \stackrel{\triangle}{=} [\hat{\sigma}_1^T, \ldots, \hat{\sigma}_n^T]^T$, then

$$\hat{\sigma} = (M \otimes I_3)\sigma + (b \otimes I_3)\sigma^r, \tag{7.12}$$

where $\sigma = [\sigma_1^T, \ldots, \sigma_n^T]^T$, $M = [m_{ij}] \in \mathbb{R}^{n \times n}$ with $m_{ij} = -a_{ij}$ and $m_{ii} = \sum_{j=1, j\neq i}^{n+1} a_{ij}$, and $b = [-a_{1(n+1)}, \ldots, -a_{n(n+1)}]^T \in \mathbb{R}^n$. Equation (7.11) can be written in matrix form as

$$H^*(\sigma)(\ddot{\hat{\sigma}} + \Lambda\dot{\hat{\sigma}}) + C^*(\sigma, \dot{\sigma})(\dot{\hat{\sigma}} + \Lambda\hat{\sigma}) + K(\dot{\hat{\sigma}} + \Lambda\hat{\sigma}) = 0, \tag{7.13}$$

where $H^*(\sigma) = \text{diag}[H_1^*(\sigma_1), \ldots, H_n^*(\sigma_n)]$,

$$C^*(\sigma, \dot{\sigma}) = \text{diag}[C_1^*(\sigma_1, \dot{\sigma}_1), \ldots, C_n^*(\sigma_n, \dot{\sigma}_n)],$$

$\Lambda = \text{diag}(\Lambda_1, \ldots, \Lambda_n)$, and $K = \text{diag}(K_1, \ldots, K_n)$. Thus according to (7.12) and (7.13), there is a unique solution for $\ddot{\sigma}$ if $H^*(\sigma)$ and $M \otimes I_3$ are invertible. Note that $H^*(\sigma)$ is invertible because all $H_i^*(\sigma_i)$ are symmetrical positive definite. Also note that M is invertible under the assumption of the theorem, according to the proof of Theorem 3.8.

Consider the Lyapunov function candidate

$$V = \frac{1}{2}\sum_{i=1}^n s_i^T H_i^*(\sigma_i) s_i,$$

where $s_i \stackrel{\triangle}{=} \dot{\hat{\sigma}}_i + \Lambda_i\hat{\sigma}_i$. Differentiating V gives

$$\dot{V} = -\sum_{i=1}^n s_i^T K_i s_i < 0,$$

where we have used the fact that (7.11) can be written as

$$H_i^*(\sigma_i)\dot{s}_i + C_i^*(\sigma_i, \dot{\sigma}_i)s_i + K_i s_i = 0 \tag{7.14}$$

and $H_i^*(\sigma_i) - 2C_i^*(\sigma_i, \dot{\sigma}_i)$ is skew symmetrical. It thus follows that V is bounded, which in turn implies that all s_i, $i = 1, \ldots, n$, are bounded. As a result, it follows that all σ_i and $\dot{\sigma}_i$, $i = 1, \ldots, n$, are bounded because σ^r and $\dot{\sigma}^r$ are bounded.

Note that (7.14) implies that all $\dot{s}_i$, $i = 1, \ldots, n$, are bounded by noting that all ${H_i^*}^{-1}(\sigma_i)$, $i = 1, \ldots, n$, are bounded. Noting that $\ddot{V} = -2\sum_{i=1}^n s_i^T K_i \dot{s}_i$, it follows that $\ddot{V}$ is bounded, which in turn implies that $\dot{V}$ is uniformly continuous. By Lemma F.6, it follows that $\dot{V}(t) \to 0$, which implies that $\dot{s}_i(t) \to 0$, $i = 1, \ldots, n$, as $t \to \infty$. Then it follows that $\tilde{\sigma}_i(t) \to 0$ and $\dot{\tilde{\sigma}}_i(t) \to 0$, as $t \to \infty$, which in turn implies from (7.12) that $(M \otimes I_3)\sigma(t) + (b \otimes I_3)\sigma^r(t) \to 0$ and $(M \otimes I_3)\dot{\sigma}(t) + (b \otimes I_3)\dot{\sigma}^r(t) \to 0$, as $t \to \infty$. Then by following the proof of Theorem 3.15, we know that $\sigma_i(t) \to \sigma^r(t)$ and $\dot{\sigma}_i(t) \to \dot{\sigma}^r(t)$, as $t \to \infty$, under the assumption of the theorem. ∎

7.3 Simulation Results

In this section, we first apply control law (7.5) to guarantee that six rigid bodies follow their team leader with a time-varying reference attitude $q^r(t)$ and angular velocity $\omega^r(t)$. We then apply control law (7.8) to guarantee that six rigid bodies follow their team leader with a time-varying reference attitude $\sigma^r(t)$ and angular velocity $\omega^r(t)$.

The rigid body specifications are chosen to be same as in Table 6.1. With control law (7.5), we let $k_{qi} = 1$ and $K_{\omega i} = 2I_3$. Also let $\tau^r = [0, 0, 0]^T$, $J^r = \text{diag}(1, 2, 1)$, $q^r(0) = [0, 0, 0, 1]^T$, and $\omega^r(0) = [0.1, 0.3, 0.5]^T$ in (7.3). We choose $q_i(0)$ and $\omega_i(0)$ randomly. With control law (7.8), we let $\Lambda_i = I_3$, $K_i = 2I_3$, and $a_{ij} = 1$ if $(j, i) \in \mathcal{E}_{n+1}$ in (7.9). Suppose that the reference attitude σ^r, reference angular velocity $\omega^r = F^{-1}(\sigma^r)\dot{\sigma}^r$, reference torque τ^r, and reference inertia J^r satisfy (D.4), where $\tau^r = [0, 0, 0]^T$, $J^r = \text{diag}(1, 2, 1)$, $\sigma^r(0) = [0, 0, 0]^T$, and $\omega^r(0) = [0.1, 0.3, 0.5]^T$. We choose $\sigma_i(0)$ and $\omega_i(0)$ randomly. In the following, a superscript (j) denotes the jth component of a quaternion or vector.

The directed graph $\mathcal{G}_7$ used in (7.5) is shown in Fig. 7.2, where node q^r denotes the team leader. Note that the directed graph can be simplified to a graph with only one node using the approach described in Remark 7.8.

Figures 7.3 and 7.4 show, respectively, the attitudes and angular velocities of rigid bodies 1, 3, and 5 with (7.5) as well as their references. Note that the attitudes and angular velocities of each rigid body converge to their reference values. Figure 7.5 shows the control torques of rigid bodies 1, 3, and 5 with (7.5).

Directed graph $\mathcal{G}_7$ used in (7.8) is shown in Fig. 7.6, where node σ^r denotes the team leader. In contrast to the directed graph shown in Fig. 7.2, the

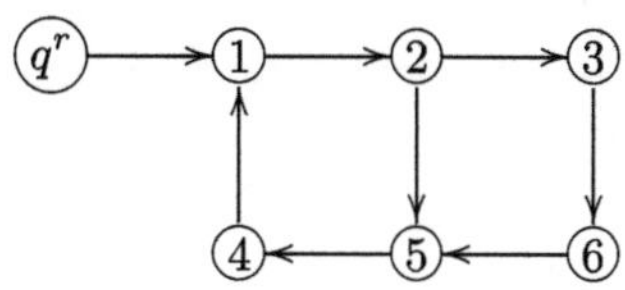

Fig. 7.2. Directed graph $\mathcal{G}_7$ used in (7.5)

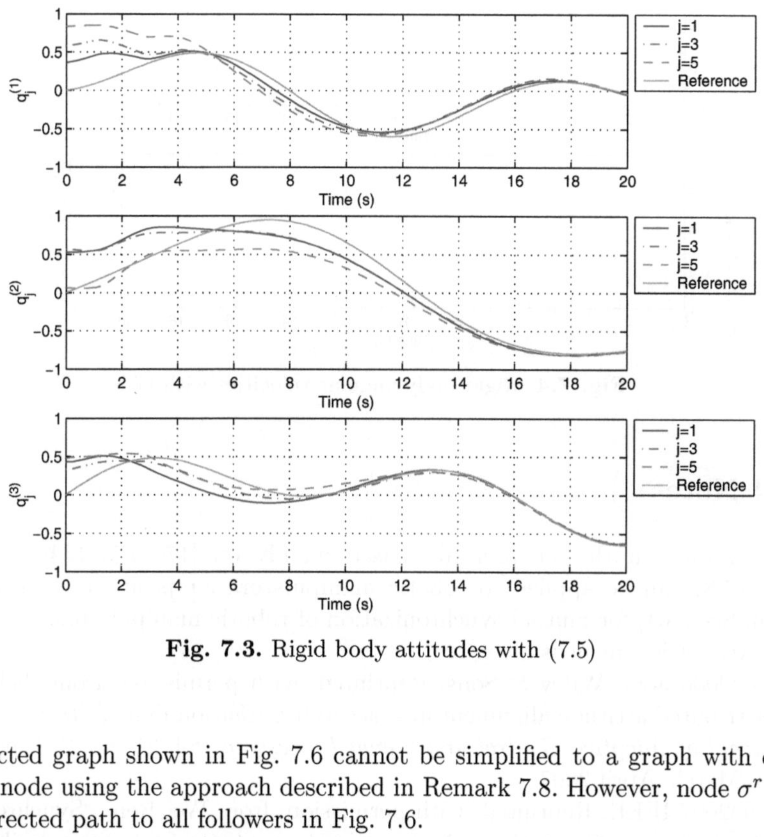

Fig. 7.3. Rigid body attitudes with (7.5)

directed graph shown in Fig. 7.6 cannot be simplified to a graph with only one node using the approach described in Remark 7.8. However, node σ^r has a directed path to all followers in Fig. 7.6.

Figures 7.7 and 7.8 show, respectively, the attitudes and angular velocities of rigid bodies 1, 3, and 5 with (7.8) as well as their references. Note that the attitudes and angular velocities track their reference values even if the directed graph shown in Fig. 7.6 cannot be simplified to a graph with only one node. Figure 7.9 shows the control torques of rigid bodies 1, 3, and 5 with (7.8).

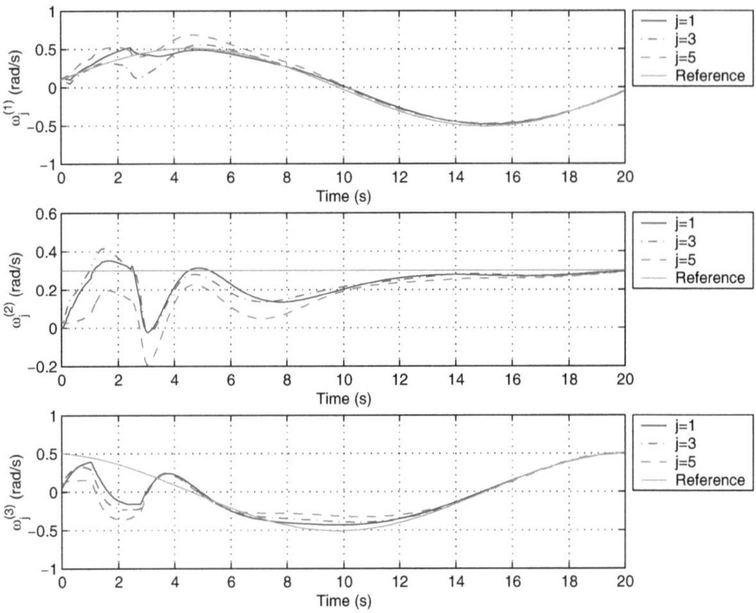

Fig. 7.4. Rigid body angular velocities with (7.5)

7.4 Notes

The results in this chapter are based mainly on [176–179, 184]. Control law (7.8) can be applied to robotic manipulators for position synchronization. See [201] for mutual synchronization of robotic manipulators.

Acknowledgment is given to

©2006 John Wiley & Sons. Reprinted, with permission, from Wei Ren, "Distributed attitude alignment in spacecraft formation flying," *International Journal of Adaptive Control and Signal Processing*, vol. 21, no. 2–3, pp. 95–113, March–April 2007.

©2007 IEEE. Reprinted, with permission, from Wei Ren, "Synchronized multiple spacecraft rotations: A revisit in the context of consensus building," *Proceedings of the American Control Conference*, pp. 3174–3179, New York City, 2007.

©2007 IEEE. Reprinted, with permission, from Wei Ren, "Distributed attitude synchronization for multiple rigid bodies with Euler-Lagrange equations of motion," *Proceedings of the IEEE Conference on Decision and Control*, New Orleans, LA, 2007 (to appear).

©2007 IEEE. Reprinted, with permission, from Wei Ren, "Distributed cooperative attitude synchronization and tracking for multiple rigid bodies," *IEEE Transactions on Control Systems Technology*, 2007 (under review).

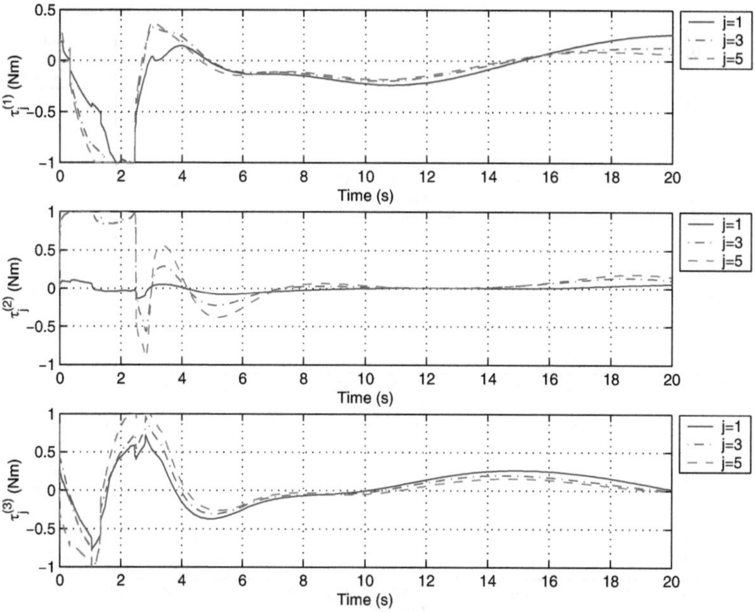

Fig. 7.5. Rigid body control torques with (7.5)

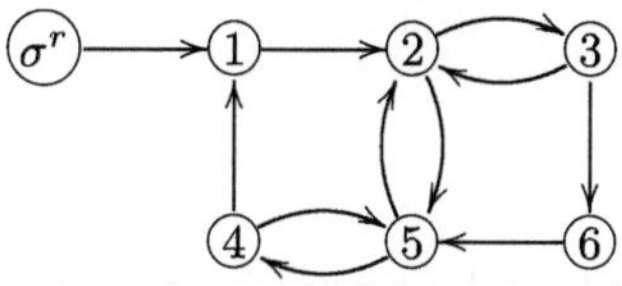

Fig. 7.6. Directed graph $\mathcal{G}_7$ used in (7.8)

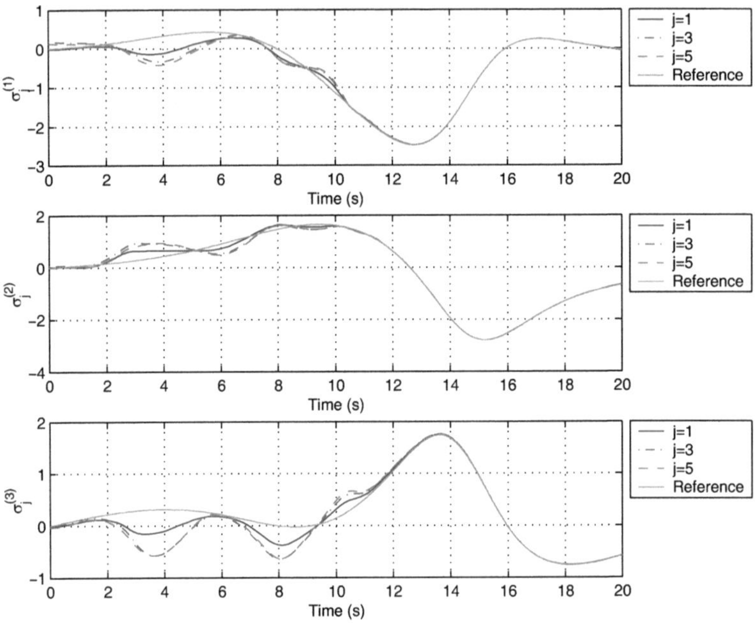

Fig. 7.7. Rigid body attitudes with (7.8)

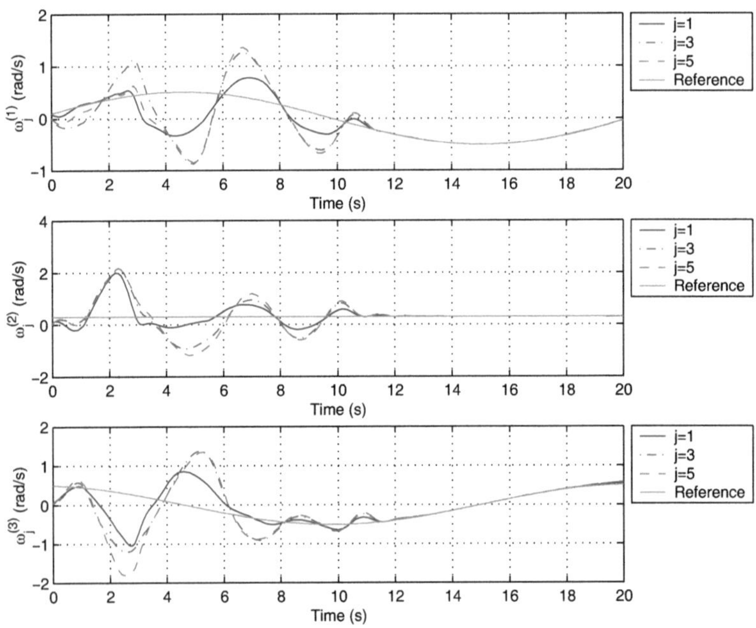

Fig. 7.8. Rigid body angular velocities with (7.8)

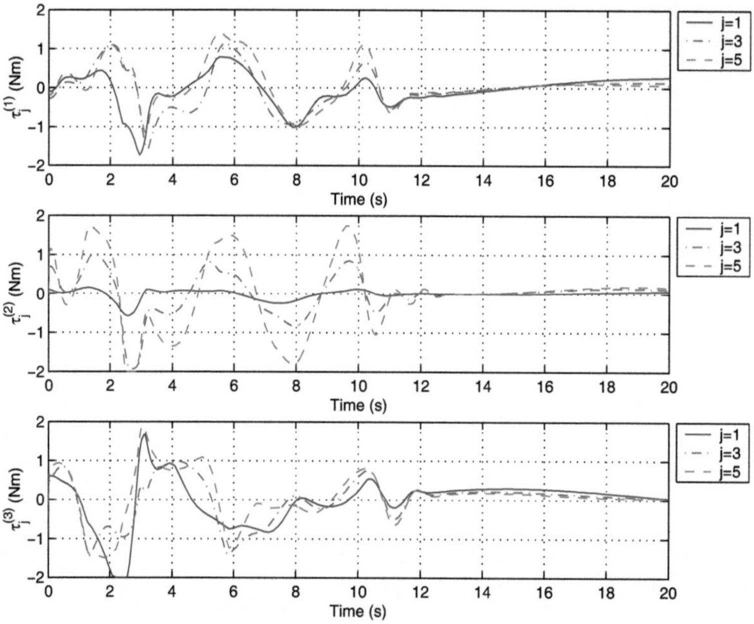

Fig. 7.9. Rigid body control torques with (7.8)

Part V

Consensus-based Design Methodologies for Distributed Multivehicle Cooperative Control

8

Consensus-based Design Methodologies for Distributed Multivehicle Cooperative Control

Contributed partly by Tim McLain

This chapter introduces a few design methodologies for distributed multivehicle cooperative control based on consensus algorithms. Our approach to distributed multivehicle cooperative control problems can be summarized in four steps: (1) the definition of a cooperation constraint and cooperation objective; (2) the definition of a coordination variable as the minimal amount of information needed to effect cooperation; (3) the design of a centralized cooperation strategy; and (4) the use of consensus algorithms to transform the centralized strategy into a distributed strategy. In addition, we overview research in formation control and UAV cooperation.

8.1 Introduction

Group cooperative behavior implies that individuals in the group share a common objective and act according to the mutual interest of the group. Effective cooperation often requires that individuals coordinate their actions. Coordination can take many forms ranging from staying out of each other's way to directly assisting another individual. In general, group cooperation is facilitated by coordinating the actions of individuals. However, each individual may not necessarily need to coordinate directly with every other individual in the group to effect group cooperative behavior. For example, fish engaged in schooling behavior react only to other fish that are close physically. We will term this type of coordination *local coordination.* At the other extreme is *global coordination*, where an individual coordinates its action with every other individual in the group. Due to communication constraints and computational feasibility, we are interested primarily in group cooperation problems where the coordination occurs locally. One of the interesting challenges in cooperative control is to design coordination strategies so that local coordination will result in group cooperation.

Although there have been numerous publications detailing specialized approaches to cooperation problems, general design methodologies are only beginning to emerge. For the most part, these methodologies assume a group of homogeneous robots with local coordination. The objective of this chapter is to describe a design approach that we have successfully applied to a variety of cooperation problems. Our approach allows a range of coordination strategies from local to global coordination. We do not claim that our approach will be appropriate for all cooperation problems. We expect that it will be many more years before the general principles underlying cooperative systems will be fully understood. However, we hope that our approach contributes toward that goal.

The essence of our approach is explained in the following steps:

Step 1. **Cooperation Objective and Constraints.** The first step is to define the cooperation objective analytically. Cooperation can often be identified when certain relationships between state variables are satisfied. These relationships are called *cooperation constraints.*

Step 2. **Coordination Variable and Coordination Function.** The next step is to identify the essential information that each vehicle must know to coordinate with the team. This information is called the *coordination variable* as mentioned in Chapter 1. Often, cooperation can be achieved through a variety of individual actions. To facilitate the selection of the individual actions that best contribute to the cooperation objective, we quantify the relationship between the coordination variable and the cooperation objective and call this function the *coordination function.*

Step 3. **Centralized Cooperation Scheme.** The next step is to derive a cooperation strategy for minimizing the team objective function assuming that each member of the team has global knowledge of the coordination variable and the coordination functions of each member of the team.

Step 4. **Consensus Building.** In a distributed situation where communication links are noisy and not persistent and where the communication topology is dynamically changing and is unknown to each team member, the centralized solution will fail. The final step of our approach is to implement a consensus algorithm that ensures that each member of the team has consistent coordination information despite the inadequacies of the communication network.

In this chapter, we describe our approach to cooperative control which is based on two main ideas. The first is the notion of coordination variables and coordination functions, which were introduced in [138, 139]. The coordination variable is the minimum amount of information that needs to be exchanged between two vehicles to effect coordination. Although it is known by different names, the notion of a coordination variable is found in many other works on cooperative control. For example, [101, 102] introduce an "action reference" which, if known by each vehicle, facilitates formation keeping. In leader-following applications [214, 244], the states of the leader constitute the coordination variable because the actions of the other vehicles in the formation are completely specified once the leader states are known. In [22,116,121,123], the notion of a virtual structure is used to derive formation control strategies. The motion of each vehicle is causally dependent on the dynamic states of the virtual structure; therefore, the states of the virtual structure are the coordination variable. In [225], a team of autonomous underwater vehicles (AUVs) is controlled to "swarm" around a desired mean location of the team with a specified standard deviation. The action of each vehicle is dependent on the location of its nearest neighbor and on the desired mean and standard deviation. This information is the coordination variable. Coordination variables may also be discrete. For example, in [27, 199], cooperative task allocation is addressed. Individual vehicle behavior is dependent on the task allocation vector which becomes the coordination variable. Similarly, in [63], the coordination variable is the dynamic role assignment in a robot soccer scenario.

The second main idea in our approach to cooperative control is the notion of consensus-seeking. Because coordination may be required between two vehicles that do not communicate directly, distributed consensus algorithms introduced in Chapters 2–7 are required to ensure that the vehicles share similar coordination variables.

8.2 Coupling in Cooperative Control Problems

One of the primary challenges in developing generalized strategies for cooperative control is the identification of broad classes of problems that are amenable to well-defined, straightforward approaches. One way to classify cooperative control problems is by the level and type of coupling involved. For example, flying a UAV in the wake of another UAV to minimize the induced drag requires tight physical coupling among the UAVs. On the other hand, a team of UAVs tasked to search a particular region cooperatively is coupled primarily through the cooperation algorithms employed by the individual UAVs. We believe that our approach is well suited to algorithmically coupled problems. In this section, we describe different types of coupling that occur in cooperative control problems.

With respect to the cooperative control of multivehicle systems, the degree and form of the coupling among the vehicles composing the system is of paramount importance to the nature and level of the cooperation that can be achieved. Cooperation implies some degree of coupling, if only through the cooperation objective or constraints involved. Generally speaking, the greater the degree of the coupling, the more challenging it is to formulate effective cooperative solutions.

Cooperative control problems can typically be formulated with a cooperation objective or cooperation constraints or both. A cooperation objective is typically optimized to increase the level of cooperation. Cooperation constraints, when satisfied, can be used to define the occurrence of cooperation.

8.2.1 Objective Coupling

Objective coupling describes the least restrictive form of coupling. It occurs when a vehicle's decisions affect only its costs and outcomes and do not influence another vehicle's costs and outcomes directly. Each vehicle's decisions affects the cooperation objective and the feasibility of cooperation constraints. Objective coupling requires vehicles to coordinate to ensure that constraints are satisfied and that the objective is optimized. An example of objective coupling is the UAV cooperative timing problem described in [139]. Suppose that two vehicles are to navigate independently through a threat field with the objective of simultaneously flying over a prescribed destination. The cooperation constraint requires the vehicles to arrive at the same instant, although the cooperation objective might be to minimize the collective power required for the mission. The trajectory taken by one vehicle (its decisions) does not affect the trajectory taken by the other vehicle directly, but it does affect the other vehicle through the cooperation constraint and objective. As this example illustrates, some degree of coupling is necessary for cooperation to occur, but the coupling is strictly through the mission objective.

8.2.2 Local Coupling

Local coupling describes a more restrictive type of coupling in cooperative systems. As with objective coupling, each vehicle's decisions influence the cooperation objective and the feasibility of cooperation constraints. Under local coupling, however, a vehicle's decisions affect not only its own costs and outcomes, but also the decisions (and hence the costs and outcomes) of its local neighbors. A simple example of local or nearest neighbor coupling is shown in the cooperative search problem described in [18], where n UAVs are assigned the task of cooperatively searching an area of interest. To provide some structure to the search task, the vehicles are required to maintain a loose row formation. To avoid collisions, the UAVs are not allowed to overlap laterally. To maintain communication, the lateral spacing of the vehicles must be kept less than the communication range. The lateral spacing constraints can be viewed as cooperation constraints. The cooperation objective is to visit as many targets as possible. The decision by one vehicle to alter its trajectory to visit a sequence of targets will directly affect the decisions of its neighbors and the costs and benefits of their decisions. The key difference between objective coupling and local coupling is that for objective coupling the cost accrued for any vehicle is a function only of that vehicle's decisions. For local coupling, the cost accrued for any vehicle is a function of its own decisions as well as the decisions of its local neighbors.

8.2.3 Full Coupling

Fully coupled systems involve vehicles whose decisions affect the costs and outcomes of all other members of the team, and thus their decisions, *i.e.*, what a single vehicle chooses to do is influenced by what all other vehicles on the team are doing. Coupling exists through the cooperation objective and cooperation constraint as before, but in this most restrictive form of coupling, the decisions of the individual vehicles are coupled directly. An example of full coupling is the wide area search munition problem described in [209]. In this problem, a team of autonomous flying munitions is tasked to search a region and identify potential targets. Once a target is identified, it must be classified by multiple passes over the target using one or more munitions. Upon identification and verification, the target is attacked removing one of the munitions from the team. The target must then be revisited for battle damage assessment. If the minimum turning radius of the munition is large in relation to the search area, then each of these tasks will likely be performed by a different vehicle. The coupling in this scenario is complex and requires that each member of the team knows the intentions and flight paths of all other members of the team.

8.2.4 Dynamic Coupling

When vehicles are coupled through physical interactions, we call it dynamic coupling. The coupling in these systems can be either local or full coupling. For example, in close-formation flight of aircraft, the aerodynamic coupling that exists is local in that it affects those aircraft in the immediate wake of a leading aircraft. Those aircraft on the outer edges of the formation are not affected by those in the center of the formation.

Although the physics of dynamic coupling can be quite complex, there is one significant advantage when the actions of one vehicle are captured directly in the equations of motion describing other vehicles. These systems can be treated as one large system by combining their equations of motion. In this way, these large systems are amenable to control theoretical approaches. With one model to characterize the behavior and interaction of multiple vehicles, conventional single-vehicle approaches can be applied to achieve cooperation.

8.3 Approach to Distributed Cooperative Control Problems with an Optimization Objective

In this section, we give an overview of our approach to cooperative control problems with an optimization objective and illustrate it with a simple example. The approach has been applied to problems with objective coupling [110,139], loose coupling [18], and dynamic coupling [188]. We will illustrate the main ideas through the use of a simple academic example.

Example 8.1. Suppose that five air vehicles are assigned virtual lanes, as shown in Fig. 8.1. Assume that the lateral positions of the vehicles are maintained inside the lanes by an onboard autopilot and that the longitudinal position y_i

of each vehicle is governed by the dynamics $\dot{y}_i = u_i$. Assume that the vehicles are initially at different longitudinal positions in the lane. The cooperation goal is to maneuver the vehicles so that they proceed along a uniform front at a constant known ground speed v, as shown in Fig. 8.1. □

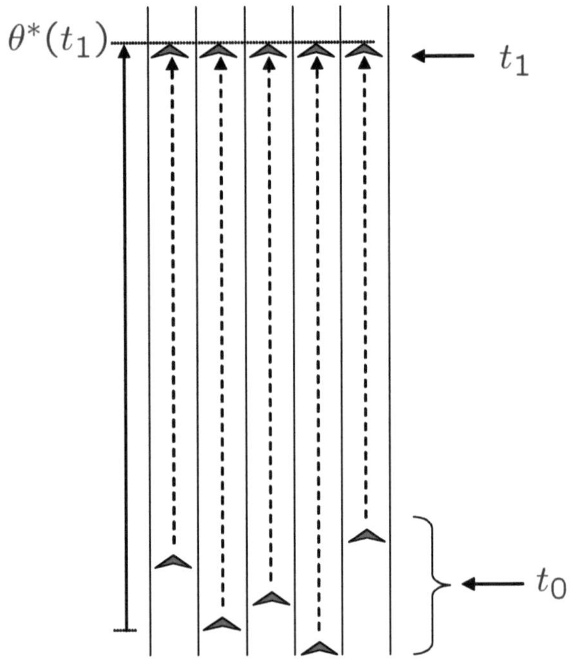

Fig. 8.1. A team of air vehicles is tasked to proceed along a uniform front at a constant ground speed

8.3.1 Cooperation Constraints and Objectives

The first step in our approach is to identify and quantify the cooperation constraint and the cooperation objective. The cooperation constraint is a formal definition of the team goal and indicates exact conditions when cooperation is achieved. More precisely, if x_i is the situational state of the ith vehicle and u_i is the decision variable, then the cooperation constraint is a positive definite mapping $J_{\text{constraint}}(x_1, u_1, x_2, u_2, \ldots, x_n, u_n)$ that is identically zero when cooperation is achieved. In Example 8.1, if n is the number of vehicles, then a possible cooperation constraint is the mapping

$$J_{\text{constraint}} = \frac{1}{2} \sum_{i=1}^{n} \sum_{j=1}^{n} (y_i - y_j)^2, \tag{8.1}$$

which is identically zero when $y_i = y_j$. In Example 8.1, note that if the team is closely but not precisely aligned, then the team may still be considered in cooperation. When $J_{\text{constraint}} < \epsilon$, we say that the team has achieved *ϵ-cooperation.*

For a given world situational state $\mathcal{X} = \{x_1, \ldots, x_n\}$, there may be many different decision variables $\mathcal{U} = \{u_1, \ldots, u_n\}$ that achieve ϵ-cooperation. In addition, there may be auxiliary objectives that we would like to minimize. For instance, in Example 8.1, we may also want to minimize overall fuel expenditure. To capture these auxiliary objectives, we introduce a positive definite function $J_{\text{objective}}(\mathcal{X},\mathcal{U})$ that quantifies these objectives. This function is called the *cooperation objective* [139]. In Example 8.1, a possible cooperation objective is given by the linear quadratic regulator equation [122]

$$J_{\text{objective}} = \sum_{i=1}^{n} \int_{t}^{\infty} q[y_i(\tau) - \theta^*(\tau)]^2 + r[u_i(\tau) - v]^2 \, d\tau,$$

where q and r are positive constants and where θ^* is the position of the uniform front.

8.3.2 Coordination Variables and Coordination Functions

The second step in our approach is to determine the information that must be shared to achieve cooperation and to organize that into a single vector called the coordination variable. We will let θ^* denote the coordination variable. The approach is to distill the essentials of the cooperation problem to a set of parameters that, if known by every vehicle in the group, can be used to select the decision variable so that the cooperation constraint is achieved. In Example 8.1, if every vehicle knows the desired position of the front, then it can regulate its position to align with the front. Therefore, let θ^* be the desired position of the uniform front. In Example 8.1, we will assume that θ^* evolves according to the equation

$$\dot{\theta}^* = v.$$

Our method assumes that the cooperation constraint can be written as a function of the coordination variable. Note that for Example 8.1, the cooperation constraint given in (8.1) can be written as a function of the coordination variable θ^*, as

$$J_{\text{constraint}} = \frac{1}{2} \sum_{i=1}^{n} \sum_{j=1}^{n} [(y_i - \theta^*) - (y_j - \theta^*)]^2 < \epsilon.$$

To facilitate minimization of the auxiliary cooperation objective subject to the cooperation constraint, we desire to write the cooperation objective in terms of θ^*. To this end, we assume that the cooperation objective can be expressed as a convex function of myopic objective functions for each vehicle. The myopic objective functions depend on the coordination variable as well as the situational state and decision variable of the individual vehicle. This myopic objective function is the coordination function [139] denoted $J_{\text{cf},i}(x_i, u_i, \theta^*)$. In Example 8.1,

$$J_{\text{objective}} = \sum_{i=1}^{n} J_{\text{cf},i},$$

where

$$J_{\text{cf},i} = \int_t^{\infty} q[y_i(\tau) - \theta^*(\tau)]^2 + r[u_i(\tau) - v]^2 \, d\tau. \tag{8.2}$$

The coordination function parameterizes the effect of the coordination variable on the objectives of each vehicle, *i.e.*, the coordination function describes how the myopic objective of each vehicle changes with changes in the coordination variable.

Posing the cooperation problem in terms of coordination variables and coordination functions will usually reduce the dimensionality of the problem. Example 8.1 is already a scalar problem and so the dimensionality has not been reduced.

8.3.3 Centralized Cooperation Scheme

Given the terminology introduced in the previous two sections, we can pose the cooperation scenario as the following optimization problem:

$$\theta^* = \arg\min \left\{ \lim_{t\to\infty} \sum_{i=1}^{n} J_{\text{cf},i}(\theta; x_i, u_i) \right\} \tag{8.3}$$
$$\text{subject to: } \lim_{t\to\infty} J_{\text{constraint}}(\theta; \mathcal{X}, \mathcal{U}) < \epsilon.$$

The next step in our approach to cooperative control is the design of a *centralized* strategy that solves this optimization problem. Centralized strategies are usually easier to design than distributed strategies. Note that the centralized algorithm will be problem dependent. In the process of solving problem (8.3), the centralized algorithm produces a decision variable for the ith vehicle denoted as [139]

$$u_i = f_i^{\dagger}(\theta^*, x_i), \tag{8.4}$$

where we assume that $f_i^{\dagger}$ is continuous in θ^*. Equations (8.3) and (8.4) represent what we term the *cooperation algorithm.*

For Example 8.1, the centralized solution requires that each vehicle knows the position of the front $\theta^*(t)$. Accordingly, the vehicles implement the control law

$$u_i = v + k(\theta^* - y_i),$$

where $k = \sqrt{q/r}$ is chosen to minimize the coordination function given in (8.2). Given that $\dot{\theta}^* = v$, it is straightforward to show that this strategy yields $y_i(t) \to \theta^*(t)$ for each $i = 1, \ldots, n$, which implies that

$$\lim_{t \to \infty} J_{\text{constraint}} \to 0,$$

i.e., the cooperation constraint is satisfied for every ϵ. Figure 8.2 shows a simulation plot of the centralized solution where $v = 0.1$ and $k = 1$.

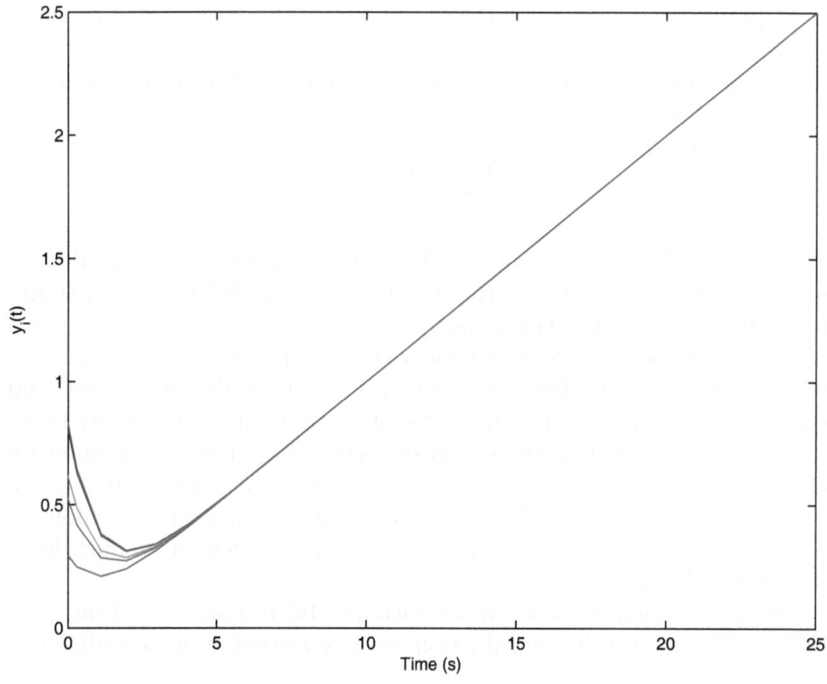

Fig. 8.2. Centralized solution of the example problem. The position of the uniform front is transmitted to all vehicles from a centralized location.

8.3.4 Consensus Building

The final step in our approach is to use consensus algorithms to decentralize the cooperation algorithm. Multivehicle cooperation requires communication among vehicles. In real-world environments, the communication links will be noisy and nonpersistent, and the communication topology will be dynamically changing and unknown to each team member. Therefore, centralized solutions are rarely feasible. The key insight is that if each vehicle instantiates the cooperation algorithm locally, and the inputs to each local instantiation are identical, then assuming that the algorithm is deterministic, it will produce identical outputs on each vehicle. However, if the local inputs are different, then each vehicle will compute a different instantiation of the coordination variable which we label θ_i. Therefore, from (8.4), we see that the decision variable for the ith vehicle is given by

$$u_i = f_i^\dagger(\theta_i, x_i).$$

Because $f_i^\dagger$ is continuous, $f_i^\dagger(\theta_i) \to f_i^\dagger(\theta^*)$ as $\theta_i \to \theta^*$. Therefore, the objective of the consensus algorithm is to ensure that $\theta_i \to \theta_j$ for every i, j. Our approach to consensus building is built on the algorithms introduced in Chapters 2–7.

If communication noise is present in the system, then we need to guarantee that ϵ-consensus is achieved, *i.e.*,

$$\lim_{t\to\infty} \sum_{ij} \|\theta_i(t) - \theta_j(t)\| < \epsilon.$$

Theorems 2.14 and 2.35 show that the consensus error $\theta_i - \theta_j$ is ISS, which implies that the consensus error is uniformly bounded by a gain times the power in the communication noise.

A fundamental result in nonlinear control theory is that the cascade of two ISS systems is also ISS (see Theorem F.10). Consider the control diagram shown in Fig. 8.3. The consensus algorithm on each vehicle is ISS from the communication noise to the consensus error. Therefore, if the cooperation algorithm is ISS from the consensus error to the cooperation constraint, then the cascade system is ISS from the communication noise to the cooperation constraint, implying that ϵ-cooperation is achieved for low enough levels of communication noise.

The application of consensus algorithm (2.10) to Example 8.1 implies that each vehicle updates its coordination variable instantiation according to

$$\dot{\theta}_i = -\sum_{j=1}^{n} a_{ij}(t)(\theta_i - \theta_j) + v,$$

where a_{ij} is the (i, j) entry of the adjacency matrix $\mathcal{A}_n \in \mathbb{R}^{n\times n}$. Each vehicle then implements the modified control law

$$u_i = f_i^\dagger(\theta_i, y_i) \stackrel{\triangle}{=} v + k(\theta_i - y_i),$$

where k can be chosen to minimize the modified coordination function

$$J_{\mathrm{cf},i} = \int_t^\infty q[y_i(\tau) - \theta_i(\tau)]^2 + r[u_i(\tau) - v]^2 \, d\tau. \qquad (8.5)$$

Because the system

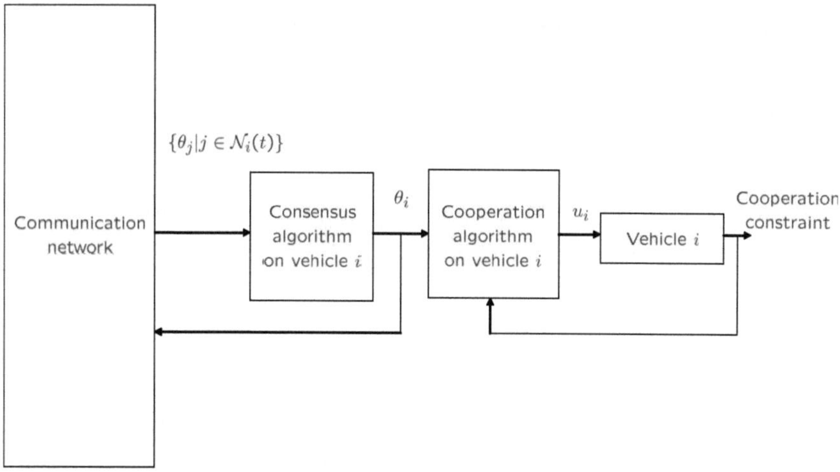

Fig. 8.3. Cascade system where the output of the consensus building mechanism drives the coordination algorithm which uses feedback from the vehicle to achieve coordination. Here $\mathcal{N}_i$ denotes the set of vehicles whose coordination variable instantiations are available to vehicle i.

$$\frac{d}{dt}(y_i - y_j) = -k(y_i - y_j) + k(\theta_i - \theta_j)$$

is ISS, we are guaranteed to achieve ϵ-cooperation. Note that because each vehicle tracks its own notion of the front in an optimal manner [with respect to (8.5)], this does not imply that the overall team has optimal performance. In fact, centralized solutions that minimize the coordination function will always have better performance than distributed solutions. Figure 8.4 shows a simulation plot of the distributed solution where $v = 0.1$ and $k = 1$ and the standard deviation of the noise on the communication channels is $\sigma = 0.1$. The first subplot shows the evolution of the coordination variables. The second subplot shows the evolution of the longitudinal position of the vehicles, and the third subplot shows the number of unidirectional communication links that are active in the network as a function of time. Note that updates of θ_i occur only when the ith vehicle is communicating with another vehicle. Despite the low levels of communication, the coordination constraint is still satisfied.

8.4 Approach to Distributed Cooperative Control Problems Without an Optimization Objective

Often, the primary goal is to effect cooperation in a team without an auxiliary cooperation objective. In that case, there is not a cooperation objective and the algorithms focus on satisfying the cooperation constraint. When there is

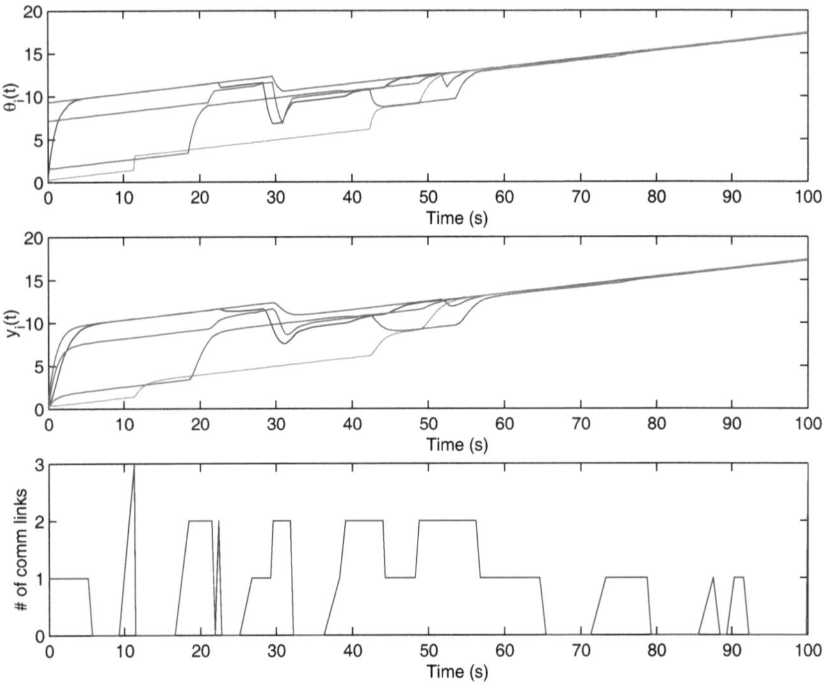

Fig. 8.4. Distributed solution of Example 8.1. A consensus algorithm is used to negotiate a common position of the uniform front despite low levels of communication among vehicles.

no need to minimize the cooperation objective, we can simplify the approach described in Section 8.3. The approach that we adopt in this case is first to identify the coordination variable and then to apply the consensus algorithms to ensure that team members have a consistent view of the coordination variable or to incorporate group behavior or dynamics even in the presence of limited/unreliable intervehicle information exchange. Two cases will be considered.

8.4.1 Coordination Variable Constituted by a Group-level Reference State

In this subsection, we consider the case where there exists an identifiable group-level reference state for each individual vehicle on the team. Here the group-level reference state constitutes the coordination variable. The group-level reference state serves as a basis for each vehicle to derive local control strategies. As a result, having the knowledge of the group-level reference state facilitates cooperation for the whole team. For example, in cooperative timing missions, the estimated team arrival time at specified destinations serves as a group-level reference for each vehicle because the motion of each vehicle can be adjusted according to the estimated team arrival time. In leader-following formation flying applications, the state of an actual or virtual team leader serves as a group-level reference for each vehicle because the actions of other vehicles in the formation are completely specified once the leader state is known. In cooperative monitoring applications, it is often desirable that multiple UAVs are evenly distributed around the perimeter of a boundary (*e.g.*, a growing forest fire) to monitor and track the perimeter as it evolves, where the equal length of coverage serves as a group-level reference for each UAV.

We let ξ denote the group-level reference state (*i.e.*, the coordination variable). Let x_i, y_i, and u_i represent, respectively, the local state, output, and control input of the ith vehicle. Suppose that the dynamics of the ith vehicle are

$$\dot{x}_i = g_i(t, x_i, u_i), \quad y_i = h_i(t, x_i), \quad i = 1, \ldots, n. \tag{8.6}$$

A centralized cooperative control strategy can be designed as

$$\dot{\xi} = f(t, \xi, y_1, \ldots, y_n), \quad u_i = \kappa_i(t, y_1, \ldots, y_n, \xi), \quad i = 1, \ldots, n,$$

where the ith control input depends on the coordination variable ξ and the outputs of all vehicles. In this centralized scheme, the coordination variable is implemented at a central location and broadcast to every vehicle on the team. However, this implementation results in a single point of failure and is not scalable to a large number of vehicles.

A remedy to these drawbacks is to instantiate a local copy of the coordination variable on each vehicle. If each vehicle implements the same local control algorithm, we might expect that a distributed scheme will achieve the same level of cooperation as the associated centralized scheme. However, due to different local situational awareness uncertainties for each vehicle, there exist discrepancies between each instantiation of the coordination variable. For example, in multivehicle simultaneous arrival missions, each vehicle's estimated team arrival time (*i.e.*, its coordination variable) may be dynamically changing and inconsistent with other vehicles' estimated team arrival time as the vehicle encounters obstacles or threats. As a result, consensus algorithms need to be applied to guarantee that each instantiation of the coordination variable converges to a common value.

Figure 8.5 shows a consensus-based distributed cooperative control scheme with a group-level reference state. The hierarchical architecture consists of three layers: consensus module, cooperation module, and physical vehicle. Each vehicle instantiates a local copy of the consensus module and cooperation module. Let ξ_i denote the ith vehicle's instantiation of the coordination variable. Let $\mathcal{N}_i(t)$ and $\mathcal{J}_i(t)$ denote, respectively, the possibly time-varying set of vehicles whose coordination variable instantiations and outputs are available to vehicle i at time t. Depending on mission requirement and communication bandwidth, ξ_i, y_i, or their combination is exchanged between neighbors. Note that the neighboring vehicles' outputs may be exchanged less frequently than the coordination variable instantiations. The objective of the consensus modules is to drive ξ_i to a consistent value that also achieves the desired mission goal. The local control law u_i for each vehicle is based on the instantiation of the coordination variable for that vehicle. Note that each layer has feedback information describing the performance of that layer to the layer above it, as denoted by z_{ci} and z_{vi}.

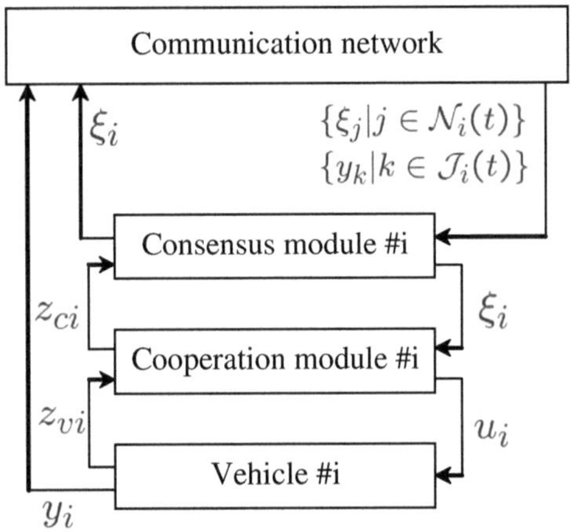

Fig. 8.5. Consensus-based distributed cooperative control scheme with a group-level reference state

We update the ith instantiation of the coordination variable according to a consensus algorithm of the form

$$\dot{\xi}_i = f_i(t, \{\xi_j | j \in \mathcal{N}_i(t)\}, \{y_k | k \in \mathcal{J}_i(t)\}), \tag{8.7}$$

where y_k, $k \in \mathcal{J}_i(t)$, introduce group feedback from vehicle i and its neighbors to the ith instantiation of the coordination variable.[1] The goal is to guarantee that $\xi_i(t) \to \xi_j(t)$, $\forall i \neq j$, and $\xi_i(t) \to \xi^{(p)}$, $i = 1, \ldots, n$, as $t \to \infty$, where $\xi^{(p)}$, $p = 1, \ldots, P$, denote a sequence of desired mission goals. We propose the local control for the ith vehicle as

$$u_i = \kappa_i(t, \{y_k | k \in \mathcal{J}_i(t)\}, \xi_i),$$

where the ith control input depends on its own and local neighbors' outputs and on the ith instantiation of the coordination variable.

8.4.2 Coordination Variable Constituted by Vehicle States

In this subsection, we consider the case where each vehicle adjusts its own state according to the states of its neighbors to achieve different group objectives.

[1] Note that (8.7) represents a generic form of the consensus algorithms described in Chapters 2–7.

Here, the states of the vehicles or their relative state deviations constitute the coordination variable. For example, to achieve a cooperative observation, attitude alignment may be required for a team of vehicles with only local information exchange of the neighboring vehicles' attitudes. In multi-vehicle flocking applications [198], each vehicle maneuvers according to its nearby neighbors' positions and velocities to achieve group behaviors such as separation, alignment, and cohesion.

When the states of the vehicles or their relative state deviations constitute the coordination variable, we can adapt the scheme shown in Fig. 8.5 to a distributed cooperative control scheme, as shown in Fig. 8.6, where each vehicle instantiates different modules such as goal seeking, consensus building, formation keeping, and collision avoidance to characterize different group behaviors. Through local communication or sensing, each vehicle specifies its local control law based on these combined modules. The functionality of the consensus module is to guarantee the alignment of velocities, attitude, altitude, and so on among multiple vehicles.

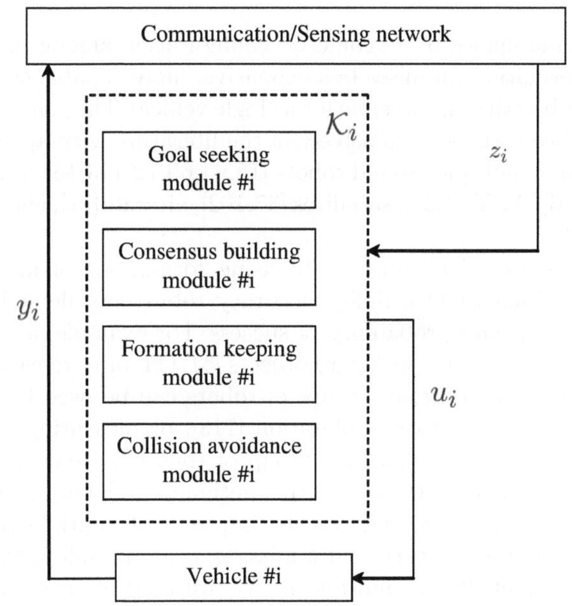

Fig. 8.6. Consensus-based distributed cooperative control scheme with different group behaviors

Suppose that the dynamics of the ith vehicle are given by (8.6). We propose the local control for the ith vehicle as

$$u_i = \kappa_i(t, z_i),$$

where $z_i \triangleq \eta_i(t, \{y_\ell | \ell \in \mathcal{J}_i(t)\})$ denotes the information sensed by and/or communicated to vehicle i and $\mathcal{J}_i(t)$ denotes the set of vehicles whose information is available to vehicle i through either communication or sensing. One example is that z_i is composed of a set of vehicle positions y_j, where $j \in \mathcal{J}_i(t)$. Another is that z_i is composed of a set of relative position measurements $y_i - y_j$, where $j \in \mathcal{J}_i(t) \setminus \{i\}$.

8.5 Literature Review

In the next several chapters, we will apply the design methodologies described in Sections 8.3 and 8.4 to formation control of multiple wheeled mobile robots and spacecraft as well as to cooperative control of multiple UAVs. Before moving on, we will overview the existing literature on formation control and multi-UAV cooperation.

8.5.1 Formation Control

Accurate maintenance of a geometric configuration among multiple vehicles moving in formation promises less expensive, more capable systems that can accomplish objectives impossible for a single vehicle. The concept of formation control has been studied extensively in the literature with applications to the coordination of multiple ground robots [15,64,69,72,123,151,152,227,233,243], UAVs [82,226], AUVs [225], satellites [37,102], aircraft [11], and spacecraft [85, 142, 200, 244].

There are several advantages to using formations of multiple vehicles. These include increased feasibility, accuracy, robustness, flexibility, lower cost, energy efficiency, and probability of success. For example, a group of robots can be used to move large awkward objects [57,111] or to move a large number of objects [240]. In addition, groups of robots can be used for terrain model acquisition [240], planetary exploration [113], or measuring radiation levels over a large area [10]. In [263], a group of robots is used for path obstruction. This could be used to impede the motion of an intruder on a battlefield. The probability of success will also be improved if multiple vehicles moving in formation are used to carry out a mission, *e.g.*, multiple UAVs are assigned to a certain target [24] or multiple AUVs are used to search an underwater object [225]. In spacecraft formation flying applications, using multiple microspacecraft instead of a monolithic spacecraft can reduce the mission cost and improve system robustness and accuracy [85].

Various strategies and approaches have been proposed for formation control. These approaches can be roughly categorized as leader-following, behavioral, and virtual leader/virtual structure approaches.

In the leader-following approach, one of the vehicles is designated as the leader, with the rest of the vehicles designated as followers. The basic idea is that the followers track the position and orientation of the leader with some prescribed (possibly time-varying) offset. There are numerous variations on this theme including designating multiple leaders, forming a chain (vehicle tracks vehicle), and other tree topologies.

One of the first studies on leader-following strategies is reported in [243], which discusses formation control laws for mobile robots. The application of these ideas to spacecraft formations is reported in [244], where explicit control laws for formation keeping and relative attitude alignment based on nearest neighbor tracking are derived. Several leader-following techniques are discussed, including leader tracking, nearest neighbor tracking, barycenter tracking, and other tree topologies. In [245], the ideas of [244] are extended to account for actuator saturation and are applied to the problem of controlling the formation to execute a continuous rotational slew. In [85], adaptive control laws are added to the control derived in [244] to reject common space disturbances. Leader-follower approaches to satellite formation keeping in earth orbit are described in [53], [103],and [265].

There have been a number of studies of leader-following techniques in the mobile robotics community. In [227], a leader-following technique is used to control a group of mobile robots to move a box cooperatively. In [56], feedback linearization techniques are used to derive tracking control laws for nonholonomic robots that are used for leader-following. In addition, the authors describe the formation configuration as a directed graph. The shape of the formation is changed as graph structures are changed. Another leader-following approach for multiple nonholonomic robots is described in [264]. A leader-following approach to the platoon problem in intelligent highways is contained in [214].

The basic idea behind the behavioral approach is to prescribe several desired behaviors for each vehicle and to make the control action of each vehicle a weighted average of the control for each behavior. Possible behaviors include collision avoidance, obstacle avoidance, goal seeking, and formation keeping. There are also numerous variations on the behavioral approach to multivehicle coordination. Most are derived by novel weights of the behaviors.

In [136], the behavioral approach is applied to the problem of maintaining a constellation of satellites in an equally distributed ring formation in earth orbit. Simple Lyapunov control functions are used to maintain distance and avoid collisions. The application of the behavioral approach to aircraft flying in formation is described in [11], where control strategies are derived to mimic the instinctive behavior of birds and fish. A paper that describes the behavioral approach to formation keeping for mobile robots is [15], where control strategies are derived by averaging several competing behaviors, including goal seeking, collision avoidance, and formation maintenance. Because competing behaviors are averaged, occasionally strange and unpredicted behaviors may occur. Unit-center tracking, leader tracking, and nearest neighbor tracking controls are also studied. In [269], the behavioral approach is used to cause a group of robots to create line and circle formations. These ideas are extended in [44] to the problem of controlling a formation of mobile robots to transport objects.

In the virtual leader/virtual structure approach, the entire formation is treated as a single structure or rigid body. In the virtual structure approach, the control is derived in three steps: first, the desired dynamics of the virtual leader/virtual structure are defined; second, the motion of the virtual leader/virtual structure is translated into the desired motion for each vehicle; and finally, tracking controls for each vehicle are derived.

The virtual structure approach is applied in [28,123] to acquire high precision formation control for mobile robots. The application to formations of spacecraft in free space is described in [19,22]. The virtual leader approach is applied in [62,121] for formation control of mobile robots. A similar approach is the action reference scheme in [101,102]. In [151], a Lyapunov formation function is used to define a formation error, and formation feedback is incorporated in the virtual leaders through parameterized trajectories. In [189,267], the virtual structure approach is used to perform elementary formation maneuvers for, respectively, mobile robots and spacecraft, where group feedback is incorporated from the followers in the virtual structure to improve group stability and robustness.

Besides the three approaches described above, approaches based on graph rigidity and artificial potential functions have been studied, respectively, in [64,153,157,268] and [61,257] for formation control. There have also been other studies of multiple satellites orbiting the earth in formation. Two spacecraft flying in a polar orbit formation are considered in [73,211], and a software package that implements their approach is described in [74]. In [241], the design of a two satellite formation flying mission for an interferometric SAR topography mission is described. Formation keeping for low earth-orbit satellites is considered in [83]. Relative formation keeping for low earth orbits using LQ regulators is discussed in [237]. There have been several studies on optimal fuel formation control, including [20,25,242].

8.5.2 Cooperation of Multiple UAVs

Cooperation among UAVs has it own set of unique challenges. For example, unlike ground robots, there is generally very little physical coupling except in the obvious cases of close formation flight. The most significant challenge that is unique to UAVs is three-dimensional flight with immediate implications on path planning algorithms. Another characteristic unique to fixed-wing UAVs is that forward motion is required. Therefore, stop-and-wait path deconfliction algorithms (*e.g.*, [5]) are not applicable. In addition, small fixed-wing UAVs are highly susceptible to wind. Therefore, cooperation strategies must incorporate feedback at the highest levels to account for objective failure modes.

However, cooperation problems for ground robots and UAVs share a number of similarities. For example, both ground and aerial robots have strict communication constraints: team members must be in close physical proximity to communicate, bandwidth is limited, and the communication topology may change unpredictably with time. Both ground and aerial robots must deal with collision avoidance constraints. Ground robots are necessarily concerned with maneuvering around each other in confined spatial environments. On the other hand, aerial robots usually have more room to maneuver, but collisions are typically catastrophic. Another similarity is that decentralized cooperation strategies are generally required for both ground and aerial robots. In addition, cooperation strategies must be robust to the failure of individual team members.

Studies specific to cooperative control of UAVs have recently appeared in the literature. Extensive efforts have been directed toward close formation flight. Early studies reported in [29] and [45] reported significant potential fuel savings that could be gained by close formation flight. In [167], the physical equations that describe a fixed-wing aircraft flying in the vortex of the leader are described, and a control system based on the linearized model is developed. The approach is extended to nonlinear aerodynamic coupling terms in [161]. Standard inner/outer loop designs are extended to close-formation flight in [210]. A behavioral approach to aircraft formation flight is given in [11]. In [163], differential flatness is used to generate group formation maneuvers. The effects of communication constraints on close formation flight are studied in [82]. Rigorous conditions for stable formation flight with limited communication are developed in [69].

Multiple UAV cooperative timing problems have also received significant attention. One version of this problem occurs where multiple UAVs are required to converge on the boundary of a radar detection area to maximize the element of surprise [24,27,40,137,138]. Cooperative timing problems also arise in re-fueling scenarios, fire and hazardous material monitoring [38], moving area of regard problems, and continuous surveillance problems [109].

Cooperative timing problems are sensitive to the assignment and ordering of tasks. One approach for handling cooperative timing is to apply timing constraints to the task assignment problem. In [7, 27, 199], mixed-integer linear programming (MILP) is used to solve tightly coupled task assignment problems with timing constraints. The advantage of this approach is that it yields the optimal solution for a given problem. The primary disadvantages are the complexity of problem formulation and the computational burden. Pruning strategies for simplifying the MILP problem have been proposed to enable near-real-time solutions.

Although path planning for single UAVs has been an active area of research for some time (*e.g.*, see [9, 66, 75, 262, 266]), cooperative path planning approaches for UAVs have only recently begun to appear. In [95], a decentralized optimization method based on a bargaining algorithm is developed and applied to a multiple aircraft coordination problem. A hybrid hierarchical control architecture is used for air traffic control in [92, 207].

Experimental work with teams of UAVs has been limited, primarily due to the practical challenges of fielding multiple vehicles simultaneously. Several researchers have demonstrated leader following with two UAVs (*e.g.*, [34, 107, 212]). Cooperative timing using a team of three UAVs has been demonstrated experimentally(*e.g.*, [150]).

8.6 The Remainder of the Book

The remainder of the book is organized as follows: Chapter 9 will apply the design scheme described in Section 8.4.1 to rendezvous and axial alignment of multiple wheeled mobile robots, where rendezvous and axial alignment can be viewed as special cases of a formation stabilization problem. Chapter 10 applies the design scheme described in Section 8.4.1 to distributed formation control of multiple wheeled mobile robots. Chapter 11 applies the design scheme described in Section 8.4.2 to decentralized formation stabilization of multiple wheeled mobile robots. Chapter 12 applies the design scheme described in Section 8.4.1 to deep space spacecraft formation flying. Chapter 13 applies the design scheme described in Section 8.4.1 to cooperative fire monitoring with multiple UAVs. Chapter 14 applies the design scheme described in Section 8.3 to cooperative surveillance with multiple UAVs.

8.7 Notes

The results in this chapter are based mainly on [26, 171]. Section 8.5.1 is mainly from [22].

Acknowledgment is given to

Part VI

Applications to Multivehicle Cooperative Control

9

Rendezvous and Axial Alignment with Multiple Wheeled Mobile Robots

In this chapter, we apply the consensus-based design scheme in Section 8.4.1 to two applications, rendezvous and axial alignment. In the rendezvous application, multiple mobile robots simultaneously arrive at a common *a priori* unknown location determined through team negotiation. In the axial alignment application, multiple mobile robots collectively align their final positions along a line. The two applications rely on the consensus algorithms introduced in Chapter 2, and are experimentally implemented and validated on a mobile actuator and sensor network platform subject to directed, possibly switching interaction topologies to explore issues and challenges in cooperative control with communication constraints. The experimental results show the effectiveness and robustness of the consensus algorithms, even in the presence of physical platform limitations, packet loss, and information delay.

9.1 Experimental Platform

The Mobile Actuator and Sensor Network (MASnet) platform in the Center for Self-Organizing and Intelligent Systems (CSOIS) at Utah State University (USU) combines wireless sensor networks with mobility, that is, a large number of robots can serve both as actuators and sensors. Although each robot has limited sensing, computation, and communication ability, they can coordinate with each other as a team to achieve challenging cooperative control tasks such as formation keeping and environment monitoring.

The MASnet platform is comprised of MASmotes, an overhead camera, and a base station PC, as shown in Fig. 9.1. MASmotes are actually two-wheel differentially steered robots that can carry sensors and actuators and are wireless networked *via* Micaz from Crossbow [1]. The functionality of MASmotes includes intermote and mote-to-base-station communication, data collecting, pulse-width modulation (PWM) signal generation, and encoder counting. An overhead CCD camera is used to identify each robot and determine its position and orientation (*i.e.*, pseudo-GPS information). Images from the camera are processed by the base station. The functionality of the base station includes image processing, serial-to-programming-board communication, pseudo-GPS information broadcasting, and decision making. The base station communicates with a gateway mote mounted on a programming board through a serial port. The gateway mote then communicates with all other MASmotes over a 2.4 GHz wireless mesh network. Note that the gateway mote serves as a gateway between wireless communication and serial port communication, and its only purpose is to forward all messages between the serial port and the RF port. Through communication, the base station can thereby send commands and pseudo-GPS information to each MASmote. All MASmotes can also communicate with each other over the 2.4 GHz wireless mesh network.

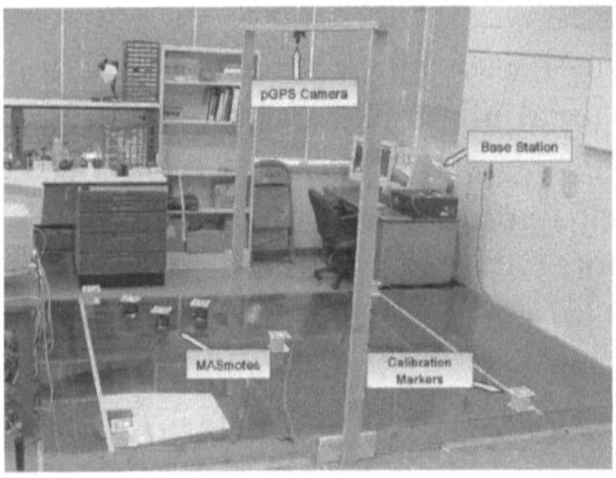

(a) MASnet testbed

(b) MASmote robot hardware

Fig. 9.1. MASnet experimental platform. Subfigure (a) shows the MASnet testbed comprised of MASmotes, an overhead camera, and a base station PC. Subfigure (b) shows the robot hardware.

9.2 Experimental Implementation

Because both intermote and mote-to-base-station communication are available, the MASnet platform can be used experimentally to test both centralized and decentralized cooperative control schemes. For a centralized scheme, each MASmote is responsible only for its low-level motor control, whereas the base station, serving as a centralized commander, broadcasts pseudo-GPS information to each MASmote robot, implements cooperative control algorithms, and sends control commands based on the information gathered from the whole team. For a decentralized cooperative control scheme, each MASmote implements its own cooperative control algorithm based on the pseudo-GPS information provided by the base station.

In the experiments, all control algorithms are implemented onboard the MASmotes, and each MASmote uses only the pseudo-GPS information of its own and its neighbors even if the pseudo-GPS information of all MASmotes is provided by the base station. By doing so, distributed cooperative control algorithms involving only neighbor-to-neighbor information exchange *via* communication or sensing can be tested for multivehicle systems. The feature of local information exchange is important in applications for the following situations: (i) communication or sensing topologies are not fully connected, (ii) vehicles have limited communication range and bandwidth, (iii) power consumption of the team is constrained, and (iv) the stealth of the team needs to be increased. Two applications are studied in the experiments, rendezvous and axial alignment. In both applications, only neighbor-to-neighbor information exchange is allowed.

One challenge of applying consensus algorithms to the MASnet platform comes from the fact that the MASmote robots are nonholonomic. Controlling the positions of the robots overcomes this issue with the compromise that the robot orientation information is lost when the robots reach their desired positions.

We apply the design scheme in Section 8.4.1 to the rendezvous and axial alignment problems. A *consensus controller* is applied to update the desired position of each MASmote robot at each instant when the robot receives the position and orientation information of its own and its neighbors from the pseudo-GPS. For the rendezvous application, the coordination variable is the team rendezvous position. For the axial alignment application, the coordination variable is the center of the line segment of alignment. For both applications, local instantiation of the coordination variable becomes the desired position of the corresponding robot. The update period for the consensus controller depends on the pseudo-GPS information update rate, which is between 0.1 and 0.2 seconds, on average. A low-level PID *position controller* is also applied to guarantee that each robot moves to its (time-varying) desired position (see [43] and references therein for the position controller). The position controller for each robot requires its current position and orientation as well as its desired position provided by its consensus controller. Each robot uses its encoders for position and orientation measurements between pseudo-GPS updates. However, a pseudo-GPS update will overwrite the encoder-based position and orientation measurements due to the inaccuracy of the low-cost encoders.

Let $r_i = [x_i, y_i]^T$ and $r_i^d = [x_i^d, y_i^d]^T$ denote, respectively, the actual and desired position of robot i. Motivated by (2.2), a possible strategy for the rendezvous problem is to update r_i^d as

$$\dot{r}_i^d = -\sum_{j=1}^{n} a_{ij}(r_i^d - r_j^d), \tag{9.1}$$

where a_{ij} is the (i,j) entry of the adjacency matrix $\mathcal{A}_n \in \mathbb{R}^{n\times n}$ associated with directed graph $\mathcal{G}_n = (\mathcal{V}_n, \mathcal{E}_n)$. Another possible strategy is to update r_i^d as

$$\dot{r}_i^d = -\sum_{j=1}^{n} a_{ij}(r_i - r_j). \tag{9.2}$$

Note that the desired team rendezvous position using (9.1) is unaffected by robot tracking performance whereas the desired team rendezvous position using (9.2) is dependent on the positions of the robots. The two strategies may be appropriate for different applications, depending on the application context.

Note that (9.2) can also be written as

$$\dot{r}_i^d = -\sum_{j=1}^{n} a_{ij}(r_i^d - r_j^d) + \sum_{j=1}^{n} a_{ij}(e_i - e_j),$$

where $e_i \stackrel{\triangle}{=} r_i^d - r_i$ denotes the ith robot's tracking error. Assuming a well-designed low-level position controller, $e_i(t)$ is bounded for all t and $e_i(t)$ approaches zero, as $t \to \infty$, $i = 1, \ldots, n$, which implies that $\sum_{j=1}^{n} a_{ij}(e_i - e_j)$ is bounded for all t and approaches zero, as $t \to \infty$. Therefore, it follows from Theorems 2.14 and 2.35 that $\left\| r_i^d(t) - r_j^d(t) \right\|$ is bounded for all t and approaches zero, as $t \to \infty$. In other words, the cascade system composed of the consensus controller and the position controller is stable, *i.e.*, $r_i^d(t) \to r_j^d(t)$ and $r_i(t) \to r_j(t)$, $\forall i \neq j$, as $t \to \infty$.

For axial alignment, motivated by (2.8), r_i^d is updated as

$$\dot{r}_i^d = -\sum_{j=1}^{n} a_{ij}[(r_i - r_j) - \Delta_{ij}], \tag{9.3}$$

where Δ_{ij} denotes the desired constant separation vector between r_i and r_j. The stability of (9.3) can be analyzed in a way similar to (9.2) by following Corollary 2.12.

In our experiments, we let $a_{ij} = 1$ if $(j,i) \in \mathcal{E}_n$ and use discrete-time versions of (9.2) and (9.3). In this case, algorithm (9.2) updates r_i^d as the average of its current position and the current positions of its neighbors at each sample time.

9.3 Experimental Results

In this section, we show experimental results for rendezvous and axial alignment on our MASnet platform.

9.3.1 Rendezvous

For the rendezvous application, rendezvous of four MASmote robots is studied under directed, time-invariant, and dynamic interaction topologies, respectively.

Figure 9.2 shows six different time-invariant interaction topologies for Cases I–VI. In particular, Case I corresponds to an undirected connected graph, Case II corresponds to a line graph, Case III corresponds to an undirected graph with two distinct subgroups, Case IV corresponds to a directed graph with multiple nodes that have only outgoing edges (vehicles 1 and 3), Case V corresponds to a cyclic pursuit graph, and Case VI corresponds to a general directed graph that has a directed spanning tree.

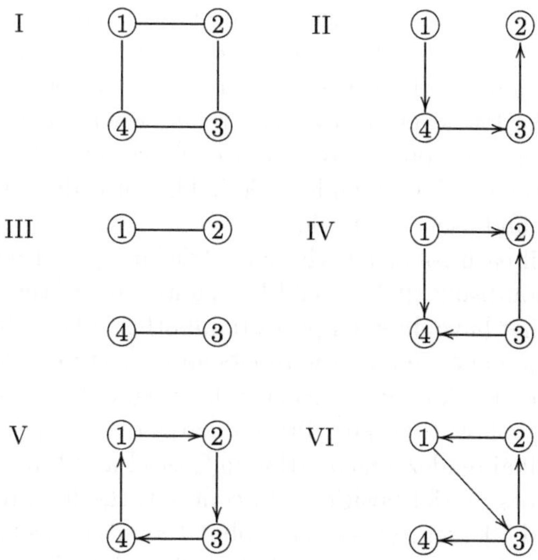

Fig. 9.2. Interaction topologies among four robots for the rendezvous application

Figure 9.3 shows the experimental results of the rendezvous captured by the overhead camera for Cases I–VI, where the circles denote the initial positions of the robots and the colored dots denote the actual trajectories of the robots identified by the overhead camera. If the vision system fails to identify a robot during a certain period, no pseudo-GPS update is available to the robot (equivalent to a pseudo-GPS packet loss), and hence no colored dot is placed on the screen during that time. The trajectories of the robots are captured between $t = 0$ and $t = t_f$ seconds.

Figure 9.3 shows that the four robots rendezvous in all cases except in Cases III and IV where only a subgroup rendezvous. This experimental result is consistent with Theorem 2.8 because only the graphs of Cases III and IV do not have directed spanning trees. In our experiments, the computationally complex task of processing the image, finding MASmote markers, and extracting position and orientation information introduces a delay of 0.1 to 0.2 seconds between image capture and position and orientation information broadcast. In addition, the hardware and software limitations of the vision system cause frequent pseudo-GPS packet loss (roughly 2–5%) and an average position measurement error of 1.32 cm. When a robot is lost from the view of the overhead camera due to glare or other issues, the robot's position and orientation information are no longer available to itself and its neighbors, implying a pseudo-GPS packet loss. Specifically, in Case VI, the upper left robot does not receive pseudo-GPS updates during several sample periods, as shown in Fig. 9.3f. Despite the existence of time delay and pseudo-GPS packet loss, the rendezvous experiments achieve results predicted by the theory. With consensus controller (9.2), the desired position for a robot simply remains constant during the period when the pseudo-GPS is not available, that is, (9.2) remains stable with time delay or pseudo-GPS loss although the algorithm will converge more slowly, which in turn results in slower rendezvous for the team. With the low-level position controller, when there is no pseudo-GPS update, a robot relies on encoder data to move toward the desired position calculated at the previous sample period. This provides a way to compensate for time delay and pseudo-GPS loss.

In theory, if each robot has the same tracking performance, the final rendezvous position using (9.2) should be the average of the initial positions of those robots that have directed paths to all other robots. In our experiments, due to the discrepancy among the robots and inaccuracy caused by the vision system, the final rendezvous position is the weighted average of the initial positions of those robots that have directed paths to all other robots, as shown in Fig. 9.3. The final rendezvous position in Cases I and V are weighted averages of all four robots' initial positions. In contrast, the final rendezvous position in Cases II and VI are, respectively, robot 1's initial position and a weighted average of the initial positions of robots 1, 2, and 3. This experimental result can be explained by noting that in Cases I and V, each robot has a directed path to every other robot; in Case II, only robot 1 has a directed path to all other robots; and in Case VI, every robot except robot 4 has a directed path to all other robots. In particular, in Case III, the final rendezvous position for the upper two robots is not the center of the two robots' initial positions but closer to the starting position of the upper left robot. We observe in the experiment that the upper left robot receives spurious orientation information during the beginning period of the experiment, which causes the robot to spin at its initial position without moving toward its desired position during the first several seconds. However, the upper right robot is able to move toward the upper left robot to achieve rendezvous. In addition, in Case IV, the overhead camera produces incorrect position and orientation measurements for the upper right robot a few seconds before the upper right and lower left robots rendezvous, which causes the upper right robot to wander around and spin.

However, the lower left robot is able to catch the upper right robot to achieve rendezvous. These two examples demonstrate the robustness of (9.2), even in the presence of incorrect output of the vision system for a certain period of time. The robustness of (9.2) is due to the fact that the desired position for each robot is dynamically determined according to its current position and the current positions of its neighbors.

Furthermore, by comparing Cases I, II, V, and VI, we can see that in Case I, the four robots rendezvous the fastest while in Case V, the four robots rendezvous the slowest. This experimental result is also predicted by theory. In particular, the fact that consensus under a line graph (Case II) converges more slowly than an undirected connected graph (Case I) can be seen by comparing the eigenvalues of the nonsymmetrical Laplacian matrix $\mathcal{L}_4$ associated with the interaction topology for the four robots in two cases. The eigenvalue $\lambda_2(\mathcal{L}_4)$ of $\mathcal{L}_4$ whose real part is the second smallest characterizes the convergence speed of the consensus algorithm. It is straightforward to show that $\lambda_2(\mathcal{L}_4) = 2$ in Case I and $\lambda_2(\mathcal{L}_4) = 1$ in Case II by noting that $a_{ij} = 1$ if $(j, i) \in \mathcal{E}_n$. By comparing Fig. 9.3a and 9.3b, we can see that in Case II, the upper right robot moves toward the lower right robot instead of moving directly toward the upper left robot due to the fact that the upper right robot can receive information only from the lower right robot, resulting in slower convergence for the team.

To test rendezvous in the case of switching interaction topologies, we force the interaction topologies for the four robots to switch randomly from the set $\bar{\mathcal{G}}_4^s = \{\mathcal{G}_{4(1)}, \mathcal{G}_{4(2)}, \mathcal{G}_{4(3)}, \mathcal{G}_{4(4)}, \mathcal{G}_{4(5)}\}$ which is shown in Fig. 9.4. Note that each directed graph in $\bar{\mathcal{G}}_4^s$ does not have a directed spanning tree but that the union of these graphs denoted by $\mathcal{G}_4^u$ does have a directed spanning tree. Because switching among the directed graphs in $\bar{\mathcal{G}}_4^s$ is random, the condition for consensus in Theorem 2.31 is generically satisfied.

Figure 9.5a shows the experimental result of rendezvous when the interaction topologies switch randomly from $\bar{\mathcal{G}}_4^s$ with switching periods randomly chosen between 2.75 and 8 seconds and Fig. 9.5b shows the experimental result under the time-invariant interaction topology $\mathcal{G}_4^u$. Note that the four robots rendezvous even when the directed topologies switch randomly with time, which validates the theory in Theorem 2.31 and demonstrates the robustness of the consensus algorithm to switching topologies, as long as the minimum connectivity condition in Theorem 2.31 is satisfied. By comparing Figs. 9.5a and 9.5b, it can be seen that convergence in the case of switching topologies is slower than in the time-invariant case because the robots simply stop when they do not receive information from their neighbors. Also, the switching topologies result in sudden drastic changes in robot directions, as shown in Fig. 9.5a. In addition, in Fig. 9.5a, when the lower left robot makes the second turn due to spurious pseudo-GPS data for several sample periods, the robot starts to wander around, as shown by the concentrated dots. However, the inherent stability of the cascade system composed of the consensus controller and the position controller causes the robot to recover, once the vision system sends out correct position and orientation data.

9.3.2 Axial Alignment

For the axial alignment application, the case where four robots are evenly distributed along a straight line is studied using a time-invariant interaction topology. We choose Δ_{ij} in (9.3) to guarantee that the robots align on a horizontal line with a separation distance of 24 cm along the x axis between two adjacent neighbors.

Figure 9.6 shows the undirected interaction topology among the four robots. Figure 9.7 shows the experimental result of the axial alignment. In theory, although each robot starts at an arbitrary initial position, their final positions should be evenly distributed along a horizontal line with a separation distance of 24 centimeters. Due to hardware and software limitations of the vision system, some robots move in the wrong direction for a period of time until the vision system correctly identifies their orientations. In addition, there exists a position tracking error with the low-level position controller due to inaccurate position and orientation measurements. Even with these limitations, the overall experiment still achieves the goal of axial alignment with an average error around 3 cm and further demonstrates the robustness of the consensus algorithm. The experimental result is consistent with the argument of Corollary 2.12.

9.3.3 Lessons Learned

Consensus algorithms have been experimentally applied to rendezvous and axial alignment. The results of both applications on the MASnet platform have demonstrated the effectiveness and robustness of the consensus algorithms to distributed cooperative control. Consensus algorithms provide a promising method for distributed multivehicle cooperative control even in the presence of physical robot limitations, packet loss, information delay, *etc.*

Despite the success of the experiments, there are limitations. One limitation is that the robots are dependent on the vision system for position and orientation measurements over long distances due to inaccuracy of the encoders in the low-cost platform. When the vision system produces an incorrect measurement or fails to identify a robot during a certain period of time, a robot may move in the wrong direction. As the time or frequency of vision failure increases, the overall team performance degrades dramatically. Another limitation is the pseudo-GPS update delay. When robots are moving slowly, the pseudo-GPS update delay has little effect, but at full speed, the difference between the actual and the broadcast position and orientation can be quite large. This is most noticeable when a robot is rotating. As a result, there is a need to improve the vision system and encoder accuracy and develop a prediction model for position and orientation. The combination of pseudo-GPS data, encoder data, and mathematical model estimates is expected to improve the accuracy of position and orientation measurements.

9.4 Notes

The results in this chapter are based mainly on [194,195]. The Mobile Actuator and Sensor Network (MASnet) platform is described in [43] and its references.

Acknowledgment is given to

©2007 IEEE. Reprinted, with permission, from Wei Ren, Haiyang Chao, William Bourgeous, Nathan Sorensen, and YangQuan Chen, "Experimental validation of consensus algorithms for multi-vehicle cooperative control," *IEEE Transactions on Control Systems Technology*, 2007 (to appear).

©2007 IEEE. Reprinted, with permission, from Wei Ren, Haiyang Chao, William Bourgeous, Nathan Sorensen, and YangQuan Chen, "Experimental implementation and validation of consensus algorithms on a mobile actuator and sensor network platform," *Proceedings of the IEEE International Conference on Systems, Man, and Cybernetics*, Montreal, Canada, 2007 (to appear).

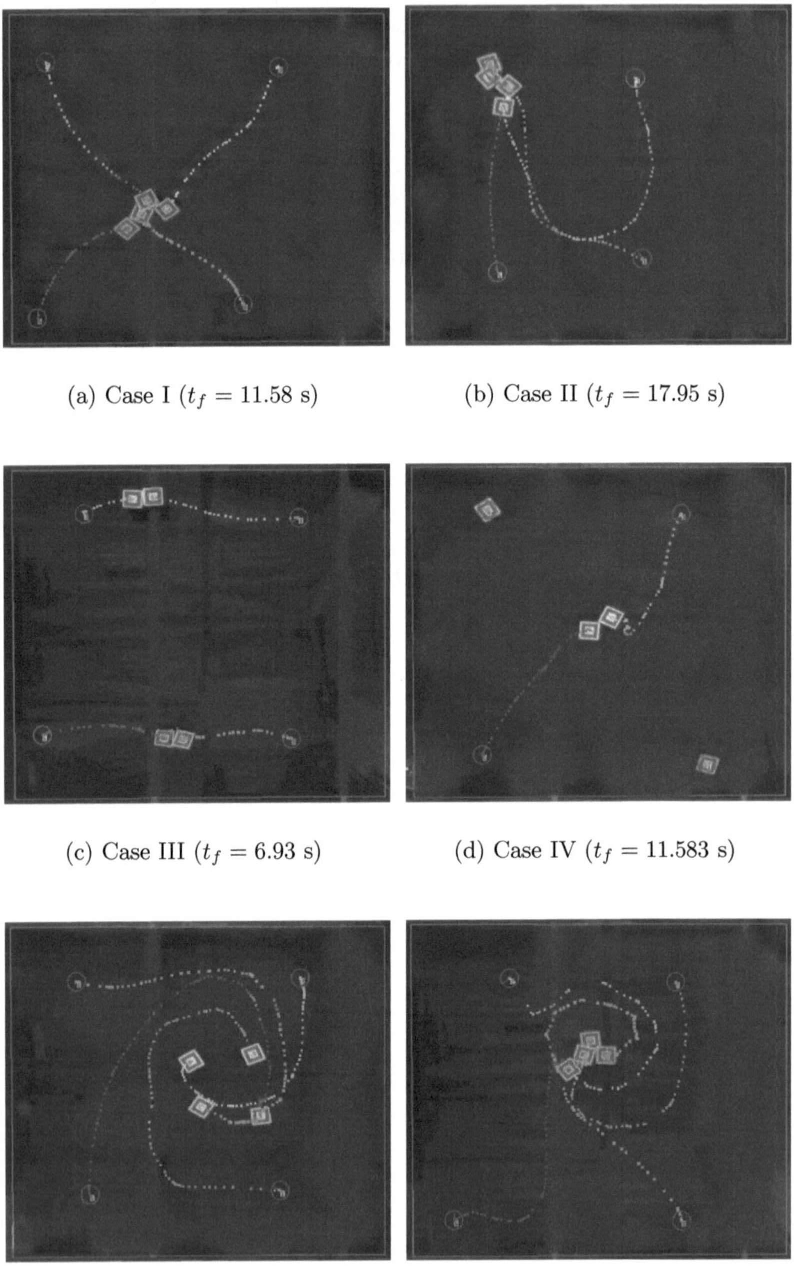

(a) Case I ($t_f = 11.58$ s)

(b) Case II ($t_f = 17.95$ s)

(c) Case III ($t_f = 6.93$ s)

(d) Case IV ($t_f = 11.583$ s)

(e) Case V ($t_f = 19.80$ s)

(f) Case VI ($t_f = 18.89$ s)

Fig. 9.3. Experimental results of rendezvous for Cases I–VI

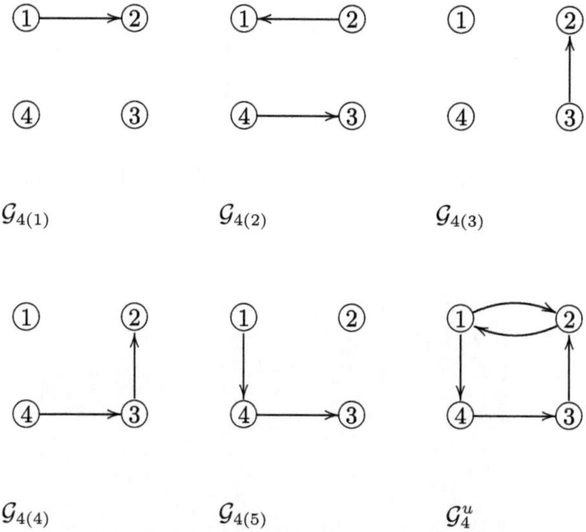

Fig. 9.4. Switching interaction topologies $\mathcal{G}_{4(1)}$–$\mathcal{G}_{4(5)}$ and their union $\mathcal{G}_4^u$ for rendezvous

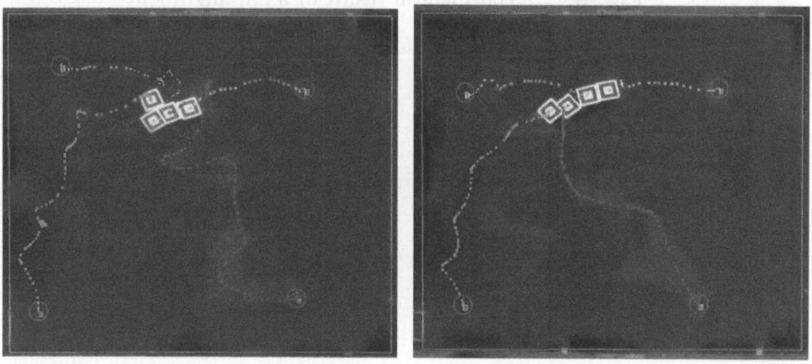

(a) Rendezvous with topologies randomly switching from $\bar{\mathcal{G}}_4^s = \{\mathcal{G}_{4(1)}, \ldots, \mathcal{G}_{4(5)}\}$ ($t_f = 45$ s)

(b) Rendezvous with a time-invariant topology $\mathcal{G}_4^u$ ($t_f = 24$ s)

Fig. 9.5. Experimental results of rendezvous with topologies randomly switching from $\bar{\mathcal{G}}_4^s$ vs a time-invariant topology $\mathcal{G}_4^u$

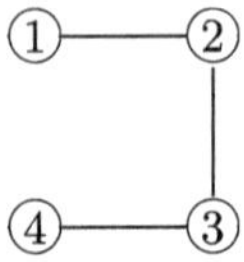

Fig. 9.6. Interaction topology for axial alignment

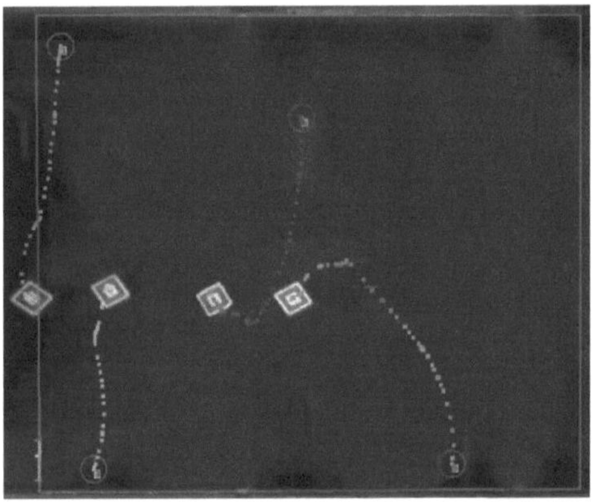

Fig. 9.7. Experimental result of axial alignment

10

Distributed Formation Control of Multiple Wheeled Mobile Robots with a Virtual Leader

In this chapter, we apply the consensus-based design scheme in Section 8.4.1 to formation control of multiple wheeled mobile robots with a virtual leader. We propose a unified, distributed formation control architecture that accommodates an arbitrary number of subgroup leaders and arbitrary information flow among the robots. The architecture requires only neighbor-to-neighbor information exchange. In particular, the consensus tracking algorithm introduced in Chapter 3 is applied on the group level to guarantee consensus on the time-varying group reference trajectory in a distributed manner. Based on the group-level consensus tracking algorithm, a consensus-based formation control strategy is then applied for vehicle level control. The proposed architecture is experimentally implemented and validated on a multirobot platform under neighbor-to-neighbor information exchange with both single and multiple subgroup leaders involved.

10.1 Distributed Formation Control Architecture

In this section, we propose a distributed formation control architecture that accommodates an arbitrary number of subgroup leaders and ensures accurate formation maintenance through information coupling between neighbors. The architecture is adapted from the design scheme in Section 8.4.1.

As described in Section 8.5.1, one solution to formation control is the virtual leader/virtual structure approach. Figure 10.1 shows an illustrative example of the virtual leader/virtual structure approach with a formation composed of four vehicles with planar motions, where C_o represents the inertial frame and C_F represents a virtual coordinate frame located at a virtual center (x_c, y_c) with an orientation θ_c relative to C_o. In Fig. 10.1, $r_j = [x_j, y_j]^T$ and $r_j^d = [x_j^d, y_j^d]^T$ represent, respectively, the jth vehicle's actual and desired position, and $r_{jF}^d = [x_{jF}^d, y_{jF}^d]^T$ represents the desired deviation of the jth vehicle relative to C_F, where

$$\begin{bmatrix} x_j^d(t) \\ y_j^d(t) \end{bmatrix} = \begin{bmatrix} x_c(t) \\ y_c(t) \end{bmatrix} + \begin{bmatrix} \cos[\theta_c(t)] & -\sin[\theta_c(t)] \\ \sin[\theta_c(t)] & \cos[\theta_c(t)] \end{bmatrix} \begin{bmatrix} x_{jF}^d(t) \\ y_{jF}^d(t) \end{bmatrix}.$$

If each vehicle can track its desired position accurately, then the desired formation shape can be preserved accurately. Note that Fig. 10.1 relies on the assumption that each vehicle knows the state of the virtual coordinate frame (*i.e.*, virtual center position and orientation), denoted as $\xi = [x_c, y_c, \theta_c]^T$. Note that here ξ is the coordination variable for the team. However, as described in Section 8.4.1, a centralized scheme is restrictive.

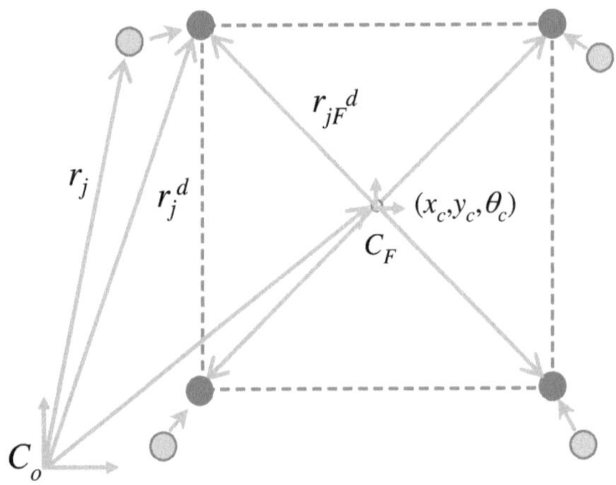

Fig. 10.1. A formation composed of four vehicles with a known virtual center

When each vehicle has inconsistent understanding or knowledge of ξ due to dynamically changing situational awareness or unreliable/limited information exchange, the desired formation geometry cannot be maintained, as shown in Fig. 10.2, where C_{Fj} represents the jth vehicle's understanding of the virtual coordinate frame with state $\xi_j = [x_{cj}, y_{cj}, \theta_{cj}]^T$. Here ξ_j denotes the jth vehicle's instantiation of the coordination variable.

Next, we propose the unified, distributed architecture for formation control that is in Fig. 10.3. The proposed architecture consists of three hierarchical layers: a consensus tracking module, a consensus-based formation control module, and the physical vehicle. In Fig. 10.3, $\mathcal{N}_i(t)$ and $\mathcal{J}_i(t)$ denote, respectively, the set of vehicles whose instantiations of the coordination variable and position tracking errors are available to vehicle i at time t. The objective of the consensus tracking module is to drive ξ_i to $\xi^r = [x_c^r, y_c^r, \theta_c^r]^T$, which represents the reference for the coordination variable (*i.e.*, the desired state of the virtual coordinate frame). As in Chapter 3, suppose that virtual vehicle $n+1$ acts as the virtual leader of the team with state $\xi_{n+1} \stackrel{\triangle}{=} \xi^r$. We use directed graph $\mathcal{G}_{n+1}^c = (\mathcal{V}_{n+1}^c, \mathcal{E}_{n+1}^c)$ to model the interaction topology for ξ_i among the $n+1$

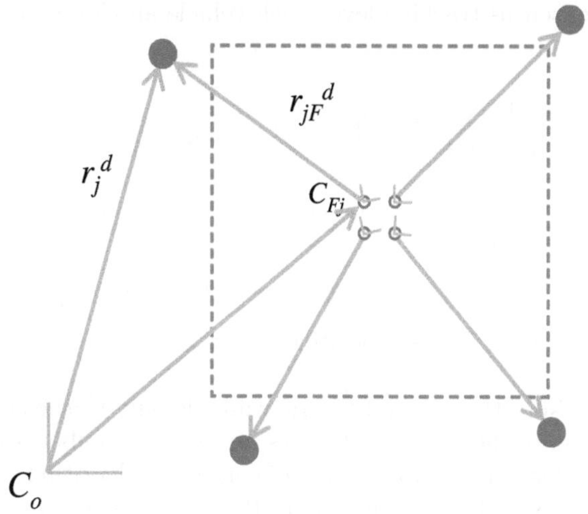

Fig. 10.2. A formation composed of four vehicles with inconsistent understanding of the virtual coordinate frame

vehicles, where $\mathcal{V}_{n+1}^c$ is the node set, $\mathcal{E}_{n+1}^c$ is the edge set, and $\mathcal{A}_{n+1}^c = [a_{ij}^c]$ is the $(n+1) \times (n+1)$ adjacency matrix. The local control law u_i, $i = 1, \ldots, n$, for each vehicle is based on its instantiation of the coordination variable and the position tracking errors of its neighbors.

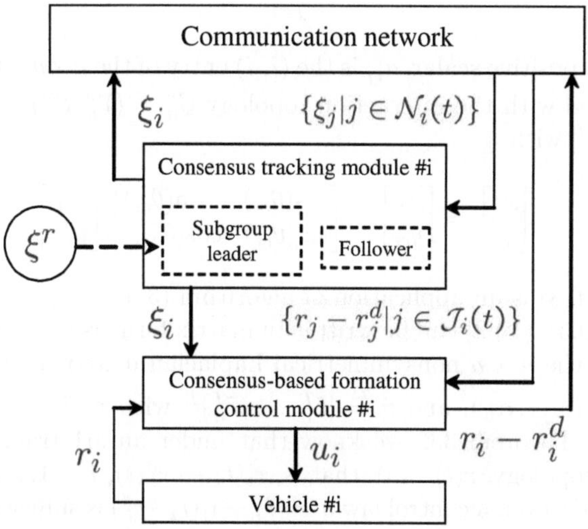

Fig. 10.3. A unified, distributed architecture for formation control

On the consensus tracking level, each vehicle applies a consensus tracking algorithm as

$$\begin{aligned} u_i =& \frac{1}{\eta_i(t)} \sum_{j=1}^{n} a_{ij}^c(t)[\dot{\xi}_j - \gamma(\xi_i - \xi_j)] \\ &+ \frac{1}{\eta_i(t)} a_{i(n+1)}^c(t)[\dot{\xi}^r - \gamma(\xi_i - \xi^r)], \quad i = 1, \ldots, n, \end{aligned} \tag{10.1}$$

where $a_{ij}^c(t)$, $i = 1, \ldots, n$, $j = 1, \ldots, n+1$, is the (i,j) entry of $\mathcal{A}_{n+1}^c \in \mathbb{R}^{(n+1)\times(n+1)}$ at time t, γ is a positive scalar, and $\eta_i(t) \stackrel{\triangle}{=} \sum_{j=1}^{n+1} a_{ij}^c(t)$.

Remark 10.1. Note that ξ^r and $\dot{\xi}^r$ are available only to the subgroup leaders and the number of the subgroup leaders can be any number from 1 to n. Note also that (10.1) is a direct application of consensus algorithm (3.4). According to Theorems 3.8 and 3.11, algorithm (10.1) guarantees that $\xi_i(t) \to \xi^r(t)$, $i = 1, \ldots, n$, as $t \to \infty$, if and only if $\mathcal{G}_{n+1}^c$ has a directed spanning tree.

Suppose that the vehicles have single-integrator dynamics given by

$$\dot{r}_i = u_i, \quad i = 1, \ldots, n, \tag{10.2}$$

where $r_i \in \mathbb{R}^m$ is the state and $u_i \in \mathbb{R}^m$ is the control input of the ith vehicle.

On the vehicle control level, we apply a consensus algorithm as

$$u_i = \dot{r}_i^d - \alpha_i(r_i - r_i^d) - \sum_{j=1}^{n} a_{ij}^v[(r_i - r_i^d) - (r_j - r_j^d)], \tag{10.3}$$

where α_i is a positive scalar, a_{ij}^v is the (i,j) entry of the $n \times n$ adjacency matrix $\mathcal{A}_n^v$ associated with the interaction topology $\mathcal{G}_n^v = (\mathcal{V}_n^v, \mathcal{E}_n^v)$ for $r_i - r_i^d$,[1] and $r_i^d = [x_i^d, y_i^d]^T$ with

$$\begin{bmatrix} x_i^d \\ y_i^d \end{bmatrix} = \begin{bmatrix} x_{ci} \\ y_{ci} \end{bmatrix} + \begin{bmatrix} \cos(\theta_{ci}) & -\sin(\theta_{ci}) \\ \sin(\theta_{ci}) & \cos(\theta_{ci}) \end{bmatrix} \begin{bmatrix} x_{iF}^d \\ y_{iF}^d \end{bmatrix}.$$

Note that (10.3) is an application of algorithm (3.3).

Using (10.3), (10.2) can be written in matrix form as $\dot{\tilde{r}} = -[(\mathcal{L}_n^v + Q) \otimes I_2]\tilde{r}$, where $\mathcal{L}_n^v$ is the $n \times n$ nonsymmetrical Laplacian matrix associated with $\mathcal{G}_n^v$, $Q \stackrel{\triangle}{=} \mathrm{diag}(\alpha_1, \ldots, \alpha_n)$, and $\tilde{r} \stackrel{\triangle}{=} [\tilde{r}_1^T, \ldots, \tilde{r}_n^T]^T$ with $\tilde{r}_i \stackrel{\triangle}{=} r_i - r_i^d$. Following the proof of Theorem 3.6, we know that under an arbitrary time-invariant interaction topology, $\tilde{r}(t) \to 0$, that is, $r_i(t) \to r_i^d(t)$, $i = 1, \ldots, n$, as $t \to \infty$. In other words, even a control law $u_i = \dot{r}_i^d - \alpha_i(r_i - r_i^d)$ is sufficient to guarantee that $r_i(t) \to r_i^d(t)$, $i = 1, \ldots, n$, as $t \to \infty$. However, the coupling between

[1] Note that the interaction topology on the consensus tracking level given by $\mathcal{G}_{n+1}^c$ is different from the interaction topology on the vehicle control level given by $\mathcal{G}_n^v$.

neighbors induced by the third term in (10.3) improves group robustness and reduces formation maintenance error.

With both (10.1) and (10.3), if $\xi_i(t) \to \xi^r(t)$ and $r_i(t) \to r_i^d(t)$, $i = 1, \ldots, n$, as $t \to \infty$, then the desired formation shape is maintained and the state of the virtual coordinate frame follows the desired reference.

Remark 10.2. Note that both (10.1) and (10.3) are distributed in the sense that only information exchange among neighbors is required. The architecture shown in Fig. 10.3 accommodates an arbitrary number of subgroup leaders and arbitrary coupling among vehicles on both the consensus tracking level and the vehicle control level. The distributed nature of (10.1) and (10.3) ensures robustness of the team to failure of follower vehicles. The introduction of multiple subgroup leaders does not make the control algorithms (10.1) and (10.3) or their convergence analysis more complicated. Also the introduction of multiple subgroup leaders reduces the risk of a single point failure that is inherent in a team with a single subgroup leader. In addition, with (10.1) and (10.3), each vehicle simply exchanges information with its neighbors without the need to identify the subgroup leaders. □

10.2 Experimental Results on a Multirobot Platform

In this section, we validate the proposed distributed formation control architecture on a multirobot experimental platform. We conduct experiments with, respectively, a single subgroup leader, multiple subgroup leaders, and directed switching interaction topologies.

10.2.1 Experimental Platform and Implementation

Experimental tests are conducted in the Mobile Robots Laboratory at USU. The mobile robot test bed consists of five AmigoBots and two Pioneer 3-DX shown in Fig. 10.4. The robots can communicate with each other through ethernet with TCP/IP protocols. The robots rely on encoder data for their position and orientation information.

In our experiments, we emulate limited interrobot information exchange by simply disallowing the use of information obtained from certain members of the group, although every robot can share information with every other robot. By doing so, we can test distributed cooperative control algorithms that involve only neighbor-to-neighbor information exchange due to limited communication or sensing.

The fact that we use differentially driven mobile robots complicates the coordination problem due to nonholonomic constraints. In the well-known paper by Brockett [33], it was shown that nonholonomic systems cannot be stabilized with continuous static state feedback. The implication for differentially driven mobile robots is that the position and orientation of the center

Fig. 10.4. Multirobot experimental platform at USU

of the robot cannot be simultaneously stabilized with a time-invariant, stabilizing control strategy. Both discontinuous control laws [14, 54] and time varying [164, 206] control laws have been found to stabilize the center of rotation and the orientation of a single robot. The multiple robot case is naturally more complex.

Define the "hand" position of the robot as the point $h \triangleq [h_x, h_y]^T$ that lies a distance L along the line that is normal to the wheel axis and intersects the wheel axis at the center point $r \triangleq [r_x, r_y]^T$, as shown in Fig. 10.5. The kinematics of the hand position are holonomic for $L \neq 0$. In our experiments, we consider the problem of coordinating the hand positions of the robots instead of coordinating their center positions. Although this assumption simplifies the control problem, it is of practical interest because the hand position may be the point of interest. For example, if the robots are equipped with a gripper located at the hand position and the coordination task is to move an object from one location to another, then the objective is to move the gripper locations in a coordinated fashion. Another example occurs when the group objective is the coordinated placement of sensors that are located at the hand position.

Let (r_{xi}, r_{yi}), θ_i, and (v_i, ω_i) denote the inertial position, orientation, and linear and angular speeds of the ith robot, respectively. The kinematic equations for the ith robot are

$$\dot{r}_{xi} = v_i \cos(\theta_i), \quad \dot{r}_{yi} = v_i \sin(\theta_i), \quad \dot{\theta}_i = \omega_i. \tag{10.4}$$

The hand position is given by

$$\begin{bmatrix} h_{xi} \\ h_{yi} \end{bmatrix} = \begin{bmatrix} r_{xi} \\ r_{yi} \end{bmatrix} + L_i \begin{bmatrix} \cos(\theta_i) \\ \sin(\theta_i) \end{bmatrix}. \tag{10.5}$$

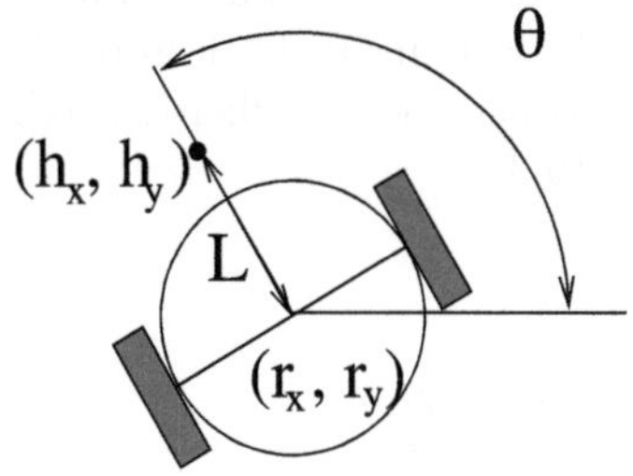

Fig. 10.5. A nonholonomic differentially driven wheeled mobile robot

Differentiating (10.5) with respect to time gives

$$\begin{bmatrix} \dot{h}_{xi} \\ \dot{h}_{yi} \end{bmatrix} = \begin{bmatrix} \cos(\theta_i) & -L_i \sin(\theta_i) \\ \sin(\theta_i) & L_i \cos(\theta_i) \end{bmatrix} \begin{bmatrix} v_i \\ \omega_i \end{bmatrix}. \tag{10.6}$$

Letting

$$\begin{bmatrix} v_i \\ \omega_i \end{bmatrix} = \begin{bmatrix} \cos(\theta_i) & \sin(\theta_i) \\ -\frac{1}{L_i}\sin(\theta_i) & \frac{1}{L_i}\cos(\theta_i) \end{bmatrix} \begin{bmatrix} u_{xi} \\ u_{yi} \end{bmatrix},$$

gives

$$\begin{bmatrix} \dot{h}_{xi} \\ \dot{h}_{yi} \end{bmatrix} = \begin{bmatrix} u_{xi} \\ u_{yi} \end{bmatrix}, \tag{10.7}$$

which is a simplified kinematic equation but is sufficient for the purpose of our application. Note that (10.7) takes in the form of single-integrator dynamics, implying that consensus algorithm (10.3) can be directly applied.

In our experiments, a team of four AmigoBots is required to maintain a square formation, and the virtual coordinate frame located at the center of the square follows a circle moving in a clockwise direction. In particular, the reference state of the virtual coordinate frame $\xi^r = [x_c^r, y_c^r, \theta_c^r]^T$ satisfies

$$\dot{x}_c^r = v_c^r \cos(\theta_c^r), \quad \dot{y}_c^r = v_c^r \sin(\theta_c^r), \quad \dot{\theta}_c^r = \omega_c^r,$$

where $v_c^r = \frac{9\pi}{500}$ m/s, $\omega_c^r = \frac{\pi}{50}$ rad/s, $[x_c^r(0), y_c^r(0)] = (0, 0)$ m, and $\theta_c^r(0) = 0$ rad. In addition, we let $L_i = 0.15$ m, $x_{jF}^d = \ell_j \cos(\phi_j)$, and $y_{jF}^d = \ell_j \sin(\phi_j)$, where $\ell_j = 0.6$ m and $\phi_j = \pi - \frac{\pi}{4} j$ rad, $j = 1, \ldots, 4$. Each robot applies (10.1) to achieves consensus tracking of the reference state ξ^r and (10.3) to compute u_{xi} and u_{yi}.

10.2.2 Formation Control with a Single Subgroup Leader

In this subsection, we consider a single subgroup leader with three followers. Figure 10.6a shows the interaction topology on the consensus tracking level, where node ξ^r denotes the virtual leader of the team, subscript L denotes a subgroup leader, subscript F denotes a follower, and an edge from node j to node i implies that $a^c_{ij} = 1$ in (10.1). Figure 10.6b shows the interaction topology on the vehicle control level, where an edge from node j to node i implies that $a^v_{ij} = 1$ in (10.3).

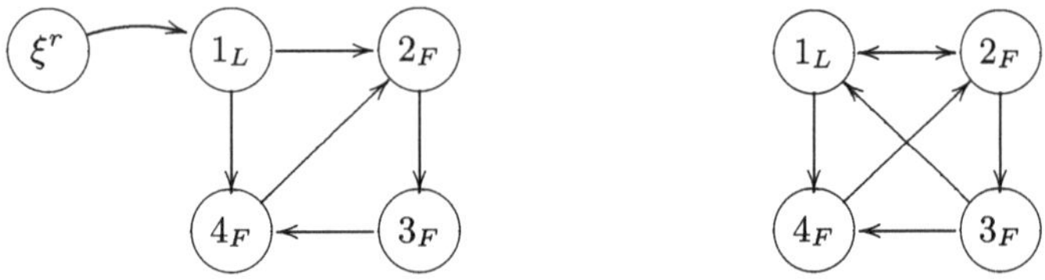

(a) Consensus tracking level topology

(b) Vehicle control level topology

Fig. 10.6. Interaction topologies with a single subgroup leader and three followers

Figure 10.7 shows the experimental result with a single subgroup leader and three followers. In particular, Fig. 10.7a shows the trajectories of the four robots at $t \in [0, t_f]$ s and snapshots at $t = 0, \frac{t_f}{3}, \frac{2t_f}{3}$ s, where t_f is the end of the experiment. Figure 10.7b shows the relative position errors, defined as the difference between the desired and actual separation distance between the robots. Figure 10.7c shows the consensus tracking errors for the virtual center position, defined as $\sqrt{(x^r_c - x_{ci})^2 + (y^r_c - y_{ci})^2}$, where (x^r_c, y^r_c) is the reference virtual center position known by the subgroup leader. Figure 10.7d shows the consensus tracking errors for the virtual center orientation, defined as $\theta^r_c - \theta_{ci}$, where θ^r_c is the reference virtual center orientation known by the subgroup leader. Note that the team is able to travel in tight formation around the circle as shown in Fig. 10.7a with relative position errors between -3 cm and 5 cm as shown in Fig. 10.7b. Also note that the consensus tracking errors for the virtual center position are below 1.6 cm, as shown in Fig. 10.7c, and the consensus tracking errors for the virtual center orientation are between 0.06° and 0.08°, as shown in Fig. 10.7d.

10.2.3 Formation Control with Multiple Subgroup Leaders

In this subsection, we consider two subgroup leaders with two followers. Figures 10.8a and 10.8b show the interaction topologies on the consensus tracking level and the vehicle control level, respectively. The experimental result is shown in Fig. 10.9. Note that good formation maintenance is also achieved with multiple subgroup leaders involved. The introduction of multiple subgroup leaders increases the robustness of the whole team in the case of failure of a certain subgroup leader. In addition, compared to the single subgroup

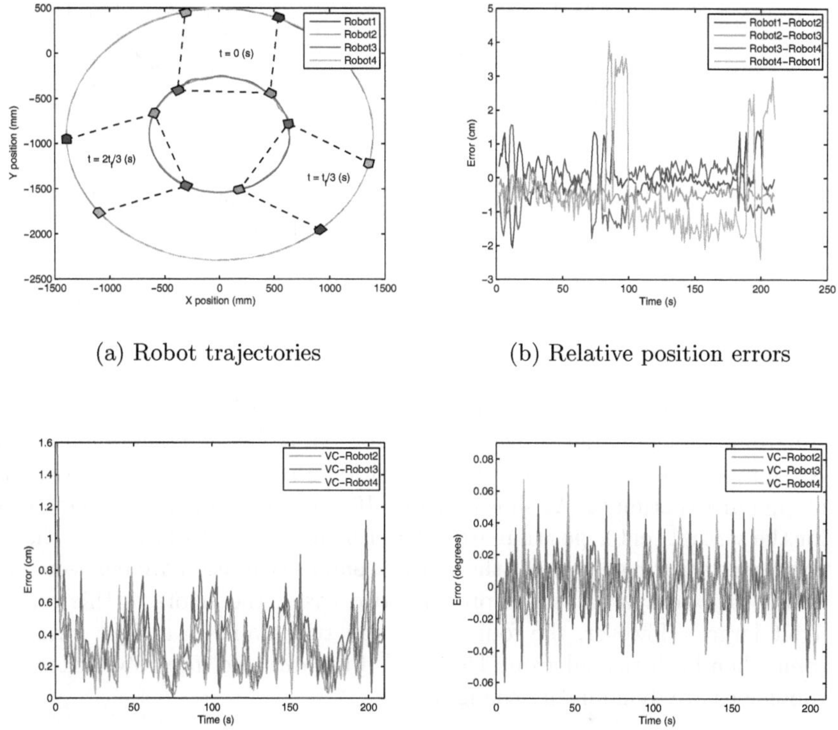

(a) Robot trajectories

(b) Relative position errors

(c) Consensus tracking errors for the virtual center position

(d) Consensus tracking errors for the virtual center orientation

Fig. 10.7. Experimental result of a square formation with a single subgroup leader and three followers

leader case, the existence of multiple subgroup leaders reduces the consensus tracking errors because more vehicles have direct access to the reference state.

10.2.4 Formation Control with Dynamically Changing Subgroup Leaders and Interrobot Interaction Topologies

In this subsection, we consider the case where the subgroup leaders are dynamically changing and the directed interrobot interaction topologies are also dynamically changing. We assume that directed graph $\mathcal{G}_5^c(t)$ switches randomly from the set $\{\mathcal{G}_{5(1)}^c, \ldots, \mathcal{G}_{5(6)}^c\}$, as shown in Fig. 10.10, with a switching time of approximately 10 seconds. Note that each $\mathcal{G}_{5(i)}^c$, $i = 1, \ldots, 6$, has a directed spanning tree with node ξ^r as the root. We assume that the directed graph $\mathcal{G}_4^v(t)$ is the same as $\mathcal{G}_5^c(t)$ with node ξ^r removed.

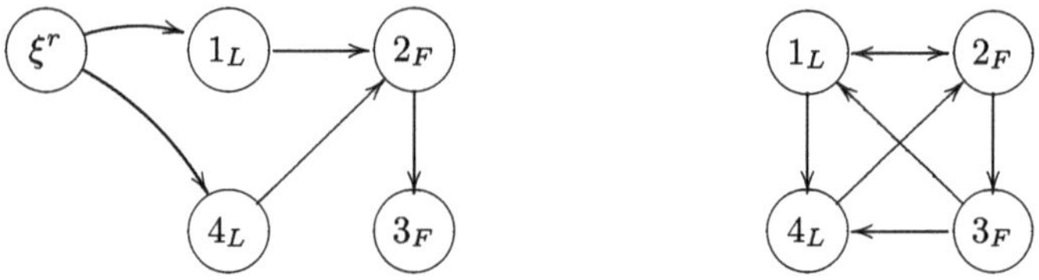

(a) Consensus tracking level topology

(b) Vehicle control level topology

Fig. 10.8. Interaction topologies with two subgroup leaders and two followers

Figure 10.11 shows the experimental result with dynamically changing subgroup leaders under directed switching interrobot interaction topologies. As shown in Fig. 10.11b, the relative position errors are between −6 cm and 5 cm in our experiment. As shown in Fig. 10.11c, the consensus tracking errors for the virtual center position are below 1.6 cm. Figure 10.11d shows that the consensus tracking errors for the virtual center orientation are between −0.1° and 0.15°. Compared to the previous two cases, the errors in this case are a bit larger. However, the four robots can still maintain a tight formation, even when both the subgroup leaders and the directed interrobot interaction topologies are dynamically changing.

10.3 Notes

The results in this chapter are based mainly on [197, 222].

Acknowledgment is given to

©2007 IEEE. Reprinted, with permission, from Nathan Sorensen and Wei Ren, "A unified formation control scheme with a single or multiple leaders," *Proceedings of the American Control Conference*, pp. 5412–5418, New York City, 2007.

©2007 Elsevier. Reprinted, with permission, from Wei Ren and Nathan Sorensen, "Distributed coordination architecture for multi-robot formation control," *Robotics and Autonomous Systems*, 2007 (to appear).

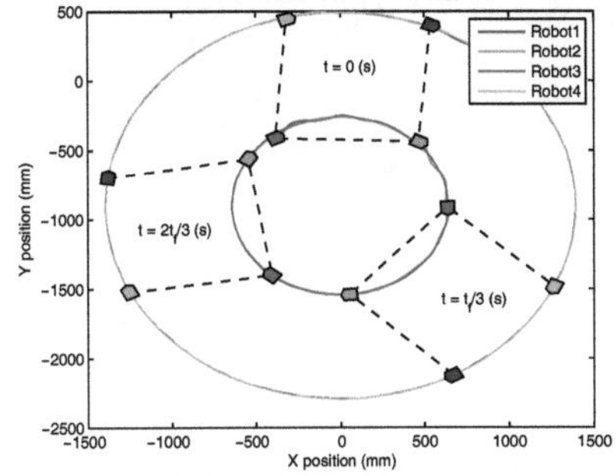

(a) Robot trajectories

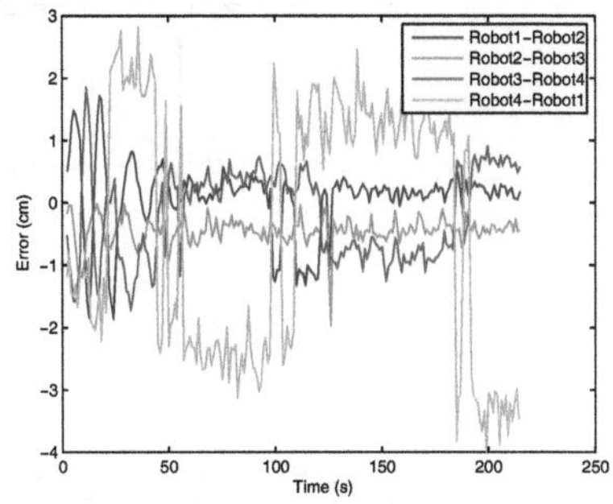

(b) Relative position errors

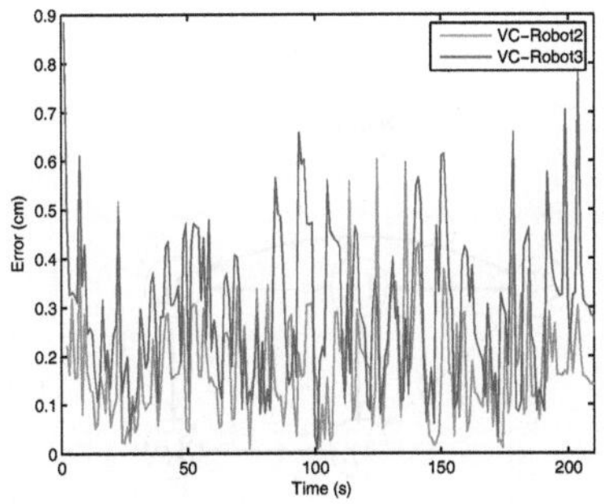

(c) Consensus tracking errors for the virtual center position

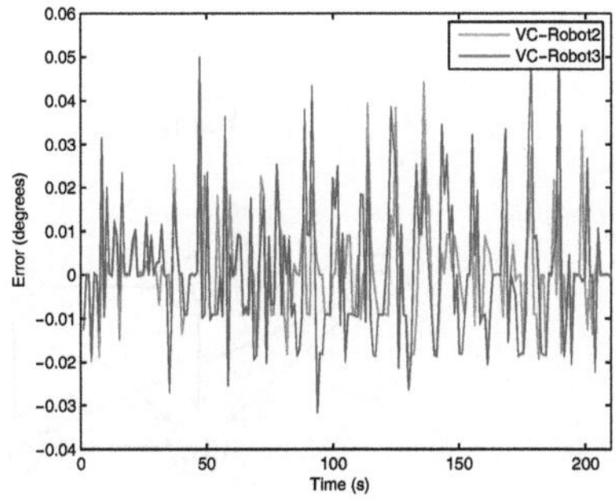

(d) Consensus tracking errors for the virtual center orientation

Fig. 10.9. Experimental result of a square formation with two subgroup leaders and two followers

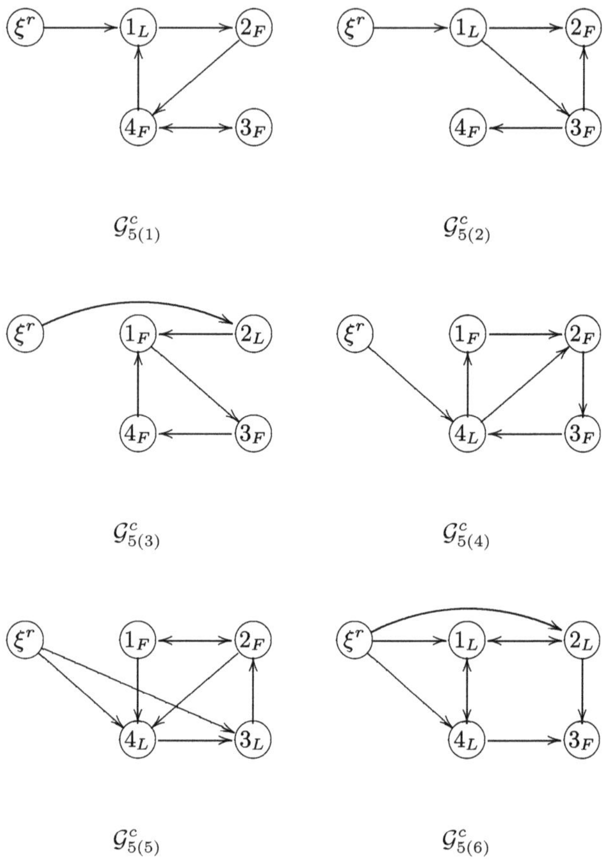

Fig. 10.10. Directed switching graphs of $\mathcal{G}^c_5(t)$

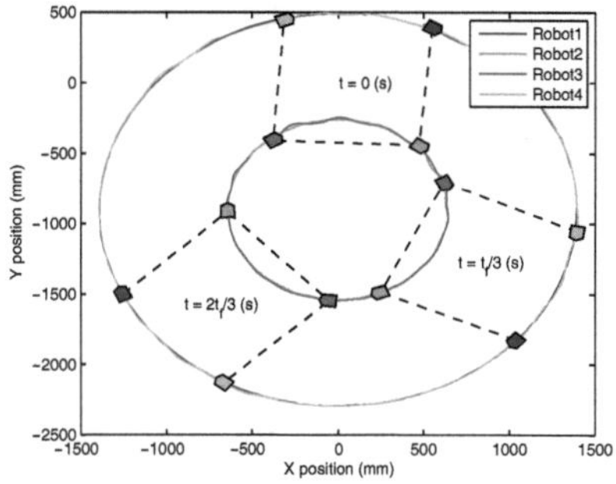

(a) Trajectories of the four robots

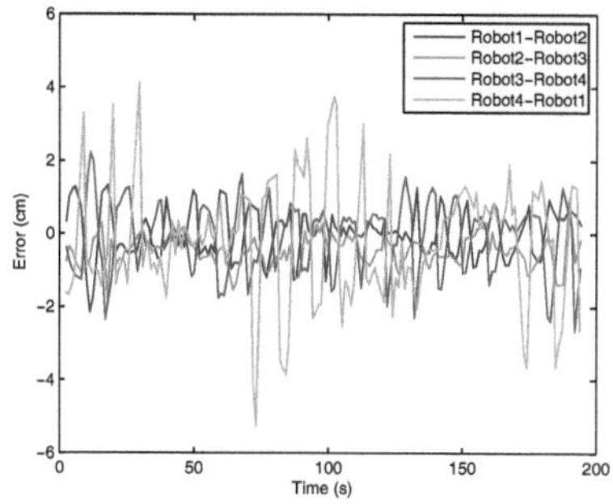

(b) Relative position errors

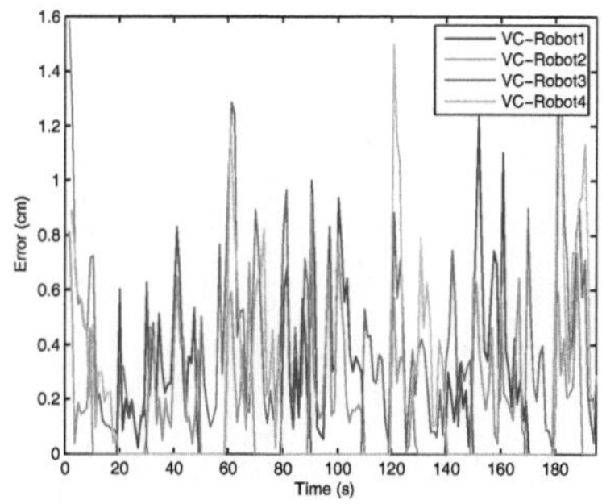

(c) Consensus tracking errors for the virtual center position

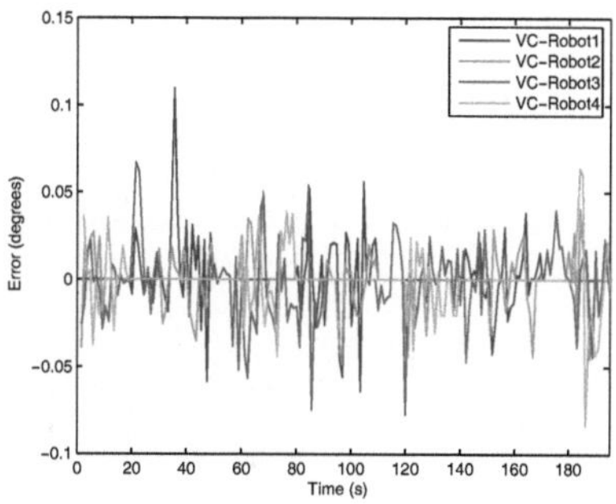

(d) Consensus tracking errors for the virtual center orientation

Fig. 10.11. Experimental result of a square formation with dynamically changing subgroup leaders under directed switching interrobot interaction topologies

11

Decentralized Behavioral Approach to Wheeled Mobile Robot Formation Maneuvers

In this chapter, we apply the design scheme in Section 8.4.2 to formation maneuvers for a group of wheeled mobile robots. Complex formation maneuvers are decomposed into a sequence of maneuvers between formation patterns. We present three decentralized behavior-based formation control strategies. The first strategy uses relative position information configured in a bidirectional ring topology to maintain the formation. The second strategy injects interrobot damping *via* passivity techniques. The third strategy accounts for actuator saturation. Hardware results on differentially driven wheeled mobile robots demonstrate the effectiveness of the proposed control strategies.

11.1 Problem Statement

The objective of this chapter is to introduce the coupled dynamics approach to formation control for a group of wheeled mobile robots, which is a behavior-based strategy. The approach is a direct application of the design scheme in Section 8.4.2. The chapter provides a rigorous analysis of formation keeping and convergence which is a contribution to behavior-based literature. Furthermore, our approach has the advantage that it can be implemented when only neighbor position information is available (*i.e.*, velocity information is not required) thereby reducing the communication overhead.

In Chapter 10, we consider kinematic models of mobile robots. In contrast, we will consider dynamic models in this chapter. Each robot has the following dynamic equations of motion:

$$\begin{bmatrix} \dot{r}_{xi} \\ \dot{r}_{yi} \\ \dot{\theta}_i \\ \dot{v}_i \\ \dot{\omega}_i \end{bmatrix} = \begin{bmatrix} v_i \cos(\theta_i) \\ v_i \sin(\theta_i) \\ \omega_i \\ 0 \\ 0 \end{bmatrix} + \begin{bmatrix} 0 & 0 \\ 0 & 0 \\ 0 & 0 \\ \frac{1}{m_i} & 0 \\ 0 & \frac{1}{J_i} \end{bmatrix} \begin{bmatrix} F_i \\ \tau_i \end{bmatrix}, \tag{11.1}$$

where $r_i = [r_{xi}, r_{yi}]^T$ is the inertial position of the ith robot, θ_i is the orientation, v_i is the linear speed, ω_i is the angular speed, τ_i is the applied torque, F_i is the applied force, m_i is the mass, and J_i is the moment of inertia. Letting $x_i \triangleq [r_{xi}, r_{yi}, \theta_i, v_i, \omega_i]^T$ and $u_i \triangleq [F_i, \tau_i]^T$, the equations of motion can be written as

$$\dot{x}_i = f(x_i) + g_i u_i, \tag{11.2}$$

where the definitions of f and g_i can be inferred from (11.1).

As in Chapter 10, we will focus on formation control of the robot hand positions $h_i \triangleq [h_{xi}, h_{yi}]^T$ defined by (10.5). The derivative of h_i is given by (10.6). The second-order derivative of h_i is given by

$$\ddot{h}_i = \begin{bmatrix} -v_i\omega_i \sin(\theta_i) - L_i\omega_i^2 \cos(\theta_i) \\ v_i\omega_i \cos(\theta_i) - L_i\omega_i^2 \sin(\theta_i) \end{bmatrix} + \begin{bmatrix} \frac{1}{m_i}\cos(\theta_i) & -\frac{L_i}{J_i}\sin(\theta_i) \\ \frac{1}{m_i}\sin(\theta_i) & \frac{L_i}{J_i}\cos(\theta_i) \end{bmatrix} \begin{bmatrix} F_i \\ \tau_i \end{bmatrix}.$$

Because

$$\det \begin{bmatrix} \frac{1}{m_i}\cos(\theta_i) & -\frac{L_i}{J_i}\sin(\theta_i) \\ \frac{1}{m_i}\sin(\theta_i) & \frac{L_i}{J_i}\cos(\theta_i) \end{bmatrix} = \frac{L_i}{m_i J_i} \neq 0,$$

system (11.2) with output (10.5) has a constant relative degree equal to two and can therefore be output feedback linearized [96] about the hand position. Toward that end, define the map $\psi : \mathbb{R}^5 \to \mathbb{R}^5$ as

$$\zeta_i = \psi(x_i) \triangleq \begin{bmatrix} r_{xi} + L_i\cos(\theta_i) \\ r_{yi} + L_i\sin(\theta_i) \\ v_i\cos(\theta_i) - L_i\omega_i\sin(\theta_i) \\ v_i\sin(\theta_i) + L_i\omega_i\cos(\theta_i) \\ \theta_i \end{bmatrix}. \tag{11.3}$$

The map ψ is a diffeomorphism whose inverse is given by

$$x_i = \psi^{-1}(\zeta_i) = \begin{bmatrix} \zeta_{1i} - L_i\cos(\zeta_{5i}) \\ \zeta_{2i} - L_i\sin(\zeta_{5i}) \\ \zeta_{5i} \\ \frac{1}{2}\zeta_{3i}\cos(\zeta_{5i}) + \frac{1}{2}\zeta_{4i}\sin(\zeta_{5i}) \\ -\frac{1}{2L_i}\zeta_{3i}\sin(\zeta_{5i}) + \frac{1}{2L_i}\zeta_{4i}\cos(\zeta_{5i}) \end{bmatrix}.$$

In transformed coordinates, (11.2) and (10.5) are given by

$$\begin{aligned}
\begin{bmatrix} \dot{\zeta}_{1i} \\ \dot{\zeta}_{2i} \end{bmatrix} &= \begin{bmatrix} \zeta_{3i} \\ \zeta_{4i} \end{bmatrix}, \\
\begin{bmatrix} \dot{\zeta}_{3i} \\ \dot{\zeta}_{4i} \end{bmatrix} &= \begin{bmatrix} -v_i\omega_i \sin(\theta_i) - L_i\omega_i^2 \cos(\theta_i) \\ v_i\omega_i \cos(\theta_i) - L_i\omega_i^2 \sin(\theta_i) \end{bmatrix} + \begin{bmatrix} \frac{1}{m_i}\cos(\theta_i) & -\frac{L_i}{J_i}\sin(\theta_i) \\ \frac{1}{m_i}\sin(\theta_i) & \frac{L_i}{J_i}\cos(\theta_i) \end{bmatrix} u_i, \\
\dot{\zeta}_{5i} &= -\frac{1}{2L_i}\zeta_{3i}\sin(\zeta_{5i}) + \frac{1}{2L_i}\zeta_{4i}\cos(\zeta_{5i}).
\end{aligned}$$

The output feedback linearizing control [96] is given by

$$u_i = \begin{bmatrix} \frac{1}{m_i}\cos(\theta_i) & -\frac{L_i}{J_i}\sin(\theta_i) \\ \frac{1}{m_i}\sin(\theta_i) & \frac{L_i}{J_i}\cos(\theta_i) \end{bmatrix}^{-1} \left\{ \nu_i - \begin{bmatrix} -v_i\omega_i\sin(\theta_i) - L_i\omega_i^2\cos(\theta_i) \\ v_i\omega_i\cos(\theta_i) - L_i\omega_i^2\sin(\theta_i) \end{bmatrix} \right\}, \tag{11.4}$$

which gives

$$\begin{aligned} \begin{bmatrix} \dot{\zeta}_{1i} \\ \dot{\zeta}_{2i} \end{bmatrix} &= \begin{bmatrix} \zeta_{3i} \\ \zeta_{4i} \end{bmatrix}, \\ \begin{bmatrix} \dot{\zeta}_{3i} \\ \dot{\zeta}_{4i} \end{bmatrix} &= \nu_i, \\ \dot{\zeta}_{5i} &= -\frac{1}{2L_i}\zeta_{3i}\sin(\zeta_{5i}) + \frac{1}{2L_i}\zeta_{4i}\cos(\zeta_{5i}), \\ h_i &= \begin{bmatrix} \zeta_{1i} \\ \zeta_{2i} \end{bmatrix}. \end{aligned}$$

The last equation represents the internal dynamics which are rendered unobservable and uncontrollable by transformation (11.3). Zero dynamics [96] are found by setting $\zeta_{1i} = \cdots = \zeta_{4i} = 0$ to get $\dot{\zeta}_{5i} = 0$. Therefore the zero dynamics are stable, but not asymptotically stable. Because $\zeta_{5i} = \theta_i$ and $[\zeta_{3i}, \zeta_{4i}]^T$ represent the velocity of the hand position, this implies that the angle θ_i will stop moving when the hand position stops moving.

In the remainder of the chapter, the input-output dynamics of each robot will be represented by the double-integrator system,

$$\ddot{h}_i = \nu_i. \tag{11.5}$$

Remark 11.1. Feedback linearization about the hand position was used in [164] for kinematic models. This section has extended the approach to dynamic models and has explicitly defined the internal dynamics. Note that (11.5) takes in the form of double-integrator dynamics, implying that the consensus algorithms in Chapters 4 and 5 can be applied directly. □

11.2 Formation Maneuvers

In this section, we describe the types of formation maneuvers that will be considered in this chapter. Let n be the number of mobile robots in the formation. A *formation pattern* is defined as a set

$$\mathcal{P} = \{h_1^d, \ldots, h_n^d\},$$

where h_i^d is the desired constant location of the hand position of the ith robot. We will consider the class of formation control problems where the group of robots is required to transition through a sequence of formation patterns $\mathcal{P}_j$,

$j = 1, \ldots, J$, where we assume that the sequence of formation patterns is designed to avoid robot collisions. We assume that during the transition from one formation pattern to another, it is desirable to maintain the robots in the same shape as the destination pattern.

There are two competing objectives. The first objective is to move the robots to their final destination, as specified in the formation pattern. The second objective is to maintain formation during the transition. Consider a simple translation, as shown in Fig. 11.1. The left triangle represents the formation pattern at the start of the maneuver, and the right triangle represents the desired formation pattern. Suppose that initially the bottom right robot leads the formation, as shown in Fig. 11.1. The robot has two conflicting objectives: move right to arrive at the final goal, and move left to regain formation. If it moves left, it will likely overshoot the formation, which is moving to the right, and if it moves right, it will take longer to regain formation as the others are required to catch up.

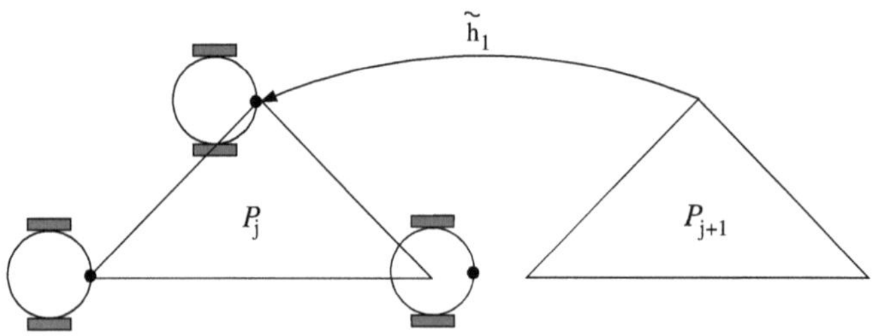

Fig. 11.1. Bottom right member of the formation is initially too far to the right

To incorporate these two competing objectives, we will define error functions for both. Let E_g be the total error between the current positions of the robots and the desired formation pattern:

$$E_g = \sum_{i=1}^{n} \tilde{h}_i^T K_g \tilde{h}_i,$$

where K_g is a symmetrical positive-definite matrix and $\tilde{h}_i \triangleq h_i - h_i^d$ (see Fig. 11.1). Similarly, define E_f as the formation error

$$E_f = \sum_{i=<n>} (\tilde{h}_i - \tilde{h}_{i+1})^T K_f (\tilde{h}_i - \tilde{h}_{i+1}),$$

where K_f is a symmetrical positive-semidefinite matrix and where the robot index is defined modulo n, *i.e.*, $\tilde{h}_{n+1} = \tilde{h}_1$ and $\tilde{h}_0 = \tilde{h}_n$. The notation $i =< n >$ is used to indicate summation around the ring defined by the formation pattern. By maintaining E_f small during the maneuver, the robots will equalize the distance that they need to go to reach the final formation pattern. Note that $E_f = 0$ if and only if $\tilde{h}_i = \tilde{h}_{i+1}$ for all i. This is equivalent to saying that $h_i - h_{i+1} = h_i^d - h_{i+1}^d$, which will be true only if the robots are in the same relative formation that they will have at the end of the maneuver. Therefore, when $E_f = 0$, the robots will be keeping formation, but they will not necessarily be in the final formation pattern.

The total error for the formation coordination problem is the sum of E_g and E_f:

$$\begin{aligned} E(t) &= E_f(t) + E_g(t) \\ &= \sum_{i=<n>} \left[\tilde{h}_i^T K_g \tilde{h}_i + (\tilde{h}_i - \tilde{h}_{i+1})^T K_f (\tilde{h}_i - \tilde{h}_{i+1}) \right], \end{aligned} \quad (11.6)$$

where K_f and K_g weigh the relative importance of formation keeping versus goal convergence. The formation control objective is to drive $E(t) \to 0$, as $t \to \infty$.

11.3 Formation Control

In this section, we propose three control strategies for driving $E(t)$ defined in (11.6) to zero, given dynamics (11.5). The first approach is *coupled dynamics formation control* derived in Section 11.3.1. The coupled dynamics approach couples the dynamics of the robots by incorporating relative position and velocity information between neighbors in the control strategy. This approach requires that each robot knows the relative position and velocity of two other robots (its neighbors in the communication ring), as well as their desired positions in the target formation pattern. The second approach, derived in Section 11.3.2, is *coupled dynamics formation control with passivity based interrobot damping.* This formation control strategy is identical to the coupled dynamics approach except that the requirement for relative velocity is removed. The third approach, derived in Section 11.3.3, is the *coupled dynamics approach with saturated control.* This strategy modifies the coupled dynamics approach so that convergence of E is guaranteed under actuator saturation constraints.

11.3.1 Coupled Dynamics Formation Control

In this section, we derive the coupled dynamics formation control strategy. The proposed control law is given by

$$\begin{aligned} \nu_i = &- K_g \tilde{h}_i - D_g \dot{h}_i \\ &- K_f(\tilde{h}_i - \tilde{h}_{i-1}) - D_f(\dot{h}_i - \dot{h}_{i-1}) \\ &- K_f(\tilde{h}_i - \tilde{h}_{i+1}) - D_f(\dot{h}_i - \dot{h}_{i+1}), \end{aligned} \quad (11.7)$$

where K_g and D_g are symmetrical positive-definite matrices, and K_f and D_f are symmetrical positive-semidefinite matrices. Note that (11.7) is an application of consensus algorithm (5.10), where the interaction topology is a bidirectional ring.

Remark 11.2. The first two terms in (11.7) drive the robot to its final position in the formation pattern (goal seeking behavior). The second two terms maintain formation with the $i-1$ robot, and the last two terms maintain formation with the $i+1$ robot (formation keeping behavior). □

Theorem 11.3. *If robot formation* (11.1) *is subject to the control strategy defined in* (11.4) *and* (11.7), *then error function* (11.6) *converges to zero asymptotically.*

Furthermore, if the formation is initially at rest, i.e., $\dot{h}(0) = 0$, *where* $\dot{h} \triangleq [\dot{h}_1^T, \dots, \dot{h}_n^T]^T$, *then the formation error is bounded by*

$$E(t) \leq E(0) - \sum_{i=1}^{n} \dot{h}_i^T \dot{h}_i. \tag{11.8}$$

The proof of this theorem and Theorem 11.7 are simplified by the use of Kronecker product notation. The following lemma serves to establish our notation and makes the derivations in the proofs of Theorems 11.3 and 11.7 more transparent.

Lemma 11.4. *Let C be the circulant matrix with the first row given by*

$$[2, -1, 0, \dots, 0, -1] \in \mathbb{R}^n.$$

Then $C \in \mathbb{R}^{n\times n}$ is symmetrical positive semidefinite. If $\xi = [\xi_1^T, \dots, \xi_n^T]^T$, where $\xi_i \in \mathbb{R}^p$, then

$$\sum_{i=<n>} (\xi_i - \xi_{i+1})^T J(\xi_i - \xi_{i+1}) = \xi^T (C \otimes J)\xi.$$

If the terms $J(\xi_i - \xi_{i-1}) + J(\xi_i - \xi_{i+1})$ are stacked in a column vector, the resulting vector can be written as $(C \otimes J)\xi$.

Proof: The first part is verified by noting that C is a (symmetrical) Laplacian matrix associated with a bidirectional ring topology whose adjacency matrix $\mathcal{A}_n = [a_{ij}] \in \mathbb{R}^{n\times n}$ is defined as $a_{ij} = 1$ for $j = i-1$ and $j = i+1$ and $a_{ij} = 0$ otherwise.

The second fact is verified as follows:

$$\begin{aligned}
\xi^T(C \otimes J)\xi &= \begin{bmatrix} \xi_1 \\ \vdots \\ \xi_n \end{bmatrix}^T \begin{bmatrix} 2J & -J & \dots & -J \\ -J & 2J & \dots & 0 \\ \vdots & & & \vdots \\ -J & \dots & -J & 2J \end{bmatrix} \begin{bmatrix} \xi_1 \\ \vdots \\ \xi_n \end{bmatrix} \\
&= 2\xi_1^T J\xi_1 - \xi_1^T J\xi_2 - \xi_1^T J\xi_n \\
&\quad - \xi_1^T J\xi_2 + 2\xi_2^T J\xi_2 - \xi_2^T J\xi_3 \\
&\quad \vdots \\
&\quad - \xi_n^T J\xi_1 - \xi_n^T J\xi_{n-1} + 2\xi_n^T J\xi_n \\
&= \xi_1^T J\xi_1 - \xi_1^T J\xi_2 - \xi_2^T J\xi_1 + \xi_2^T J\xi_2 \\
&\quad + \xi_2^T J\xi_2 - \xi_2^T J\xi_3 - \xi_3^T J\xi_2 + \xi_3^T J\xi_3 \\
&\quad \vdots \\
&\quad + \xi_n^T J\xi_n - \xi_n^T J\xi_1 - \xi_1^T J\xi_n + \xi_1^T J\xi_1 \\
&= (\xi_1 - \xi_2)^T J(\xi_1 - \xi_2) \\
&\quad \vdots \\
&\quad + (\xi_n - \xi_1)^T J(\xi_n - \xi_1) \\
&= \sum_{i=< n >} (\xi_i - \xi_{i+1})^T J(\xi_i - \xi_{i+1}).
\end{aligned}$$

The third claim is again shown by direct manipulation:

$$\begin{bmatrix} J(\xi_1 - \xi_n) + J(\xi_1 - \xi_2) \\ J(\xi_2 - \xi_1) + J(\xi_2 - \xi_3) \\ \vdots \\ J(\xi_n - \xi_{n-1}) + J(\xi_n - \xi_1) \end{bmatrix} = \begin{bmatrix} 2J & -J & \dots & -J \\ -J & 2J & -J & \dots \\ \vdots & & & \vdots \\ -J & \dots & -J & 2J \end{bmatrix} \begin{bmatrix} \xi_1 \\ \xi_2 \\ \vdots \\ \xi_n \end{bmatrix} = (C \otimes J)\xi.$$

■

Proof of Theorem 11.3: Letting

$$\tilde{h} \triangleq [\tilde{h}_1^T, \dots, \tilde{h}_n^T]^T,$$

Lemma 11.4 can be used to write E as

$$E = \tilde{h}^T(I_n \otimes K_g + C \otimes K_f)\tilde{h}.$$

Consider the Lyapunov function candidate

$$V = \frac{1}{2}E + \frac{1}{2}\sum_{i=1}^{n} \dot{h}_i^T \dot{h}_i, \tag{11.9}$$

which is positive definite and radially unbounded with respect to $\tilde{h}$ and $\dot{h}$ by part (v) of Lemma C.8. Note that V can be written as

$$V = \frac{1}{2}\tilde{h}^T(I_n \otimes K_g + C \otimes K_f)\tilde{h} + \frac{1}{2}\dot{h}^T\dot{h}.$$

The time derivative of V is

$$\dot{V} = \dot{h}^T\left[(I_n \otimes K_g + C \otimes K_f)\tilde{h} + \nu\right],$$

where $\nu \triangleq [\nu_1^T, \ldots, \nu_n^T]^T$. Noting that control law (11.7) can be written in stacked form as

$$\nu = -(I_n \otimes K_g + C \otimes K_f)\tilde{h} - (I_n \otimes D_g + C \otimes D_f)\dot{h}, \tag{11.10}$$

$$\dot{V} = -\dot{h}^T(I_n \otimes D_g + C \otimes D_f)\dot{h},$$

which is negative semidefinite by part (v) of Lemma C.8.

Let $\Omega = \{(\tilde{h}, \dot{h}) | \dot{V} = 0\}$. Note that $\dot{V} \equiv 0$ implies that $\dot{h} \equiv 0$, which in turn implies that $\ddot{h} \equiv 0$. It thus follows from (11.5) and (11.10) that

$$(I_n \otimes K_g + C \otimes K_f)\tilde{h} \equiv 0,$$

which implies $\tilde{h} \equiv 0$ by part (v) of Lemma C.8. The proof then follows from Theorem F.4.

The second statement follows by noting that $\dot{V} \leq 0$ implies that $V(t) \leq V(0)$ and

$$E(t) + \sum_{i=1}^{n} \dot{h}_i^T \dot{h}_i = 2V(t) \leq 2V(0) = E(0)$$

$$\Rightarrow E(t) \leq E(0) - \sum_{i=1}^{n} \dot{h}_i^T \dot{h}_i,$$

where the third equality follows from the fact that the formation is initially at rest. ■

Remark 11.5. Equation (11.8) provides a bound on the error function $E(t)$. Although the bound is conservative, it implies that during a maneuver, the error will never be worse than the initial error. It is interesting to note that the bound becomes tighter as the velocity of the robots increases. Hardware results demonstrating the effectiveness of (11.7) are contained in Section 11.4. □

11.3.2 Coupled Dynamics Formation Control with Passivity-based Interrobot Damping

The control strategy given in (11.7) requires that the relative velocity between neighbors is known. If relative velocity information is not known, then one possible strategy is to set $D_f = 0$. Unfortunately, this choice results in relative motion that is oscillatory, despite a smooth transition of each of the individual robots to its desired formation pattern. To eliminate this oscillation, we use the passivity techniques in Section 4.3 to inject relative damping into the system. The proposed control strategy is given by

$$\begin{aligned}\dot{\hat{x}}_i &= A\hat{x}_i + \tilde{h}_i,\\ \nu_i &= -(K_g + P)\tilde{h}_i - D\dot{h}_i - K_f(\tilde{h}_i - \tilde{h}_{i-1}) - K_f(\tilde{h}_i - \tilde{h}_{i+1}) - PA\hat{x}_i,\end{aligned} \tag{11.11}$$

where K_g and D are positive-definite matrices, K_f is a positive-semidefinite matrix, A is Hurwitz, and P is the symmetrical positive-definite solution of the Lyapunov equation $A^T P + PA^T = -Q$, where Q is a symmetrical positive-definite matrix. Note that (11.11) is a variant of (4.26).

Remark 11.6. The state of the dynamic controller is $\hat{x}$ and represents, in a sense, the estimate of the relative velocities between neighbors. Note the presence of robot velocity $\dot{h}$. Because this information is required for feedback linearization, we assume that it is also available to the controller. □

Theorem 11.7. *If robot formation* (11.1) *is subject to the control strategy defined in* (11.4) *and* (11.11)*, then error function* (11.6) *converges to zero asymptotically.*

Furthermore, if the formation is initially at rest, i.e., $\dot{h}(0) = 0$*, and the passivity filter is initialized as*

$$\hat{x}_i(0) = -A^{-1}\tilde{h}_i(0),$$

then the formation error is bounded by

$$E(t) \leq E(0) - \sum_{i=1}^{n} \dot{h}_i^T \dot{h}_i - \sum_{i=1}^{n} \hat{x}_i^T P \hat{x}_i. \tag{11.12}$$

Proof: Defining $\hat{x} = [\hat{x}_1^T, \ldots, \hat{x}_n^T]^T$, and using Lemma 11.4, control strategy (11.11) can be written as

$$\dot{\hat{x}} = (I_n \otimes A)\hat{x} + \tilde{h}, \tag{11.13a}$$

$$\begin{aligned}\nu = &-\big[(I_n \otimes K_g) + (C \otimes K_f) + (I_n \otimes P)\big]\tilde{h}\\ &- (I_n \otimes D)\dot{h} - (I_n \otimes P)(I_n \otimes A)\hat{x}.\end{aligned} \tag{11.13b}$$

Consider the Lyapunov function candidate

$$V = \frac{1}{2}\tilde{h}^T(I_n \otimes K_g + C \otimes K_f)\tilde{h} + \frac{1}{2}\dot{h}^T\dot{h} + \frac{1}{2}\dot{\hat{x}}(I_n \otimes P)\dot{\hat{x}},$$

which is positive definite and radially unbounded with respect to $\tilde{h}$, $\dot{h}$, and $\dot{\hat{x}}$ by part (v) of Lemma (C.8). The time derivative of V is given by

$$\dot{V} = \dot{h}\big[(I_n \otimes K_g + C \otimes K_f)\tilde{h} + \nu\big] + \frac{1}{2}\ddot{\hat{x}}(I_n \otimes P)\dot{\hat{x}} + \frac{1}{2}\dot{\hat{x}}(I_n \otimes P)\ddot{\hat{x}},$$

where $\ddot{x} = (I_n \otimes A)\dot{x} + \dot{h}$. Using the fact that $(I_n \otimes P)(I_n \otimes A) + (I_n \otimes A)^T(I_n \otimes P) = -(I_n \otimes Q)$,

$$\begin{aligned}\dot{V} = \dot{h}\big[&(I_n \otimes K_g + C \otimes K_f + I_n \otimes P)\tilde{h} + (I_n \otimes P)(I_n \otimes A)\dot{\hat{x}} + \nu\big] \\ &- \dot{\hat{x}}(I_n \otimes Q)\dot{\hat{x}}.\end{aligned}$$

Application of control law (11.11) gives

$$\dot{V} = -\dot{h}^T(I_n \otimes D)\dot{h} - \dot{\hat{x}}(I_n \otimes Q)\dot{\hat{x}},$$

which is negative semidefinite.

Let $\Omega = \{(\tilde{h}, \dot{h}, \dot{\hat{x}}) | \dot{V} = 0\}$. Note that $\dot{V} \equiv 0$ implies $\dot{h} \equiv 0$ and $\dot{\hat{x}} \equiv 0$, which in turn implies that $\ddot{h} \equiv 0$. It thus follows from (11.5) and (11.13) that the following two equalities hold:

$$\begin{aligned}&(I_n \otimes A)\hat{x} + \tilde{h} \equiv 0, \\ &\big[(I_n \otimes K_g) + (C \otimes K_f) + (I_n \otimes P)\big]\tilde{h} + (I_n \otimes P)(I_n \otimes A)\hat{x} \equiv 0.\end{aligned}$$

Combining these two equations gives

$$\big[(I_n \otimes K_g) + (C \otimes K_f)\big]\tilde{h} \equiv 0,$$

which implies $\tilde{h} \equiv 0$. Therefore, the proof then follows from Theorem F.4.

The second statement follows from the same argument used in Theorem 11.3. ■

11.3.3 Saturated Control

Our experience is that (11.7) and (11.11) work well in the presence of actuator saturation, but convergence is not necessarily guaranteed. In this section, we derive a coupled dynamics strategy that explicitly accounts for actuator saturation.

The saturation control problem appends the additional constraints, $|F| \leq F_{\max}$ and $|\tau| \leq \tau_{\max}$, to dynamics (11.1). For the robots used in our testbed, these bounds are $F_{\max} = 30$ N and $\tau_{\max} = 230$ Nm. Unfortunately, force and torque bounds cannot be applied directly to feedback linearized dynamics (11.4) because the feedback linearization explicitly depends on the tangential and angular velocity of each robot. However, because each formation maneuver prescribes a finite motion, we can assume that, given the bounded acceleration of the system, the robot velocity will also be bounded, and thereby derive bounds

$$\|\nu\| \leq \nu_{\max}$$

on the feedback linearized forces. The saturation problem is solved by modifying error function (11.6) to have linear, rather than quadratic, growth. Accordingly, let

$$E_g = \sum_{i=1}^{n} \{\frac{1}{k}\log[\cosh(k\tilde{h}_{xi})] + \frac{1}{k}\log[\cosh(k\tilde{h}_{yi})]\}$$

$$E_f = \sum_{i=<n>} \Big(\frac{1}{k}\log\{\cosh[k(\tilde{h}_{xi} - \tilde{h}_{x,i+1})]\} + \frac{1}{k}\log\{\cosh[k(\tilde{h}_{yi} - \tilde{h}_{y,i+1})]\}\Big).$$

Similar to (11.6), the total error function is defined as

$$E = k_f E_f + k_g E_g. \tag{11.14}$$

The proposed control strategy is

$$\begin{aligned} \nu_i = &-k_g \tanh(k\tilde{h}_i) - d\tanh(k\dot{h}_i), \\ &- k_f \tanh[k(\tilde{h}_i - \tilde{h}_{i-1})] \\ &- k_f \tanh[k(\tilde{h}_i - \tilde{h}_{i+1})], \end{aligned} \tag{11.15}$$

where k_g, k, and d are positive scalars, k_f is a nonnegative scalar, and $\tanh(\cdot)$ is defined componentwise. Note that (11.15) is a variant of (4.13).

Remark 11.8. The first two terms in (11.15) move the robot toward the desired formation pattern; the last two terms cause the robots to move into, and maintain, formation. □

Theorem 11.9. *If robot formation* (11.1) *is subject to the control strategy defined in* (11.4) *and* (11.15), *then the control satisfies the saturation constraint*

$$\|\nu\|_\infty \leq k_g + d + 2k_f,$$

and error function (11.14) *converges to zero asymptotically.*

Furthermore, if the formation is initially at rest, then formation error (11.14) *satisfies*

$$E(t) \leq E(0) - \frac{1}{2}\sum_{i=1}^{n} \dot{h}_i^T \dot{h}_i.$$

Proof: Because each component of tanh is bounded by one, each component of the control law will be bounded by $k_g + d + 2k_f$.

Consider the Lyapunov function candidate

$$V = E + \frac{1}{2}\sum_{i=1}^{n} \dot{h}_i^T \dot{h}_i,$$

where E is given by (11.14). Note that V is positive definite and radially unbounded with respect to $\tilde{h}$ and $\dot{h}$. Using the fact that $\frac{d}{dt}\log[\cosh(x)] = \dot{x}\tanh(x)$, where $x \in \mathbb{R}$, the time derivative of V is given by

$$\dot{V} = k_g \sum_{i=1}^{n} \dot{h}_i^T \tanh(k\tilde{h}_i) + k_f \sum_{i=<n>} (\dot{h}_i - \dot{h}_{i+1})^T \tanh[k(\tilde{h}_i - \tilde{h}_{i+1})] + \sum_{i=1}^{n} \dot{h}_i^T \nu_i,$$

which, using the fact that tanh is an odd function, can be rearranged as

$$\dot{V} = \sum_{i=<n>} \dot{h}_i^T \{k_g \tanh(k\tilde{h}_i) + k_f \tanh[k(\tilde{h}_i - \tilde{h}_{i+1})] + k_f \tanh[k(\tilde{h}_i - \tilde{h}_{i-1})] + \nu_i\}.$$

Substituting from (11.15),

$$\dot{V} = -d \sum_{i=1}^{n} \dot{h}_i^T \tanh(k\dot{h}_i),$$

which implies that V is a valid Lyapunov function candidate. Let $\Omega = \{(\tilde{h}_i, \dot{h}_i) | \dot{V} = 0\}$. Note that $\dot{V} \equiv 0$ implies that $\dot{h}_i \equiv 0$, $i = 1, \ldots, n$, which in turn implies that $\ddot{h}_i \equiv 0$, $i = 1, \ldots, n$. It thus follows from (11.5) and (11.15) that

$$k_g \tanh(k\tilde{h}_i) + k_f \tanh[k(\tilde{h}_i - \tilde{h}_{i+1})] + k_f \tanh[k(\tilde{h}_i - \tilde{h}_{i-1})] \equiv 0. \quad (11.16)$$

Note that (11.16) implies that

$$k_g \sum_{i=1}^{n} \tilde{h}_i^T \tanh(k\tilde{h}_i) + k_f \sum_{i=1}^{n} \tilde{h}_i^T \{\tanh[k(\tilde{h}_i - \tilde{h}_{i+1})] + \tanh[k(\tilde{h}_i - \tilde{h}_{i-1})]\} \equiv 0, \quad (11.17)$$

which can be written as

$$k_g \sum_{i=1}^{n} \tilde{h}_i^T \tanh(k\tilde{h}_i) - k_f \sum_{i=1}^{n} \sum_{j=1}^{n} c_{ij} \tilde{h}_i^T \tanh[k(\tilde{h}_i - \tilde{h}_j)] \equiv 0, \quad (11.18)$$

where c_{ij} is the (i, j) entry of the circulant matrix C in Lemma 11.4. From (4.14) in Lemma 4.18, (11.18) can be written as

$$k_g \sum_{i=1}^{n} \tilde{h}_i^T \tanh(k\tilde{h}_i) - \frac{1}{2} k_f \sum_{i=1}^{n} \sum_{j=1}^{n} c_{ij} (\tilde{h}_i - \tilde{h}_j)^T \tanh[k(\tilde{h}_i - \tilde{h}_j)] \equiv 0, \quad (11.19)$$

where the second term is nonpositive by noting that $c_{ij} \leq 0$, $\forall i \neq j$. It thus follows that $\tilde{h}_i \equiv 0$, $i = 1, \ldots, n$. Therefore, the proof then follows from Theorem F.4.

The second statement follows from the same argument used in Theorem 11.3. ■

11.4 Hardware Results

This section describes experimental results using the control strategies described in Section 11.3. Experimental results were obtained on three robots in the Multiple AGent Intelligent Coordination and Control (MAGICC) laboratory at Brigham Young University (BYU). The experimental platform is shown in Fig. 11.2. The position of each robot is measured using a combination of dead reckoning and an overhead camera system. The physical parameters of the robots shown in Fig. 11.2 are listed in Table 11.1.

Fig. 11.2. Robots from the BYU MAGICC laboratory used to obtain experimental results

The robots are commanded to transition through the series of formation patterns given by

$$\mathcal{P}_1 = \left\{ \begin{bmatrix} 0 \\ -3 \end{bmatrix}, \begin{bmatrix} 1 \\ -4 \end{bmatrix}, \begin{bmatrix} -1 \\ -4 \end{bmatrix} \right\},$$

$$\mathcal{P}_2 = \left\{ \begin{bmatrix} 0 \\ 4 \end{bmatrix}, \begin{bmatrix} 2 \\ 2 \end{bmatrix}, \begin{bmatrix} -2 \\ 2 \end{bmatrix} \right\},$$

$$\mathcal{P}_3 = \left\{ \begin{bmatrix} 5 \\ 6 \end{bmatrix}, \begin{bmatrix} 5 \\ 4.5 \end{bmatrix}, \begin{bmatrix} 5 \\ 3 \end{bmatrix} \right\},$$

where the units are given in feet. The test facility at the BYU MAGICC lab is a 15 foot square. The robots are initially in the formation pattern given by

$$\mathcal{P}_0 = \left\{ \begin{bmatrix} 0 \\ -6 \end{bmatrix}, \begin{bmatrix} 1.5 \\ -6 \end{bmatrix}, \begin{bmatrix} -1.5 \\ -6 \end{bmatrix} \right\}.$$

Table 11.1. Parameters used to obtain simulation and experimental results

Parameter	Value
m_i	10.1 kg
J_i	0.13 kg m^2
L_i	0.12 m
K_g in (11.7)	$0.5I_2$
D_g in (11.7)	I_2
K_f in (11.7)	$5I_2$
D_f in (11.7)	I_2
A in (11.11)	$-5I_2$
Q in (11.11)	I_2
K_g in (11.11)	$10I_2$
D_g in (11.11)	I_2
K_f in (11.11)	$5I_2$
k_g in (11.15)	1
d_g in (11.15)	5
k_f in (11.15)	5
k in (11.15)	1

Figure 11.3 shows the robots transitioning between the formation pattern using control strategy (11.7) and the gains given in Table 11.1. The desired formation patterns are shown by "x" marks. Notice that the robots move into the formation specified by the target formation pattern at the beginning of each maneuver. The formation maneuver is defined to be complete when $E(t) < \epsilon$, where $\epsilon = 0.5$ m for all results shown in this chapter. When the formation maneuver is complete, the robots begin to maneuver to the next formation pattern. This accounts for the fact that the robots do not exactly reach the formation patterns shown in Fig. 11.3.

Figure 11.4 shows the robots transitioning between the formation patterns using control strategy (11.11) and the gains given in Table 11.1.

Figure 11.5 shows the robots transitioning between the formation patterns using control strategy (11.15) and the gains given in Table 11.1. For this experiment, the formation gains have been tuned to cause the robots to move into formation quickly before transitioning to the desired formation pattern.

11.5 Notes

The results in this chapter are based mainly on [118].

Acknowledgment is given to

©2003 IEEE. Reprinted, with permission, from Jonathan R. Lawton, Randal W. Beard, and Brett Young, "A decentralized approach to formation maneuvers," *IEEE Transactions on Robotics and Automation*, vol. 19, no. 6, pp. 933–941, December 2003.

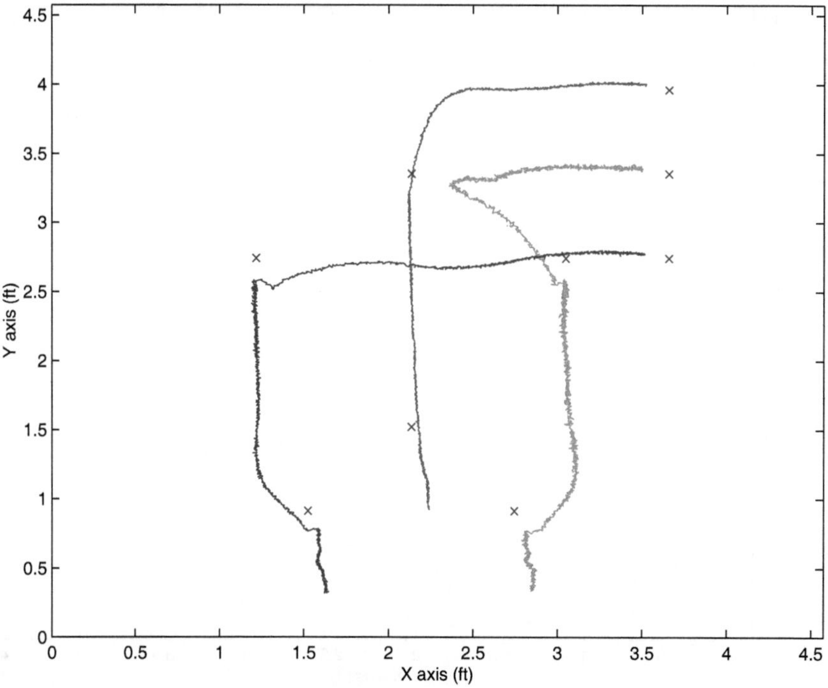

Fig. 11.3. Robot hand positions with control strategy (11.7)

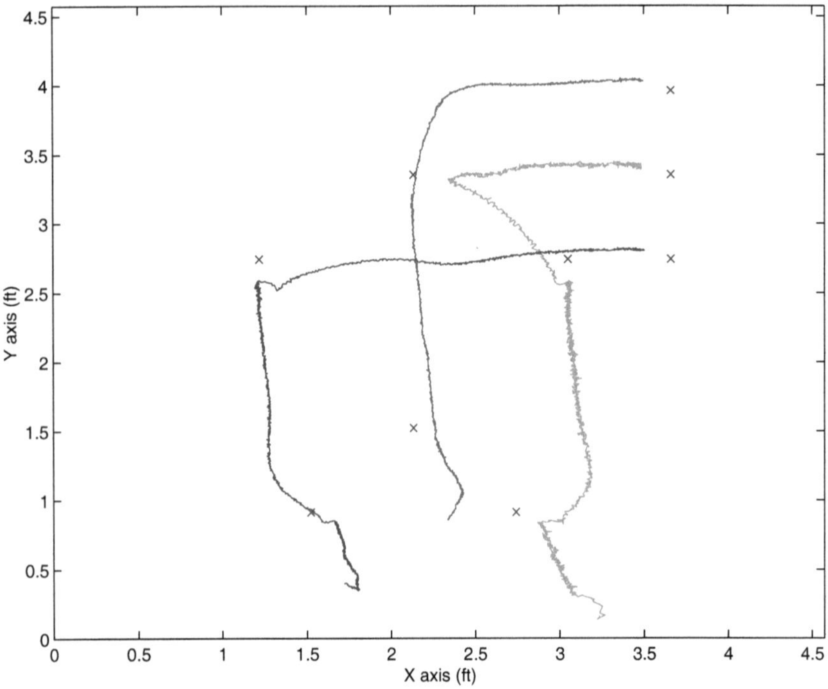

Fig. 11.4. Robot hand positions with passivity control strategy (11.11)

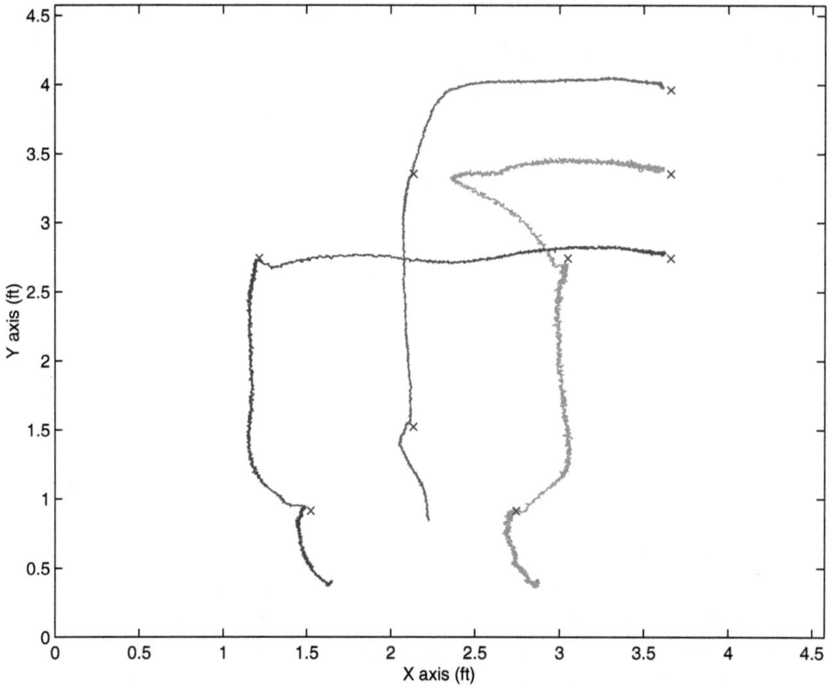

Fig. 11.5. Robot hand positions with input saturation control strategy (11.15)

12

Deep Space Spacecraft Formation Flying

This chapter applies the design scheme in Section 8.4.1 to deep space spacecraft formation flying. We propose a decentralized formation control scheme that is built on the combined strength of consensus algorithms and the virtual structure approach. By following a decentralized coordination architecture *via* the virtual structure approach, we introduce decentralized formation control strategies, which are appropriate when a large number of spacecraft are involved and/or stringent interspacecraft communication limitations are applied. In our decentralized scheme, each spacecraft in the formation instantiates a local copy of the state of the virtual structure, which is the coordination variable using the virtual structure framework. The local instantiation of the coordination variable in each spacecraft is then driven into consensus by communication with its neighbors following a bidirectional ring topology. The effectiveness of the proposed control strategies is demonstrated through simulation results.

12.1 Problem Statement

The concept of multiple spacecraft flying in formation is emerging as an attractive alternative to traditional monolithic spacecraft for both scientific and military applications. The multiple spacecraft approach has several advantages, including increased instrument resolution, reduced cost, reconfigurability, and overall system robustness. Some of the potential applications for formation flying include space-based interferometers and military surveillance instruments. Both NASA and the Air Force have identified spacecraft formation flying as a key technology for the twenty first century. In this section, we introduce some preliminary notation and properties for spacecraft formation flying, including reference frames, desired states for each spacecraft, and spacecraft dynamics.

12.1.1 Reference Frames

As shown in Fig. 12.1, four coordinate frames are used in this chapter. Reference frame $\mathcal{F}_O$ is used as an inertial frame. The formation frame, or the virtual structure frame $\mathcal{F}_F$ is fixed at the virtual center of the formation. Reference frame $\mathcal{F}_i$ is embedded at the center of mass of each spacecraft as a body frame, which rotates with the spacecraft and represents its orientation. Reference frame $\mathcal{F}_i^d$ represents the desired configuration for each spacecraft. Given any vector p, the representation of p in terms of its components in $\mathcal{F}_O$, $\mathcal{F}_F$, and $\mathcal{F}_i$ are represented by $[p]_O$, $[p]_F$, and $[p]_i$, respectively.

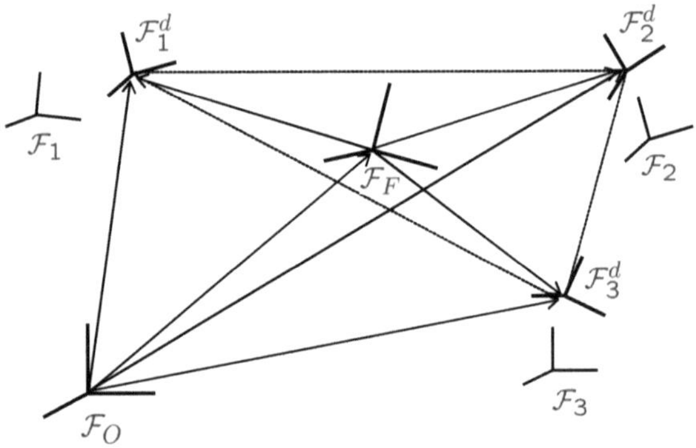

Fig. 12.1. Coordinate frame geometry

Let the direction cosine matrix C_{ab} denote the orientation of frame $\mathcal{F}_a$ with respect to $\mathcal{F}_b$; then $[p]_a = C_{ab}[p]_b$, where $[p]_a$ and $[p]_b$ are the coordinate representations of vector p in $\mathcal{F}_a$ and $\mathcal{F}_b$, respectively. We use unit quaternions to represent spacecraft attitudes (see Appendix D).

12.1.2 Desired States for Each Spacecraft

As described in Section 8.5.1, in the virtual structure approach, the entire desired formation is treated as a single structure with a formation frame $\mathcal{F}_F$ located at its virtual center of mass to represent its configuration. The virtual structure then has position $r_F \in \mathbb{R}^3$, velocity $v_F \in \mathbb{R}^3$, attitude $q_F \in \mathbb{R}^4$, and angular velocity $\omega_F \in \mathbb{R}^3$ relative to $\mathcal{F}_O$.

Let $r_i \in \mathbb{R}^3$, $v_i \in \mathbb{R}^3$, $q_i \in \mathbb{R}^4$, and $\omega_i \in \mathbb{R}^3$ represent the position, velocity, attitude, and angular velocity of the ith spacecraft relative to the inertial frame $\mathcal{F}_O$. Similarly, let r_{iF}, v_{iF}, q_{iF}, and ω_{iF} represent the position, velocity, attitude, and angular velocity of the ith spacecraft relative to formation frame $\mathcal{F}_F$. A superscript "d" is also used to represent the corresponding desired state of each spacecraft relative to either $\mathcal{F}_O$ or $\mathcal{F}_F$.

Conceptually, we can think that place holders corresponding to each spacecraft are embedded in the virtual structure to represent the desired position and attitude of each spacecraft. As the virtual structure as a whole evolves in time, the place holders trace out trajectories for each corresponding spacecraft to track. As a result, the actual states of the ith place holder represent the desired states of the ith spacecraft. With $\mathcal{F}_F$ as a reference frame, these states can be denoted by $r_{iF}^d \in \mathbb{R}^3$, $v_{iF}^d \in \mathbb{R}^3$, $q_{iF}^d \in \mathbb{R}^4$, and $\omega_{iF}^d \in \mathbb{R}^3$.

Generally, r_{iF}^d, q_{iF}^d, v_{iF}^d, and ω_{iF}^d can vary with time, which means that the desired formation shape is time-varying. However, if we are concerned with formation maneuvers that preserve the overall formation shape, that is, each place holder needs to preserve a fixed relative position and orientation in the virtual structure, r_{iF}^d and q_{iF}^d should be constant, and v_{iF}^d and ω_{iF}^d should be zero. This requirement can be loosened to make the formation shape more flexible by allowing the place holders to expand or contract while still keeping a fixed relative orientation. We will focus on this scenario in this latter section. Of course, the approach here can be readily extended to the general case.

Let $\lambda_F = [\lambda_1, \lambda_2, \lambda_3]$, where the components represent the expansion/contraction rates of the virtual structure along each $\mathcal{F}_F$ axis. The state of the virtual structure can be defined as

$$\xi = [r_F^T, v_F^T, q_F^T, \omega_F^T, \lambda_F^T, \dot{\lambda}_F^T]^T. \tag{12.1}$$

We note that if each spacecraft has knowledge of ξ and of its own desired position and orientation with respect to the virtual structure, then formation keeping is transformed into an individual tracking problem. Therefore, the vector ξ represents the minimum amount of information needed by each spacecraft to coordinate its motion with the group. Therefore, ξ is the coordination variable for the team.

Given ξ, the desired states for the ith spacecraft are given by

$$\begin{aligned}
[r_i^d]_O &= [r_F]_O + C_{OF}\Lambda[r_{iF}^d]_F, \\
[v_i^d]_O &= [v_F]_O + C_{OF}\dot{\Lambda}[r_{iF}^d]_F + [\omega_F]_O \times (C_{OF}\Lambda[r_{iF}^d]_F), \\
[q_i^d]_O &= [q_F]_O[q_{iF}^d]_F, \\
[\omega_i^d]_O &= [\omega_F]_O,
\end{aligned} \tag{12.2}$$

where $C_{OF}(q_F)$ is the rotational matrix of frame $\mathcal{F}_O$ with respect to $\mathcal{F}_F$ and $\Lambda = \text{diag}(\lambda_F)$. Note that unlike the constant desired states r_{iF}^d, v_{iF}^d, q_{iF}^d, and ω_{iF}^d relative to $\mathcal{F}_F$, the desired states r_i^d, v_i^d, q_i^d, and ω_i^d relative to $\mathcal{F}_O$ are time-varying because ξ is time-varying. The evolution equations of the desired states are given by

$$
\begin{aligned}
[\dot{r}_i^d]_O =& [v_i^d]_O, \\
[\dot{v}_i^d]_O =& [\dot{v}_F]_O + 2[\omega_F]_O \times (C_{OF}\dot{\Lambda}[r_{iF}^d]_F) \\
& + C_{OF}\ddot{\Lambda}[r_{iF}^d]_F + [\dot{\omega}_F]_O \times (C_{OF}\Lambda[r_{iF}^d]_F), \\
[\dot{q}_i^d]_O =& [\dot{q}_F]_O [q_{iF}^d]_F, \\
[\dot{\omega}_i^d]_O =& [\dot{\omega}_F]_O.
\end{aligned}
\tag{12.3}
$$

12.1.3 Spacecraft Dynamics

The translational dynamics of each spacecraft relative to $\mathcal{F}_O$ are

$$
\frac{dr_i}{dt_o} = v_i, \quad m_i \frac{dv_i}{dt_o} = f_i, \tag{12.4}
$$

where m_i and $f_i \in \mathbb{R}^3$ are, respectively, the mass and control force associated with the ith spacecraft.

The rotational dynamics of each spacecraft relative to $\mathcal{F}_O$ (see [244]) are

$$
\begin{aligned}
\frac{d\hat{q}_i}{dt_o} &= -\frac{1}{2}\omega_i \times \hat{q}_i + \frac{1}{2}\bar{q}_i\omega_i, \quad \frac{d\bar{q}_i}{dt_o} = -\frac{1}{2}\omega_i \cdot \hat{q}_i, \\
J_i \frac{d\omega_i}{dt_o} &= -\omega_i \times (J_i\omega_i) + \tau_i,
\end{aligned}
\tag{12.5}
$$

where $\hat{q}_i$ and $\bar{q}_i$ are, respectively, the vector and scalar parts of q_i; and $J_i \in \mathbb{R}^{3\times 3}$ and $\tau_i \in \mathbb{R}^3$ are, respectively, the inertial tensor and control torque of the ith spacecraft.

12.2 Decentralized Architecture *via* the Virtual Structure Approach

In this section, we propose a decentralized architecture for deep space spacecraft formation flying *via* the virtual structure approach. To demonstrate the salient features of our decentralized scheme, we first introduce a centralized architecture *via* the virtual structure approach.

12.2.1 Centralized Architecture

Figure 12.2 shows a centralized coordination architecture *via* the virtual structure approach.

System **G** is a discrete event supervisor, which evolves a series of formation patterns by outputting its current formation pattern y_G. System **F** is the formation control module, which produces and broadcasts coordination variable ξ. System $\mathbf{K}_i$ is the local spacecraft controller for the ith spacecraft, which receives coordination variable ξ from the formation control module, converts

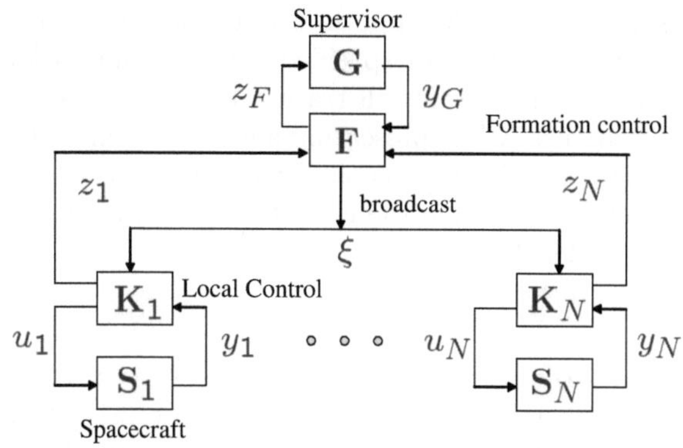

Fig. 12.2. A centralized architecture *via* the virtual structure approach

ξ to the desired states for the ith spacecraft, and then controls the actual state for the ith spacecraft to track its desired state. System $\mathbf{S}_i$ is the ith spacecraft, with control input u_i representing the control force and torque, and output y_i representing the measurable outputs from the ith spacecraft. In this centralized architecture, **G** and **F** are implemented at a centralized location (*e.g.*, spacecraft #1), and then coordination variable ξ is broadcast to the local controllers of the other spacecraft. Note that there is formation feedback from each local spacecraft controller to formation control module **F** through the performance measure z_i. Also, there is formation feedback from **F** to **G** through performance measure z_F.

Remark 12.1. The strength of this centralized architecture is that formation algorithms are fairly easy to realize. The weakness is that a heavy communication and computational burden is concentrated on the centralized location, which may degrade overall system performance. Also the centralized location results in a single point of failure for the whole system. □

12.2.2 Decentralized Architecture

In this section, instead of using a set of desired locations for each agent as a formation pattern as in Chapter 11, we take advantage of the virtual structure approach to define the formation pattern by $P = \xi^d$, where $\xi^d = [r_F^{d^T}, v_F^{d^T}, q_F^{d^T}, \omega_F^{d^T}, \lambda_F^{d^T}, \dot{\lambda}_F^{d^T}]^T$ is the desired constant reference for the coordination variable representing the desired states of the virtual structure. We will assume piecewise rigid formations which implies that $v_F^d = \omega_F^d = \dot{\lambda}_F^d \equiv 0$. By specifying the formation pattern for the group, the movements of each spacecraft can be completely defined. Through a sequence of formation patterns $P^{(k)} = \xi^{d(k)}, k = 1, \cdots, K$, the group can achieve a class of formation maneuver goals. In Chapter 11, the formation pattern is defined so that each vehicle knows only its final location in the formation, whereas its trajectory throughout the maneuver is not specified. Here the formation pattern is defined such that each spacecraft will track a trajectory specified by the state of the virtual structure, while preserving a certain formation shape.

A decentralized architecture is shown in Fig. 12.3. In the decentralized architecture, each spacecraft in the formation instantiates a local copy of the coordination variable. We use

$$\xi_i = [r_{Fi}^T, v_{Fi}^T, q_{Fi}^T, \omega_{Fi}^T, \lambda_{Fi}^T, \dot{\lambda}_{Fi}^T]^T$$

to represent the coordination variable instantiated in the ith spacecraft corresponding to the coordination variable ξ defined in (12.1). A bidirectional ring topology is used to communicate the coordination variable instantiations, instead of the position or attitude information among the spacecraft, to bring each local instantiation into consensus.

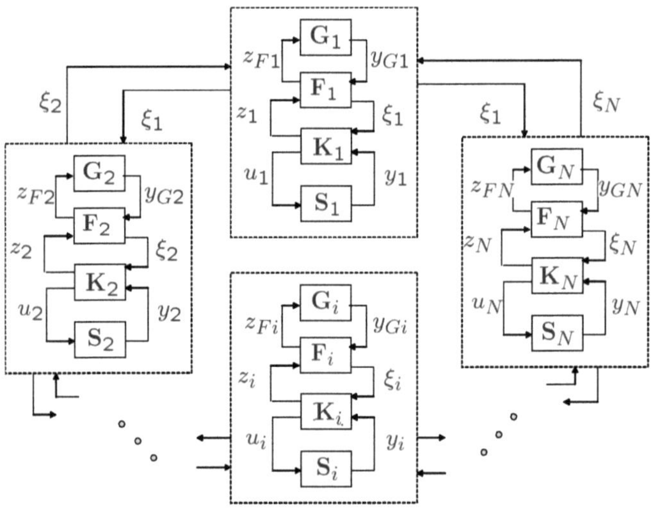

Fig. 12.3. Decentralized architecture *via* the virtual structure approach

In Fig. 12.3, instead of implementing the discrete event supervisor and formation control module at a centralized location, each spacecraft has a local copy of the discrete event supervisor $\mathbf{G}$ and formation control module $\mathbf{F}$, denoted by $\mathbf{G}_i$ and $\mathbf{F}_i$ for the ith spacecraft, respectively. As in Fig. 12.2, $\mathbf{K}_i$ and $\mathbf{S}_i$ represent the ith local spacecraft controller and the ith spacecraft, respectively.

Before the group maneuver starts, a sequence of formation patterns has been preset in each discrete event supervisor $\mathbf{G}_i$. The goal of $\mathbf{G}_i$ is to transition through the sequence of formation patterns so that a class of group maneuver goals can be achieved sequentially. Certain mechanisms need to be applied to coordinate and drive the group starting time into consensus, *e.g.*, simple semaphores. When the group maneuver starts, each discrete event supervisor $\mathbf{G}_i$ outputs the current formation pattern, $y_{Gi} = \xi^{d(1)}$, to the formation control module $\mathbf{F}_i$. Each formation control module $\mathbf{F}_i$ implements a coordination variable instantiation ξ_i. The goal of $\mathbf{F}_i$ is to evolve ξ_i to its current desired formation pattern $\xi^{d(k)}$ and drive ξ_i into consensus with coordination variable instantiations implemented on other spacecraft. Here we use a bidirectional ring topology, which means that coordination variable ξ_i instantiated in the ith spacecraft is driven into consensus with its two neighbors, that is, instantiations ξ_{i-1} and ξ_{i+1} implemented in the $(i-1)$th and the $(i+1)$th spacecraft, respectively. Communications between the ith spacecraft and the $(i-1)$th and $(i+1)$th spacecraft need to be established to transmit and receive the coordination variable instantiations. Formation control module $\mathbf{F}_i$ then sends its coordination variable instantiation ξ_i to the local spacecraft controller $\mathbf{K}_i$. Based on ξ_i, the local controller $\mathbf{K}_i$ can derive the desired states and the corresponding derivatives for the ith spacecraft from (12.2) and (12.3). A local controller $\mathbf{K}_i$ is designed to guarantee that the ith spacecraft tracks its desired states asymptotically. Formation feedback is also included from the ith spacecraft controller $\mathbf{K}_i$ to the ith formation control module $\mathbf{F}_i$ through the performance measure z_i indicating the ith spacecraft's tracking performance. Accordingly, as we will see in Section 12.3, control law for ξ_i implemented in $\mathbf{F}_i$ depends on the performance measure z_i, the current desired formation pattern $y_{Gi} = \xi^{d(k)}$, and the corresponding coordination variable instantiations ξ_{i-1} and ξ_{i+1} from the ith spacecraft's neighbors. Of course, formation feedback can also be included from other spacecraft to the ith formation control module $\mathbf{F}_i$ at the cost of additional communication. Formation feedback from ith formation control module $\mathbf{F}_i$ to ith discrete event supervisor $\mathbf{G}_i$ is also included through the performance measure z_{Fi}, which indicates how far the ith instantiation ξ_i is from its current maneuver goal $\xi^{d(k)}$ and consensus performance between ξ_i and its neighbors. Like the coordination and consensus of the first group maneuver starting time, similar mechanisms can be applied to indicate accomplishment of the current formation pattern and coordinate and drive the starting time into consensus for the next formation pattern among spacecraft. Then the same procedure described above repeats so that a sequence of formation patterns can be achieved.

The communication requirement for each spacecraft during the maneuver can be estimated as follows. We know that r_{Fi}, v_{Fi}, ω_{Fi}, λ_{Fi}, and $\dot{\lambda}_{Fi}$ all have three components and q_{Fi} has four components. Thus the coordination variable ξ_i has 19 components. Assume that each component is encoded as B bits and the sample rate of the system is given by L Hz. By communicating with its two adjacent neighbors, the required bandwidth for each spacecraft can be estimated as $38BL$ bits/s. Note that this is the case when group translation, group rotation, and group expansion/contraction are all involved. If only one group maneuver is involved, the bandwidth can be further reduced to almost one-third of the above estimate.

Remark 12.2. In [37], a decentralized architecture is proposed for autonomous establishment and maintenance of satellite formations, where each satellite processes only local measurement information and transmission vectors from other nodes so that a local Kalman filter can be implemented to obtain a local control. It is also shown that the decentralized framework generates a neighboring optimal control if the planned maneuvers and trajectories are themselves optimal. Compared with the architecture in [37], which is based on a fully interconnected network, the architecture proposed here imposes fewer communication requirements. Even if the compression of data transmission is realized in [37], each vehicle still needs extensive data transmitted from all other vehicles, which causes additional intervehicle communications especially when a large number of vehicles are involved. The architecture proposed here requires only communication between adjacent neighbors during the maneuver. □

Remark 12.3. Compared to its centralized alternative, there is no master in the loop, and each spacecraft evolves in a parallel manner so that a single point of failure existing in any centralized implementation can be eliminated and the total system performance will not degrade catastrophically under failure. As a result, decentralized implementation offers more flexibility, reliability, and robustness than the corresponding centralized alternative. The weakness is that each local instantiation must be driven into consensus, which accounts for additional complexity and intervehicle communications. Due to the ring topology and the implementation of the coordination variable, information exchange among spacecraft can be reduced in the above decentralized architecture. Therefore, this weakness can be somewhat mitigated although the disadvantage of increased intervehicle communication requirements is a typical concern for decentralized systems. Of course, there may exist discrepancies between the starting time of each instantiation of the coordination variable dynamics. This starting time discrepancy can be mitigated through the control law for each coordination variable, which will drive neighboring coordination variable instantiations into consensus. Also, there may exist a time delay when neighboring spacecraft exchange information. This issue is not modeled in the above decentralized architecture. □

12.3 Decentralized Formation Control Strategies

Two major tasks need to be carried out in the decentralized formation control scheme *via* the virtual structure approach. One is to propose suitable control laws for each spacecraft to track its desired states defined by the virtual structure. The other is to control and drive each virtual structure instantiation into consensus to achieve the desired formation patterns in a decentralized manner. In Sections 12.3.1 and 12.3.2, we present control strategies for each spacecraft and each virtual structure instantiation, respectively. In Section 12.3.3, we provide convergence analysis for the system composed of the coupled dynamics of n spacecraft and n coordination variable instantiations.

12.3.1 Formation Control Strategies for Each Spacecraft

For the ith spacecraft, define

$$X_i = [r_i^T, v_i^T, q_i^T, \omega_i^T]^T$$

and

$$X_i^d = [r_i^{d^T}, v_i^{d^T}, q_i^{d^T}, \omega_i^{d^T}]^T$$

as the actual state and desired state, respectively. Define

$$\tilde{X}_i = X_i - X_i^d = [\tilde{r}_i^T, \tilde{v}_i^T, \tilde{q}_i^T, \tilde{\omega}_i^T]^T$$

as the error state for the ith spacecraft.

We know that the desired states for each spacecraft also satisfy translational and rotational dynamics (12.4) and (12.5), respectively, that is,

$$\begin{aligned} \frac{dr_i^d}{dt_o} &= v_i^d, \\ m_i \frac{dv_i^d}{dt_o} &= f_i^d, \\ \frac{d\hat{q}_i^d}{dt_o} &= -\frac{1}{2}\omega_i^d \times \hat{q}_i^d + \frac{1}{2}\bar{q}_i^d \omega_i^d, \\ \frac{d\bar{q}_i^d}{dt_o} &= -\frac{1}{2}\omega_i^d \cdot \hat{q}_i^d, \\ J_i \frac{d\omega_i^d}{dt_o} &= -\omega_i^d \times (J_i \omega_i^d) + \tau_i^d. \end{aligned} \tag{12.6}$$

This is valid because the desired states for each spacecraft are the same as the actual states for each corresponding place holder, which satisfies translational and rotational dynamics.

The proposed control force for the ith spacecraft is given by

$$f_i = m_i[\dot{v}_i^d - K_{ri}(r_i - r_i^d) - K_{vi}(v_i - v_i^d)], \tag{12.7}$$

where K_{ri} and K_{vi} are symmetrical positive-definite matrices.

The proposed control torque for the ith spacecraft is given by

$$\tau_i = J_i \dot{\omega}_i^d + \frac{1}{2}\omega_i \times J_i(\omega_i + \omega_i^d) - k_{qi}\widehat{q_i^{d*} q_i} - K_{\omega i}(\omega_i - \omega_i^d), \tag{12.8}$$

where k_{qi} is a positive scalar, $K_{\omega i}$ is a symmetrical positive-definite matrix, and $\widehat{q_i^{d*} q_i}$ represents the vector part of the unit quaternion $q_i^{d*} q_i$.

Note that (12.7) and (12.8) require both X_i^d and $\dot{X}_i^d$ which are obtained from ξ_i and $\dot{\xi}_i$ using (12.2) and (12.3).

12.3.2 Formation Control Strategies for Each Virtual Structure Instantiation

As in Section 12.2.2, ξ_i is the ith coordination variable instantiation and $\xi^{d(k)}$ is the current kth desired constant goal for the coordination variable instantiations, *i.e.*, the current formation pattern. For notation simplicity, hereafter, we use ξ^d instead of $\xi^{d(k)}$ to represent a certain formation pattern to be achieved. Define

$$\tilde{\xi}_i = \xi_i - \xi^d = [\tilde{r}_{Fi}^T, \tilde{v}_{Fi}^T, \tilde{q}_{Fi}^T, \tilde{\omega}_{Fi}^T, \tilde{\lambda}_{Fi}^T, \dot{\tilde{\lambda}}_{Fi}^T]^T$$

as the error state for the ith coordination variable instantiation. There are two objectives for the instantiation of the coordination variable implemented in each spacecraft. The first objective is to reach its desired constant goal ξ^d defined by the formation pattern set. The second objective is to drive each instantiation into consensus, *i.e.*, $\xi_1 = \xi_2 = \cdots = \xi_n$. We apply the ideas introduced in Chapters 5 and 6 to drive the coordination variable instantiations into consensus during the maneuver, as well as evolve it to its desired goal at the end of the maneuver.

Define E_G as the goal seeking error between the current coordination variable instantiation ξ_i and the desired goal ξ^d:

$$E_G(t) = \sum_{i=1}^{n} \left\| \xi_i - \xi^d \right\|^2 .$$

Also define E_S as the total consensus error between neighboring coordination variable instantiations:

$$E_S(t) = \sum_{i=1}^{n} \left\| \xi_i - \xi_{i+1} \right\|^2 ,$$

where the summation index i is defined modulo n, *i.e.*, $\xi_{n+1} = \xi_1$ and $\xi_0 = \xi_n$. Defining $E(t) = E_G(t) + E_S(t)$, then the control objective is to drive $E(t)$ to zero asymptotically.

Because the coordination variable represents the states of the virtual structure, the ith coordination variable instantiation satisfies the following rigid body dynamics:

$$\begin{bmatrix} \dot{r}_{Fi} \\ m_F \dot{v}_{Fi} \\ \dot{\widehat{q}}_{Fi} \\ \dot{\overline{q}}_{Fi} \\ J_F \dot{\omega}_{Fi} \\ \dot{\lambda}_{Fi} \\ \ddot{\lambda}_{Fi} \end{bmatrix} = \begin{bmatrix} v_{Fi} \\ f_{Fi} \\ -\frac{1}{2}\omega_{Fi} \times \widehat{q}_{Fi} + \frac{1}{2}\overline{q}_{Fi}\omega_{Fi} \\ -\frac{1}{2}\omega_{Fi} \cdot \widehat{q}_{Fi} \\ -\omega_{Fi} \times J_F \omega_{Fi} + \tau_{Fi} \\ \dot{\lambda}_{Fi} \\ \nu_{Fi} \end{bmatrix}, \tag{12.9}$$

where m_F and J_F are the virtual mass and virtual inertia of the virtual structure, f_{Fi} and τ_{Fi} are the virtual force and virtual torque exerted on the ith implementation of the virtual structure, and ν_{Fi} is the virtual control effort used to expand or contract the formation.

The tracking performance for the ith spacecraft is defined as $e_{Ti} \triangleq \left\|\tilde{X}_i\right\|^2$. Also define

$$\Gamma_{Gi} = D_G + K_F e_{Ti} \tag{12.10}$$

to incorporate formation feedback from the ith spacecraft to the ith coordination variable implementation, where D_G and K_F are symmetrical positive-definite matrices. Obviously, Γ_{Gi} is also a positive-definite matrix. If we let $K_F = 0$, there is no formation feedback.

The proposed control force f_{Fi} is given by

$$\begin{aligned} f_{Fi} =& m_F\{-K_G(r_{Fi} - r_F^d) - \Gamma_{Gi} v_{Fi} \\ & - K_S[r_{Fi} - r_{F(i+1)}] - D_S[v_{Fi} - v_{F(i+1)}] \\ & - K_S[r_{Fi} - r_{F(i-1)}] - D_S[v_{Fi} - v_{F(i-1)}]\}, \end{aligned} \tag{12.11}$$

where K_G is a symmetrical positive-definite matrix, Γ_{Gi} is defined in (12.10), and K_S and D_S are symmetrical positive semidefinite matrices. Note that (12.11) is a direct application of (5.10).

The proposed control torque τ_{Fi} is given by

$$\begin{aligned} \tau_{Fi} =& - k_G \widehat{q_F^{d*} q_{Fi}} - \Gamma_{Gi} \omega_{Fi} \\ & - k_S \widehat{q_{F(i+1)}^{*} q_{Fi}} - D_S[\omega_{Fi} - \omega_{F(i+1)}] \\ & - k_S \widehat{q_{F(i-1)}^{*} q_{Fi}} - D_S[\omega_{Fi} - \omega_{F(i-1)}], \end{aligned} \tag{12.12}$$

where $k_G > 0$ and $k_S \geq 0$ are scalars, Γ_{Gi} is defined in (12.10), D_S is a symmetrical positive-semidefinite matrix, and $\hat{q}$ represents the vector part of the unit quaternion. Note that (12.12) is a direct application of (6.1).

Similar to (12.11), the proposed control effort ν_{Fi} is given by

$$\begin{aligned} \nu_{Fi} =& - K_G(\lambda_{Fi} - \lambda_F^d) - \Gamma_{Gi} \dot{\lambda}_{Fi} \\ & - K_S[\lambda_{Fi} - \lambda_{F(i+1)}] - D_S[\dot{\lambda}_{Fi} - \dot{\lambda}_{F(i+1)}] \\ & - K_S[\lambda_{Fi} - \lambda_{F(i-1)}] - D_S[\dot{\lambda}_{Fi} - \dot{\lambda}_{F(i-1)}], \end{aligned} \tag{12.13}$$

where K_G is a symmetrical positive-definite matrix, Γ_{Gi} is defined in (12.10), and K_S and D_S are symmetrical positive-semidefinite matrices. Note that (12.13) is a direct application of (5.10).

Remark 12.4. Note that the matrices in (12.11), (12.12), and (12.13) can be chosen differently based on specific requirements to change the effective weights of translation, rotation, and expansion/contraction. In (12.11), (12.12), and (12.13), the first two terms are used to drive E_G to zero, the third and fourth terms are used to drive the ith and $(i+1)$th coordination variable instantiations into consensus, and the fifth and sixth terms are used to drive the ith and $(i-1)$th coordination variable instantiations into consensus. The second term, that is, the formation feedback term is also used to slow down the ith virtual structure when the ith spacecraft has a large tracking error. This strategy requires that each spacecraft know the coordination variable instantiations of its neighbors, which can be accomplished by nearest neighbor communication. From (12.11), (12.12), and (12.13), we can also see that besides ξ_{i-1}, ξ_i, and ξ_{i+1}, the control laws for the ith coordination variable instantiation also require the current constant formation pattern ξ^d and $\tilde{X}_i$ through the formation feedback gain matrix Γ_{Gi}. □

12.3.3 Convergence Analysis

From (12.4), (12.5), (12.7), and (12.8), the dynamics for the ith spacecraft can be represented by $\dot{\tilde{X}}_i = f(\tilde{X}_i, \xi_i)$, where $f(\cdot,\cdot)$ can be determined from those equations. From (12.9), (12.11), (12.12), and (12.13), the dynamics for the ith coordination variable instantiation can be represented by $\dot{\xi}_i = g(\xi_{i-1}, \xi_i, \xi_{i+1}, \tilde{X}_i)$, where $g(\cdot,\cdot,\cdot,\cdot)$ can also be determined from those equations. Therefore, the coupled dynamics of the whole system composed of n spacecraft and n coordination variable instantiations are time-invariant with states $\tilde{X}_i$ and ξ_i, $i = 1, \cdots, n$. Theorem F.3 will be used to prove the main theorem for convergence of the whole system.

Theorem 12.5. *If the control strategy for each spacecraft is given by* (12.7) *and* (12.8) *and the control strategy for each coordination variable instantiation is given by* (12.11), (12.12), *where* $k_G > 4k_S$, *and* (12.13), *then* $\sum_{i=1}^n e_{Ti}(t) + E(t) \to 0$, *as* $t \to \infty$.

Proof: For the system consisting of n spacecraft and n coordination variable instantiations, consider the Lyapunov function candidate:

$$V = V_{sp} + V_{Ft} + V_{Fr} + V_{Fe}, \tag{12.14}$$

where

$$
\begin{aligned}
V_{sp} &= \sum_{i=1}^{n} \left(\frac{1}{2}\tilde{r}_i^T K_{ri}\tilde{r}_i + \frac{1}{2}\tilde{v}_i^T \tilde{v}_i + k_{qi}\tilde{q}_i^T \tilde{q}_i + \frac{1}{2}\tilde{\omega}_i^T J_i\tilde{\omega}_i \right), \\
V_{Ft} &= \frac{1}{2}\sum_{i=1}^{n} [r_{Fi} - r_{F(i+1)}]^T K_S [r_{Fi} - r_{F(i+1)}] \\
&\quad + \frac{1}{2}\sum_{i=1}^{n} \left(\tilde{r}_{Fi}^T K_G \tilde{r}_{Fi} + v_{Fi}^T v_{Fi}\right), \\
V_{Fr} &= \sum_{i=1}^{n} k_S [q_{Fi} - q_{F(i+1)}]^T [q_{Fi} - q_{F(i+1)}] \\
&\quad + \sum_{i=1}^{n} \left(k_G \tilde{q}_{Fi}^T \tilde{q}_{Fi} + \frac{1}{2}\omega_{Fi}^T J_F \omega_{Fi} \right), \\
V_{Fe} &= \frac{1}{2}\sum_{i=1}^{n} [\lambda_{Fi} - \lambda_{F(i+1)}]^T K_S [\lambda_{Fi} - \lambda_{F(i+1)}] \\
&\quad + \frac{1}{2}\sum_{i=1}^{n} \left(\tilde{\lambda}_{Fi}^T K_G \tilde{\lambda}_{Fi} + \dot{\lambda}_{Fi}^T \dot{\lambda}_{Fi} \right).
\end{aligned}
$$

With proposed control force (12.7) for each spacecraft, the second equation in translational dynamics (12.4) for the ith spacecraft can be rewritten as $\dot{\tilde{v}}_i = -K_{ri}\tilde{r}_i - K_{vi}\tilde{v}_i$. Applying Lemma D.1, the derivative of V_{sp} is

$$
\begin{aligned}
\dot{V}_{sp} &= \sum_{i=1}^{n} (-\tilde{v}_i^T K_{vi}\tilde{v}_i) \\
&\quad + \sum_{i=1}^{n} \tilde{\omega}_i^T \left[k_{qi}\widehat{q_i^{d*} q_i} + \tau_i - \tau_i^d - \frac{1}{2}(\omega_i \times J_i\tilde{\omega}_i) \right].
\end{aligned}
$$

From (12.6), $\tau_i^d = J_i\dot{\omega}_i^d + \omega_i^d \times (J_i\omega_i^d)$. With proposed control torque (12.8) for each spacecraft, after some manipulation, we obtain

$$
\dot{V}_{sp} = \sum_{i=1}^{n} \left(-\tilde{v}_i^T K_{vi}\tilde{v}_i - \tilde{\omega}_i^T K_{\omega i}\tilde{\omega}_i \right) \le 0. \tag{12.15}
$$

Differentiating V_{Ft} gives

$$
\begin{aligned}
\dot{V}_{Ft} &= \sum_{i=1}^{n} v_{Fi}^T \{ K_S [r_{Fi} - r_{F(i+1)}] \\
&\quad + K_S [r_{Fi} - r_{F(i-1)}] + K_G \tilde{r}_{Fi} + \frac{f_{Fi}}{m_F} \}.
\end{aligned}
$$

With control force (12.11) for each coordination variable instantiation,

$$\begin{aligned}\dot{V}_{Ft} = &-\sum_{i=1}^{n}\{v_{Fi}^T\Gamma_{Gi}v_{Fi}\\&+[v_{Fi}-v_{F(i+1)}]^T D_S[v_{Fi}-v_{F(i+1)}]\}\leq 0.\end{aligned}\tag{12.16}$$

Applying Lemma D.1, the derivative of V_{Fr} is

$$\begin{aligned}\dot{V}_{Fr} = &\sum_{i=1}^{n}[\omega_{Fi}-\omega_{F(i+1)}]^T k_S\widehat{q_{F(i+1)}^* q_{Fi}}\\&+\sum_{i=1}^{n}\omega_{Fi}^T(k_G\widehat{q_F^{d*}q_{Fi}}+\tau_{Fi}-\frac{1}{2}\omega_{Fi}\times J_F\omega_{Fi}),\end{aligned}$$

and after some manipulation,

$$\dot{V}_{Fr} = \sum_{i=1}^{n}\omega_{Fi}^T(k_S\widehat{q_{F(i+1)}^* q_{Fi}} - k_S\widehat{q_{Fi}^* q_{F(i-1)}} + k_G\widehat{q_F^{d*}q_{Fi}}+\tau_{Fi}).$$

With proposed control torque (12.12), for each coordination variable instantiation,

$$\begin{aligned}\dot{V}_{Fr} = &-\sum_{i=1}^{n}[\omega_{Fi}^T\Gamma_{Gi}\omega_{Fi}\\&+(\omega_{Fi}-\omega_{F(i+1)})^T D_S(\omega_{Fi}-\omega_{F(i+1)})]\leq 0.\end{aligned}\tag{12.17}$$

Similar to $\dot{V}_{Ft}$, with proposed control effort (12.13) for each coordination variable instantiation, the derivative of V_{Fe} is

$$\begin{aligned}\dot{V}_{Fe} = &-\sum_{i=1}^{n}\{\dot{\lambda}_{Fi}^T\Gamma_{Gi}\dot{\lambda}_{Fi}\\&+[\dot{\lambda}_{Fi}-\dot{\lambda}_{F(i+1)}]^T D_S[\dot{\lambda}_{Fi}-\dot{\lambda}_{F(i+1)}]\}\leq 0.\end{aligned}\tag{12.18}$$

From (12.15), (12.16), (12.17), and (12.18), it is obvious that $\dot{V} = \dot{V}_{sp} + \dot{V}_{Ft} + \dot{V}_{Fr} + \dot{V}_{Fe} \leq 0$. Let $\Omega = \{(\tilde{X}_1,\cdots,\tilde{X}_n,\tilde{\xi}_1,\cdots,\tilde{\xi}_n)|\dot{V}=0\}$. Note that $\dot{V}\equiv 0$, *i.e.*, $\dot{V}_{sp} = \dot{V}_{Ft} = \dot{V}_{Fr} = \dot{V}_{Fe} \equiv 0$, implies that $\tilde{v}_i\equiv 0$, $\tilde{\omega}_i\equiv 0$, $v_{Fi}\equiv 0$, $\omega_{Fi}\equiv 0$, $\dot{\lambda}_{Fi}\equiv 0$, $i=1,\cdots,n$.

Because $\tilde{v}_i \equiv 0$, we know that $\tilde{r}_i \equiv 0$ from (12.4) and (12.7). Because $\tilde{\omega}_i \equiv 0$, we also know that $\widehat{q_i^{d*}q_i} \equiv 0$ from (12.5) and (12.8), which then implies that $q_i \equiv q_i^d$, *i.e.*, $\tilde{q}_i \equiv 0$.

Then, following $v_{Fi}\equiv 0$, from (12.11) and the second equation in (12.9), it can be seen that

$$\begin{aligned}&K_G\tilde{r}_{Fi} + K_S[r_{Fi}-r_{F(i+1)}]\\&+K_S[r_{Fi}-r_{F(i-1)}]\equiv 0,\quad i=1,\cdots,n,\end{aligned}$$

which is equivalent to

$$\begin{aligned} K_G\tilde{r}_{Fi} + K_S[\tilde{r}_{Fi} - \tilde{r}_{F(i+1)}] \\ +K_S[\tilde{r}_{Fi} - \tilde{r}_{F(i-1)}] \equiv 0, \quad i = 1, \cdots, n. \end{aligned} \tag{12.19}$$

From Lemma 11.4, (12.19) can also be written as $(I_n \otimes K_G + C \otimes K_S)\tilde{r}_F \equiv 0$, where $\tilde{r}_F \triangleq [\tilde{r}_{F1}^T, \cdots, \tilde{r}_{Fn}^T]^T$ and C is the circulant matrix defined in Lemma 11.4. Based on Lemmas C.8 and 11.4, $I_n \otimes K_G$ is positive definite and $C \otimes K_S$ is positive semidefinite. Thus we know that $\tilde{r}_F \equiv 0$.

Following a procedure similar to the above, we can also show that $\tilde{\lambda}_{Fi} \equiv 0$ because $\dot{\lambda}_{Fi} \equiv 0$.

Also following $\omega_{Fi} \equiv 0$, from (12.12) and the fourth equation in (12.9), we know that

$$k_G\widehat{q_F^{d*}q_{Fi}} + k_S\widehat{q_{F(i+1)}^*q_{Fi}} + k_S\widehat{q_{F(i-1)}^*q_{Fi}} \equiv 0, \quad i = 1, \cdots, n. \tag{12.20}$$

Note that (12.20) is a special case of (6.5), where the interaction topology is a bidirectional ring. Then following the proof of Theorem 6.2, we can show that $\widehat{q_F^{d*}q_{Fi}} \equiv 0$, which implies that $q_{Fi} \equiv q_F^d$, *i.e.*, $\tilde{q}_{Fi} \equiv 0$, if $k_G > 4k_S$.

Therefore, by Theorem F.3, $\left\|\tilde{X}_i(t)\right\| \to 0$, $\left\|\tilde{\xi}_i(t)\right\| \to 0$, and $\|\xi_i(t) - \xi_{i+1}(t)\| \to 0$, $i = 1, \cdots, n$, as $t \to \infty$. Accordingly, $\sum_{i=1}^n e_{Ti} + E(t) \to 0$, as $t \to \infty$. ■

From Theorem 12.5, we can see that each virtual structure instantiation will achieve its final goal asymptotically and each spacecraft will also track its desired state specified by the virtual structure asymptotically during the maneuver. Therefore, the formation maneuver can be achieved asymptotically.

12.3.4 Discussion

Note that different performance measure functions e_{Ti} may be chosen to measure formation maintenance. For example, $e_{Ti} = \tilde{X}_i^T P \tilde{X}_i$, where P is symmetrical positive definite. The matrix P can be designed to adjust the relative weights of translational and rotational formation error based on certain requirements. The motivation for the design of the nonlinear gain matrices Γ_{Gi} is to meet the following requirements. When a spacecraft is out of its desired configuration, that is, e_{Ti} is large, its coordination variable instantiation will slow down or even stop, allowing the spacecraft to regain formation. When a spacecraft is maintaining its desired configuration, that is, e_{Ti} is small, its coordination variable instantiation will keep moving toward its final goal at a reasonable speed. By this design, each coordination variable instantiation will be aimed at performing reasonably fast formation maneuvers as well as preserving tight formation shape during the maneuver, despite control saturation, disturbances, and malfunctions. In this chapter, we choose a candidate for such gain matrices as $\Gamma_{Gi} = D_G + K_F e_{Ti}$, where D_G is the symmetrical positive-definite gain matrix, which corresponds to the nominal formation speed when the formation is preserved tightly, and K_F is the symmetrical positive-definite formation gain matrix which weights the performance measure e_{Ti}. Of course, other choices are also feasible. If $K_F = 0$, no formation feedback is introduced. We will see that a formation gain matrix with larger entries results in better formation maintenance but slower convergence speed. Note that

$$e_{Ti} \to 0 \Rightarrow \Gamma_{Gi} \to D_G,$$
$$e_{Ti} \to \infty \Rightarrow \Gamma_{Gi} \to \infty.$$

As a result, nonlinear gains slow down or speed up the coordination variable instantiation based on how far each spacecraft is out of the desired configuration.

Because PD-like control laws are used for each spacecraft and each coordination variable instantiation, the transient specifications for each spacecraft and each coordination variable instantiation can be satisfied by designing corresponding gain matrices in control laws following the design procedure for coefficients of a second-order system. For a second-order system $s^2 + k_1 s + k_2 = 0$, if we define rise time t_r and damping ratio ζ, then natural frequency ω_n is approximately $1.8/t_r$. Therefore, if we let $k_2 = \omega_n^2 = (1.8/t_r)^2$ and $k_1 = 2\zeta\omega_n = 2\zeta(1.8/t_r)$, the transient specifications for the system are satisfied. We can design K_{ri}, k_{qi}, K_G, and k_G according to k_2, and design K_{vi}, $K_{\omega i}$, and D_G according to k_1. For example, K_{ri} and K_{vi} can be defined as $k_2 I_3$ and $k_1 I_3$, respectively, where I_3 is a 3×3 identity matrix.

Example 12.6. Let $K_G = I_3$ and $K_S = D_S = 0$ in (12.11). Note that the translational dynamics of the ith coordination variable instantiation can be rewritten as $\ddot{\tilde{r}}_{Fi} + \Gamma_{Gi}\dot{\tilde{r}}_{Fi} + K_G \tilde{r}_{Fi} = 0$, where $\tilde{r}_{Fi} = r_{Fi} - r_{Fi}^d = [\tilde{r}_{Fxi}, \tilde{r}_{Fyi}, \tilde{r}_{Fzi}]^T$. Figure 12.4 shows a plot of $\tilde{r}_{Fxi}$ for different choices of Γ_{Gi}. We can see that the dynamics of the ith coordination variable instantiation evolve more slowly as the elements of Γ_{Gi} are increased to be sufficiently large, that is, it takes a longer time for the ith coordination variable instantiation to achieve its desired state. When $\Gamma_{Gi} \to \infty$, the coordination variable instantiation will stop evolving. □

Moreover, for each spacecraft, if we define a translational tracking error for the ith spacecraft as $E_{ti} = \frac{1}{2}\tilde{r}_i^T K_{ri}\tilde{r}_i + \frac{1}{2}\|\tilde{v}_i\|^2$, E_{ti} decreases during the maneuver and $\tilde{r}_i^T K_{ri}\tilde{r}_i$ is bounded by $2E_{ti}(0) - \|\tilde{v}_i\|^2$. Similarly, if we define a rotational tracking error as $E_{ri} = k_{qi}\|\tilde{q}_i\|^2 + \frac{1}{2}\tilde{\omega}_i J_i \tilde{\omega}_i$, E_{ri} decreases during the maneuver, and $\|\tilde{q}_i\|^2$ is bounded by $\frac{1}{k_{qi}}(E_{ri}(0) - \frac{1}{2}\tilde{\omega}_i J_i \tilde{\omega}_i)$. For each coordination variable instantiation, following the proof for $\dot{V}_{Ft}$, $\dot{V}_{Fr}$, and $\dot{V}_{Fe}$, we know that V_{Ft}, V_{Fr}, and V_{Fe} are bounded by $V_{Ft}(0)$, $V_{Fr}(0)$, and $V_{Fe}(0)$, respectively. Therefore, $\sum_{i=1}^n [r_{Fi} - r_{F(i+1)}]^T K_S [r_{Fi} - r_{F(i+1)}] \le 2V_{Ft}(0)$, $\sum_{i=1}^n \tilde{r}_{Fi}^T K_G \tilde{r}_{Fi} \le 2V_{Ft}(0)$, $\sum_{i=1}^n \|q_{Fi} - q_{F(i+1)}\|^2 \le \frac{1}{k_S} V_{Fr}(0)$,

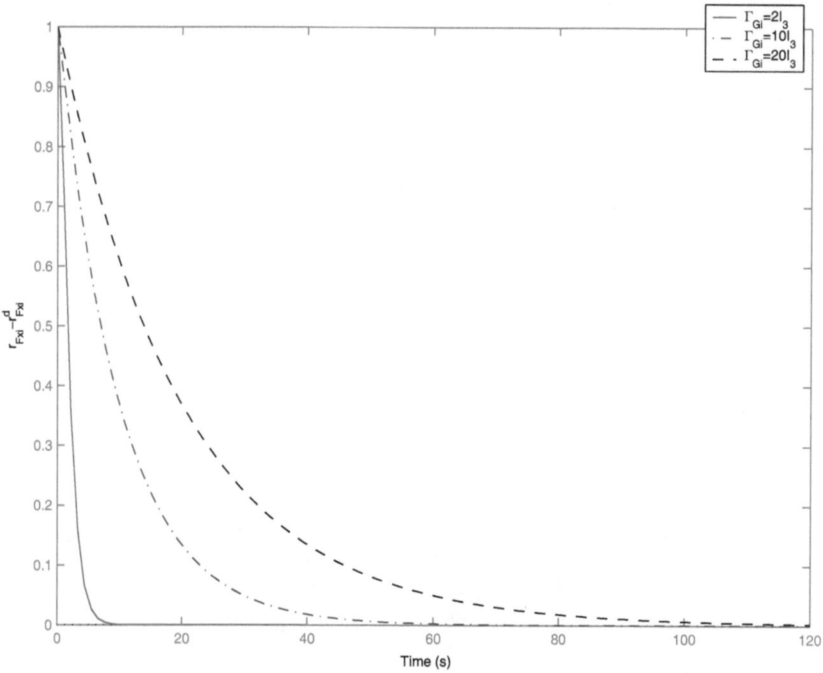

Fig. 12.4. Plot of $\tilde{r}_{Fxi}$ with initial conditions $\tilde{r}_{Fxi}(0) = 1$ and $\dot{\tilde{r}}_{Fxi}(0) = 0$ for different choices of Γ_{Gi}

$\sum_{i=1}^{n} \|\tilde{q}_{Fi}\|^2 \leq \frac{1}{k_G} V_{Fr}(0)$, $\sum_{i=1}^{n} [\lambda_{Fi} - \lambda_{F(i+1)}]^T K_S [\lambda_{Fi} - \lambda_{F(i+1)}] \leq 2V_{Fe}(0)$, and $\sum_{i=1}^{n} \tilde{\lambda}_{Fi}^T K_G \tilde{\lambda}_{Fi} \leq 2V_{Fe}(0)$.

12.4 Simulation Results

In this section, we consider a scenario with nine spacecraft. In the scenario, a mothership with mass equal to 1500 kg is located one kilometer away from a plane where eight daughter spacecraft each with mass 150 kg are distributed equally along a circle with a diameter of 1 kilometer in the plane. The configuration of the nine spacecraft is shown in Fig. 12.5. We assume that the nine spacecraft formation evolves like a rigid structure, that is, the formation shape is preserved, and each spacecraft preserves a fixed relative orientation within the formation throughout the maneuvers.

We simulate a scenario where the nine spacecraft start from rest with some initial position and attitude errors and then perform a group rotation of 45° about the inertial z axis. Here we assume that each place holder in the formation has the same orientation, that is, q_{iF}^d is the same for each spacecraft. In simulation, we instantiate a local copy of the coordination variable ξ in each

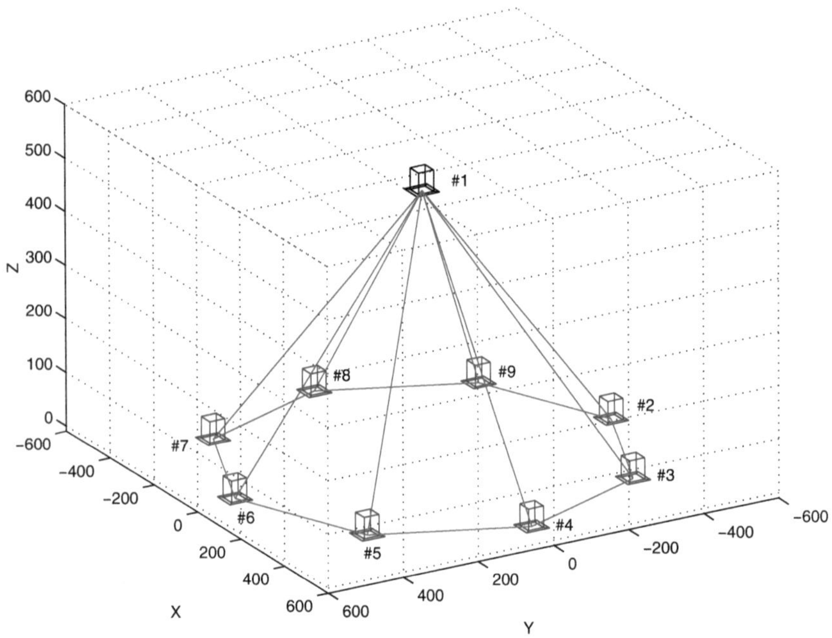

Fig. 12.5. The geometric configuration of nine spacecraft

spacecraft and drive them into consensus using the control strategy introduced in Section 12.3.2. To show the robustness of the control strategy, we start the coordination variable implementation in each spacecraft at a different time and introduce a different sample time, varying from 0.4 seconds to 0.6 seconds for each coordination variable instantiation. Various communication delays are also added among spacecraft. Three cases will be compared in this section. These include cases without actuator saturation and formation feedback (Case I), with actuator saturation but without formation feedback (Case II), with both actuator saturation and formation feedback (Case III). Here, we assume that the control force and control torque for spacecraft #1 are saturated at $|f_x|, |f_y|, |f_z| = 2$ N and $|\tau_x|, |\tau_y|, |\tau_z| = 0.0006$ Nm, respectively, and the control force and control torque for all other spacecraft are saturated at $|f_x|, |f_y|, |f_z| = 1$ N and $|\tau_x|, |\tau_y|, |\tau_z| = 0.0003$ Nm, respectively.

For these simulations, the average coordination error is defined as $\frac{1}{n}\sum_{i=1}^{n} \|\xi_i - \bar{\xi}\|$, where $\bar{\xi} = \frac{1}{n}\sum_{i=1}^{n} \xi_i$. The average coordination error in these three cases is plotted in Fig. 12.6. We can see that in all three cases, each instantiation of the coordination variable is driven into consensus asymptotically. Also, the average coordination error is large during the initial time interval because each local instantiation starts at a different time. Cases I and II are identical because the actuator saturation for each spacecraft does not affect the dynamics of the virtual structure when there is no formation feedback from each spacecraft to its coordination variable instantiation. The maximum average coordination error in Case III is larger than that in the other two cases because formation feedback is introduced for each coordination variable instantiation, which may add some dissimilarities between different instantiations.

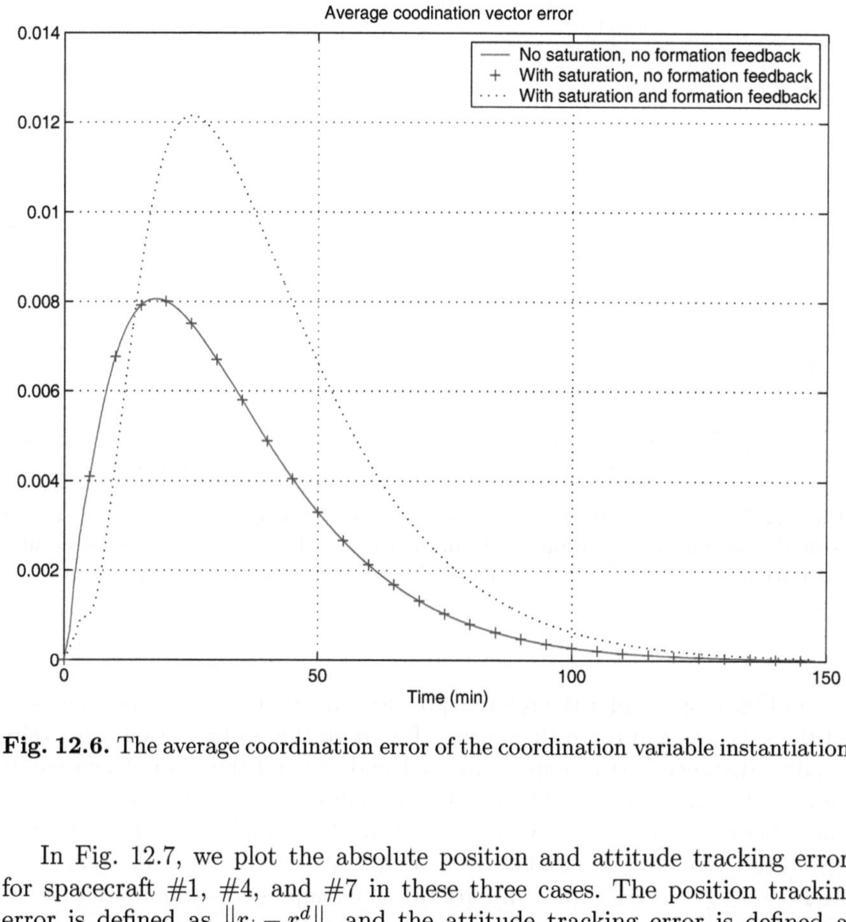

Fig. 12.6. The average coordination error of the coordination variable instantiations

In Fig. 12.7, we plot the absolute position and attitude tracking errors for spacecraft #1, #4, and #7 in these three cases. The position tracking error is defined as $\left\|r_i - r_i^d\right\|$, and the attitude tracking error is defined as $\left\|q_i - q_i^d\right\|$. We can see that the tracking errors in each case will decrease to zero asymptotically by using the control law given in Section 12.3.1. The absolute position and attitude tracking errors in Case II are much larger than those in the other two cases due to the actuator saturation. In Case III, with formation feedback, the absolute position and attitude tracking errors are similar to those in Case I, even if there is actuator saturation. When we increase the entries in the gain matrix K_F to increase formation feedback, the absolute tracking errors can be decreased further, but the system convergence time will increase accordingly.

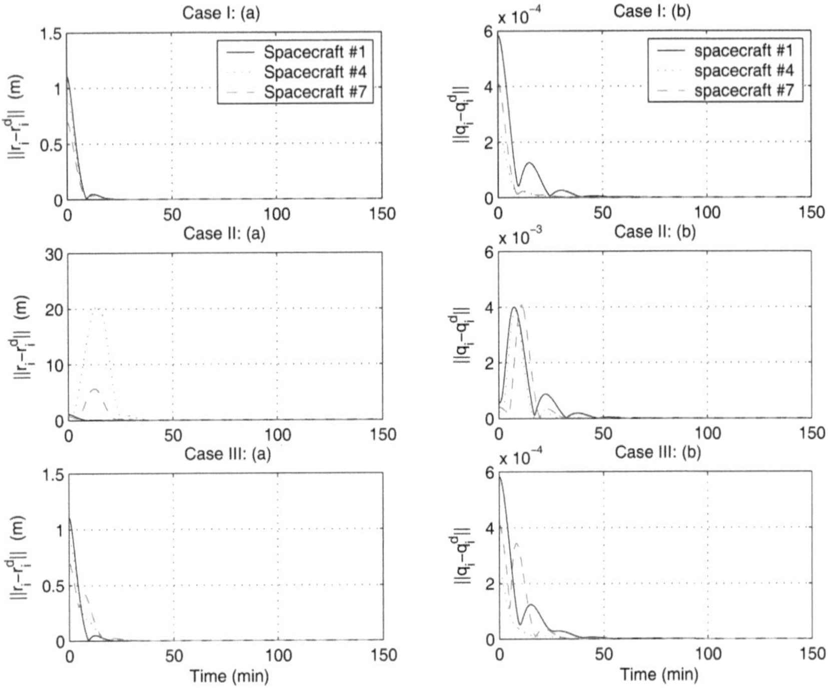

Fig. 12.7. The absolute position and attitude tracking errors. In Case I, there is neither actuator saturation nor formation feedback. In Case II, there is actuator saturation but no formation feedback. In Case III, there are both actuator saturation and formation feedback.

In Fig. 12.8, we plot the relative position and attitude errors between some of the spacecraft in these three cases. Based on the configuration, the desired relative distance between spacecraft #1 and #2 and the desired relative distance between spacecraft #1 and #6 should be equal. The desired relative distance between spacecraft #3 and #7 and the desired relative distance between spacecraft #5 and #9 should also be equal. We plot $|\,\|r_1 - r_2\| - \|r_1 - r_6\|\,|$ and $|\,\|r_3 - r_7\| - \|r_5 - r_9\|\,|$ in part (a) as examples to see how well the formation shape is preserved. The desired relative attitude between each spacecraft should be equal based on our previous assumption. We plot $\|q_1 - q_4\|$, $\|q_4 - q_7\|$, and $\|q_7 - q_1\|$ in part (b) as examples to see how well the relative orientation relationships between these spacecraft are preserved. Similarly, the relative position tracking errors in Case II are larger than those in the other two cases due to control force saturation. In Case III, with formation feedback, the relative position errors are smaller than those in Case II. The relative attitude errors in Case III are even smaller than those in the other two cases due to formation feedback.

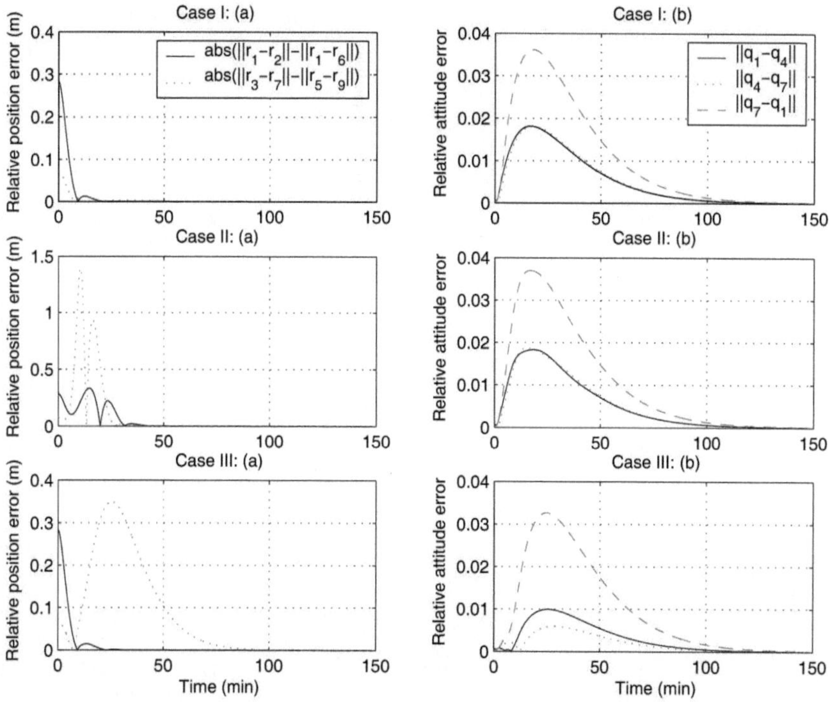

Fig. 12.8. The relative position and attitude errors. In Case I, there is neither actuator saturation nor formation feedback. In Case II, there is actuator saturation but no formation feedback. In Case III, there are both actuator saturation and formation feedback.

In Fig. 12.9, we plot the control effort for spacecraft #1 in these three cases. We can see that both the control force and control torque approach zero asymptotically. We can also see that τ_z saturates in Case II during the initial time period, whereas this saturation is mitigated with formation feedback introduced in Case III.

12.5 Notes

The results in this chapter are based mainly on [188]. Section 12.2.1 is from [22], and Section 12.3.4 is from [189]. For further results in spacecraft formation flying, see [39, 94, 103, 117, 142, 200, 244, 245].

Acknowledgment is given to

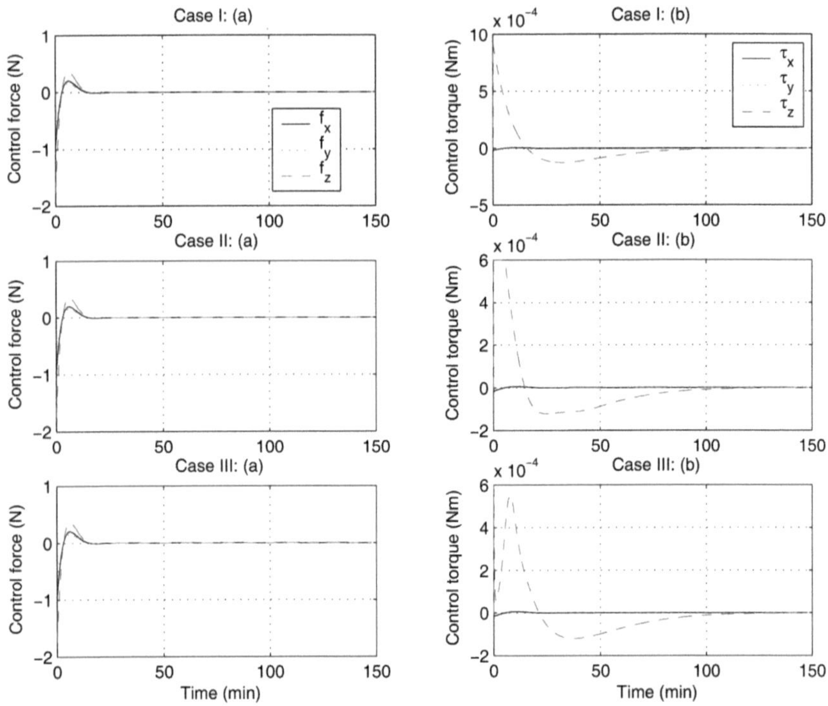

Fig. 12.9. The control effort for spacecraft #1. In Case I, there is neither actuator saturation nor formation feedback. In Case II, there is actuator saturation but no formation feedback. In Case III, there are both actuator saturation and formation feedback.

13

Cooperative Fire Monitoring with Multiple UAVs

The objective of this chapter is to explore the feasibility of using multiple low-altitude, short endurance (LASE) UAVs to monitor and track cooperatively the propagation of large forest fires. A real-time algorithm is described for tracking the perimeters of fires with an onboard infrared sensor. Using this algorithm, we apply the design scheme in Section 8.4.1 to develop a distributed multi-UAV approach to monitoring the perimeter of a fire. We propose a distributed load balancing algorithm that extends the discrete-time consensus algorithm in Chapter 2. The UAVs are assumed to have limited communication and sensing range. The effectiveness of the approach is demonstrated in simulation using a six degree-of-freedom dynamic model for the UAV and a numerical propagation model for the forest fire. Salient features of the approach include the ability to monitor a changing fire perimeter, the ability to add and remove UAVs systematically from the team, and the ability to supply time-critical information to firefighters.

13.1 Problem Statement

Forest fires cause billions of dollars in damage to property and the environment every year. To combat forest fires effectively, their early detection and continuous tracking is vital. With the help of advanced image processing techniques, many methods have been developed to detect forest fires in remote regions using satellite images (see [77, 112]). Such images are typically captured by low earth-orbiting satellites with an orbital period of about 10 hours and with a resolution that is sufficient only for fire detection. However, firefighters need frequent and high-quality information updates on the progress of fires to fight them effectively and safely. Because current forest fire monitoring techniques are deficient, firefighters are often required to enter fire regions with little knowledge of how and where the fire is propagating, placing their lives at risk. For these reasons, there is a need to develop more effective fire monitoring technologies.

High-altitude, long-endurance (HALE) UAVs such as the ALTAIR have the potential to increase image resolution and update rates over satellite based systems (see [8]). However, the limited availability of HALE systems during peak fire season may limit their overall effectiveness and emphasizes the need for lower cost, locally implementable systems.

Low-altitude, short-endurance (LASE) UAVs are expected to be a key technology for enhanced fire monitoring. Flying at low altitude, these UAVs can capture high-resolution imagery and broadcast frequent updates to fire crews. NASA is actively pursuing this possibility with ongoing research projects aimed at tracking the growth of fires using LASE UAVs (see [248]). However, a number of challenges have to be solved before LASE UAVs can be used for fire monitoring. First, with the fire growing and changing directions, UAVs need to be able to plan their path using limited real-time information. Second, LASE UAVs typically cannot carry enough fuel to endure a long fire-fighting mission, which means that the UAV must have the intelligence to return to the home base for refueling. Furthermore, for large forest fires, the information update rate may still be too low if only a single LASE UAV is employed. Finally, we note that fires generate tremendous heat and turbulence directly above the burning region. Therefore, crossing directly over the fire will place low-altitude UAVs at significant risk. As a consequence, LASE UAVs are effectively restricted to the air space over the unburned region of the fire.

The objective of this chapter is to explore the feasibility of using multiple LASE UAVs to monitor and track cooperatively the propagation of large forest fires. By using teams of inexpensive, rapidly deployable LASE UAVs, the complexity of the system will shift from the hardware platform to the cooperative control strategies employed to coordinate fire monitoring operations. Although teams of LASE UAVs will be more robust to single-point failures than a single satellite or a HALE UAV, several technical challenges must be addressed to enable their successful implementation. Issues addressed in this chapter include overcoming limited communication range and flight duration, developing a suitable coordination strategy for fire monitoring, and forming team consensus in the presence of noisy or uncertain information.

Figure 13.1 shows the forest fire monitoring scenario considered in this chapter. The objective is to capture images along the perimeter of the fire and upload the location of the fire perimeter (with associated imagery) to the base station as frequently and with as little latency as possible.

We make the following assumptions: First, we assume that each UAV can collect or receive sufficient information onboard to plan and adjust its path autonomously. This allows the UAV to adapt its path according to the fire perimeter. In particular, each UAV is assumed to be equipped with an infrared camera that captures images of a small region beneath it. An infrared camera is particularly suitable for fire monitoring because it detects the regions of the ground with the highest temperature. Second, each UAV is assumed to have limited communication range, which means that it cannot upload data to the

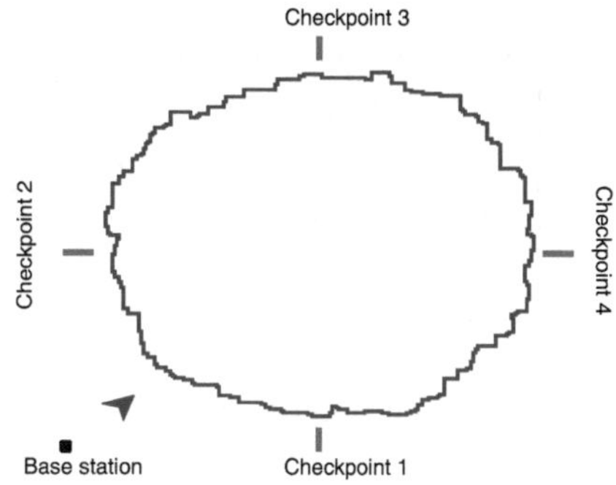

Fig. 13.1. Fire monitoring scenario. The line represents the fire perimeter. A base station deploys multiple UAVs to monitor the propagation of the fire.

base station unless it is within a certain range of the station, and it cannot communicate with other UAVs unless they are within a specified proximity. Finally, each UAV is assumed to have limited fuel, which implies that it must periodically return to the base station for refueling.

The delay between the times when images are collected and when they are transmitted to the base station, can serve as a measure of the quality of the fire monitoring algorithm. Let $\delta(x, t)$ denote the latency associated with information about the position x along the perimeter at time t. As time passes, the information at the base station grows older (more latent) until a UAV arrives to transmit the latest information it has gathered. For a particular position x along the perimeter, $\delta(x, t)$ will simply increase with time until it is replaced by the data downloaded from a UAV. Figure 13.2 gives a typical depiction of latency evolution for a particular point x_0 on the perimeter of the fire. The vertical edges of the sawtooth waveform represent the transmission of data from the UAV to the base station, and the linearly increasing portion of the waveform represents the increase in latency between UAV updates. The minimum latency $\delta_{\min}$ corresponds to the time of flight between the point of interest and the base station. The maximum latency depends on the total time required to make an observation at x_0 and to deliver that data to the base station.

In this chapter we develop a cooperative surveillance strategy that minimizes the latency of fire perimeter measurements delivered to the base station. This is done by minimizing the time of flight between points on the perimeter and the base station and by maximizing the frequency of measurement updates delivered to the base station.

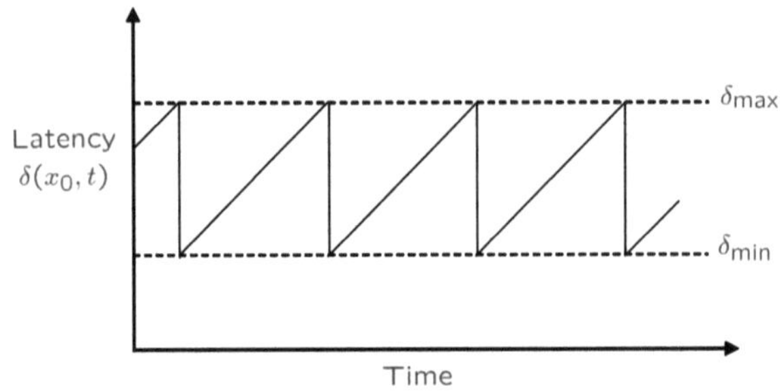

Fig. 13.2. Latency between updates for a point x_0 on the perimeter

13.2 Fire Perimeter Tracking for a Single UAV

Cooperative fire monitoring relies upon the ability of each individual UAV to track the fire perimeter independently. In this section, we will briefly describe a robust fire tracking algorithm that we have implemented in simulation. We assume that each UAV is equipped with an infrared camera on a pan and tilt gimbal and an autopilot with functionality similar to the one described in [21, 108]. Our approach to fire perimeter tracking can be summarized in six steps that are each performed at the frame rate of the onboard infrared camera.

1. Scan through the infrared image, labeling each pixel as (a) burned (or currently burning) and (b) unburned.
2. Use a linear classifier to segment the image into two partitions with burned elements in one partition and unburned in the other.
3. Project the segmentation line from the camera frame into the world coordinates of the UAV.
4. Construct the set of reachable points T seconds into the future and parameterize this set as a function of the roll angle. Details about this construction have been reported in [23].
5. Command the roll angle that corresponds to the point in the reachable set that is a distance d (in world coordinates) on the unburned side of the fire perimeter.
6. Command the pan and tilt angles of the camera gimbal so that the segmentation line of the linear classifier divides the image into two equal parts.

This approach is well suited to tracking the perimeter of a fire because it usually evolves in a noncontiguous way. Fires will jump over roads and streams and spread around rocky terrain. In addition, fire spreads faster uphill than downhill. In mountainous environments, this characteristic leads to fingering phenomena where fires spread in finger patterns along ridges in the terrain. The linear classifier used in Step 2 effectively smooths through noncontiguous boundaries. The distance d in Step 5 ensures that the UAV does not fly directly over the burning fire, but rather flies at a safe offset distance d. Maneuvering the gimbal in Step 6 ensures the continued effectiveness of the linear classifier in Step 2.

The fire tracking scheme described in this section has been successfully demonstrated in computer simulation. Results will be briefly described in Section 13.4.

13.3 Cooperative Team Tracking

For firefighters on the perimeter of a fire, frequent updates about the progression of the fire are critical for safety. Motivated by the design scheme in Section 8.4.1, this section develops a distributed monitoring scheme that allows transmitting perimeter information as often as possible to the base station. We will assume that all UAVs fly at identical constant velocities. We will also assume that the fire perimeter is homeomorphic to a circle. However, the algorithm presented in this section will function correctly if the perimeter is homeomorphic to a line, and the UAVs are commanded to reverse direction when the end of the perimeter is reached.

13.3.1 Latency Minimization

When a UAV transmits its information to the base station, an associated latency profile (in terms of time-stamped images) accompanies the data. Let the latency of a point x at the time of the base station update be denoted $\rho(x)$. Note that $\delta(x,t) = \rho(x)$ when a UAV updates the base station and $\delta(x,t) = \rho(x) + (t - t_{\text{update}})$ between updates. The objective is to design a cooperative monitoring scheme that minimizes $\rho(x)$ for every x and updates the base station as often as possible.

Figure 13.3a shows the latency of the perimeter of the fire when a clockwise flying UAV arrives back at the base station after a complete tour of the perimeter. The thickness of the path denotes the latency of the base station update of that point. Because the state of the fire is transmitted only after the UAV has traversed the entire fire perimeter, the greatest latency is associated with data gathered at the beginning of the flight near the base station. Because the UAV is traveling at constant velocity, the latency profile is a linear function of the distance traveled: $\rho(x) = (P - x)/v$, where v is the velocity of the UAV and P is the perimeter. The base station receives updates only as fast as the UAV can traverse the entire fire perimeter. The total latency of one traversal is given by $\int_0^P \rho(x)\, dx = 0.5P^2/v$.

Let one UAV be assigned to survey the upper half of the perimeter, and a second is assigned to the lower half. If the UAVs depart from the base station simultaneously and fly at the same speed, the update rate will be the same as the single UAV case (because both UAVs arrive back at the base station at the same time), but the latency of the information on both sides of the base

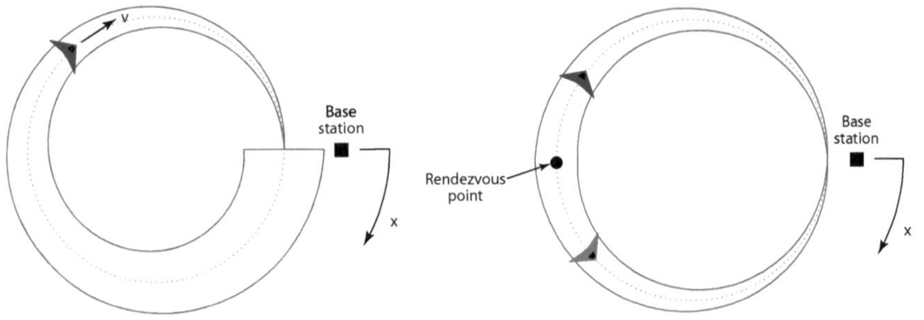

(a) Single UAV (b) Two UAVs

Fig. 13.3. (a) Latency profile for a single UAV monitoring a static circular fire. The thickness of the path denotes the latency of information at that point when it is transmitted to the base station. (b) Latency profile with a pair of UAVs monitoring a static circular fire in opposite directions.

station will be symmetrical and reduced, as can be seen in Fig. 13.3b. Here, the latency profile is given by

$$\rho(x) = \begin{cases} \frac{x}{v}, & \text{for } 0 \le x \le P/2, \\ \frac{P-x}{v}, & \text{for } P/2 < x \le P, \end{cases}$$

and the overall latency of the scheme is $\int_0^P \rho(x)\,dx = 0.25P^2/v$, which is half that of the single UAV case.

For UAVs that follow the perimeter and fly at constant velocity, the latency profile shown in Fig. 13.3b is the minimum possible latency for every x along the perimeter of the fire. To see that this is true, note that the minimum latency of data gathered at x on the perimeter is the time needed to travel from x to the base station along the shortest path. Dividing the perimeter equally between two UAVs ensures that the distance traveled between any point on the perimeter and the base station is minimal. The consequence is that adding more than two UAVs will not improve the latency profile $\rho(x)$ for any particular point on the perimeter. However, the rate at which updates occur will increase linearly with the number of UAV pairs employed and the effective latency will be reduced. This is illustrated in Fig. 13.4 where four UAV pairs are used to survey the perimeter. To maintain the minimum latency profile of Fig. 13.3b and to maximize the frequency of updates at the base station, UAVs should be distributed equally around the perimeter with each UAV assigned to monitor a segment of length P/n where n is the number of UAVs.

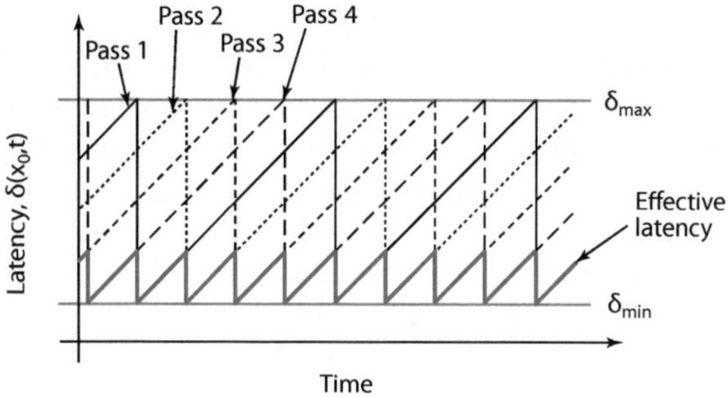

Fig. 13.4. Effective latency for a point on a perimeter surveyed by four UAV pairs

13.3.2 Distributed Fire Monitoring Algorithm

For a fixed perimeter length and a fixed number of UAVs, the minimum latency configuration occurs when pairs of UAVs are uniformly spread along the perimeter of the fire in both directions (*i.e.*, for every pair of UAVs, one is headed in the clockwise direction and the other in the counterclockwise direction). Pairs of UAVs will meet, transmit gathered information, and then each UAV will reverse its direction to meet its neighbor in the other direction. To facilitate refueling, UAVs can exchange roles at a rendezvous so that team members with the least fuel are nearest to the base station.

The aim of our distributed algorithm is to converge to this minimum latency configuration. The algorithm must converge for any perimeter size and must readjust when the perimeter length or the number of pairs of UAVs changes. By designing the algorithm so that changes in the system parameters are propagated across the team quickly, we will be able to address insertion and deletion of UAVs, as well as expansion and contraction of the perimeter.

The fundamental idea is for each UAV to take the action that will allow neighboring pairs to "share" the perimeter between them. When two UAVs meet, each has knowledge of the length just traveled from its previous rendezvous. The sum of these lengths can then be divided equally between them: the UAV that has traveled the least will loiter at the midpoint of this segment to wait for its neighbor the next time the two are to meet.

To illustrate the idea, consider a simple line segment $\overline{ab}$ with two UAVs tasked to gather information cooperatively along the segment, as shown in Fig. 13.5. Let ℓ_i be the distance traveled by the ith UAV from the last end-

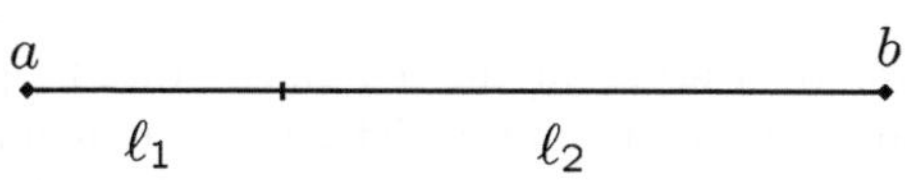

Fig. 13.5. Simple segment perimeter tracking

point visited. In Fig. 13.5, UAV 1 has traveled the least, so after returning to endpoint a, it will travel $\frac{\ell_1+\ell_2}{2}$ and then begin to loiter. UAV 2 will return to endpoint b and then reverse direction until it encounters UAV 1. Because UAV 1 traveled the shortest distance, it will arrive at the midpoint of $\overline{ab}$ first, and when both arrive, each will have traveled the same distance (*i.e.*, $\ell_1 = \ell_2$) from the endpoints which implies that the UAV pair has achieved the minimum latency configuration. An example of two UAV pairs monitoring a fixed perimeter of size $2P$ is shown in Fig. 13.6.

Any change in the size of the segment will be tracked because the pair effectively measures the current perimeter length by summing the distances traveled from the endpoints. In other words, because the UAVs have memory of the state of the perimeter from only one previous iteration, a continuous load balancing algorithm will track finite changes in the perimeter. To enable information on the growth of the perimeter to be accounted for more rapidly, the UAV assigned to wait for its neighbor will loiter at the point a distance $\frac{\ell_1+\ell_2}{2}$ from where the endpoint was previously. By measuring the discrepancy of ℓ_1 and the new distance back to the endpoint, UAV 1 can update the loiter distance to negate the effect of the growth in that region.

Adding UAVs to the perimeter is equivalent to stringing perimeter segments together that have changing endpoints: the endpoints for a segment shared by one pair of UAVs are the outside neighbors of these UAVs. We will use Monte Carlo simulations to verify that pairwise load balancing will lead to team convergence. We maintain that by balancing the length shared by every pair of UAVs, the team as a whole will spread itself evenly around the perimeter and so achieve the minimum latency configuration. If the algorithm can be shown to converge for arbitrary initial conditions with an arbitrary number of team members, then insertion/deletion can be analyzed by considering the modified system (after the insertion/deletion) with initial conditions given from the state of the original system at the time of the insertion/deletion. Each UAV implements the following algorithm, which is an extension of discrete-time consensus algorithm (2.4):

Load Balancing Algorithm

1. Maintain an estimate of the distance traveled from the last rendezvous in each direction (each UAV shares a segment with its clockwise neighbor and its counterclockwise neighbor).
2. At a rendezvous, the UAV that has traveled the smallest distance since its last rendezvous agrees to loiter at the midpoint of the shared segment the next time it is tracking the perimeter in that direction (clockwise/counterclockwise).
3. If the endpoint of a segment has changed (due to perimeter growth or neighbor actions), then the loiter distance is augmented by the change in distance of the endpoint. This keeps the loiter point at the same position relative to the segment length as communicated during the rendezvous.

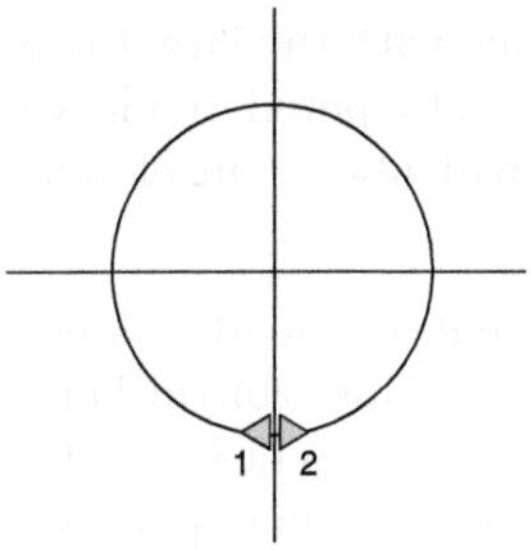

(a) UAVs 1 and 2 are launched

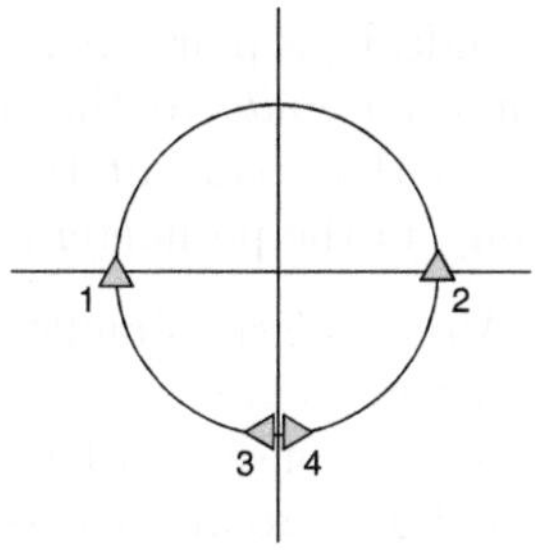

(b) UAVs 3 and 4 are launched

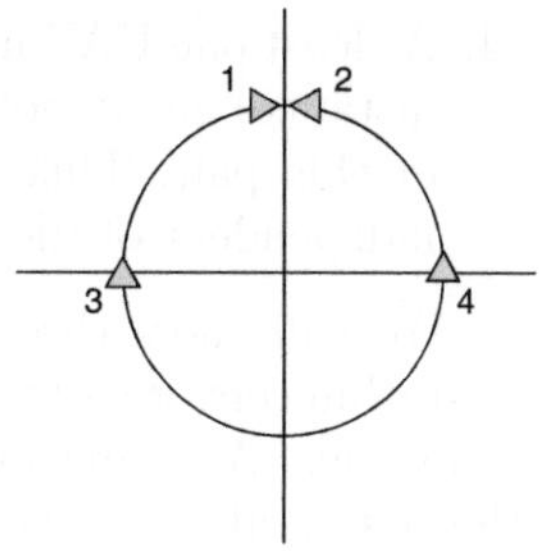

(c) $\ell_1 = \ell_2 = P$, so neither will loiter

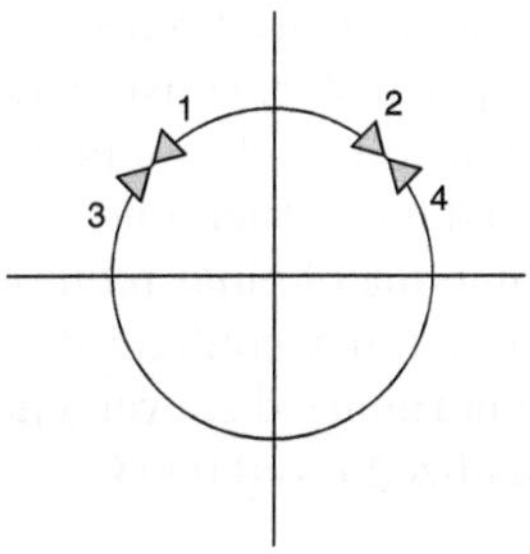

(d) $\ell_1 = 0.25P$, $\ell_3 = 0.75P$, so UAV 1 will loiter at $P/2$

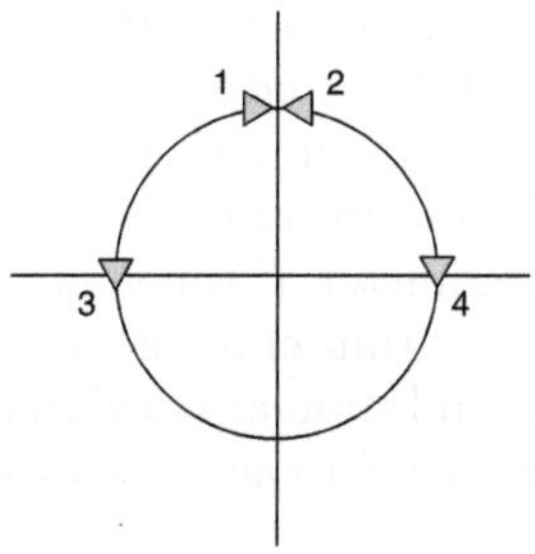

(e) UAVs 1 and 2 return to their previous rendezvous point

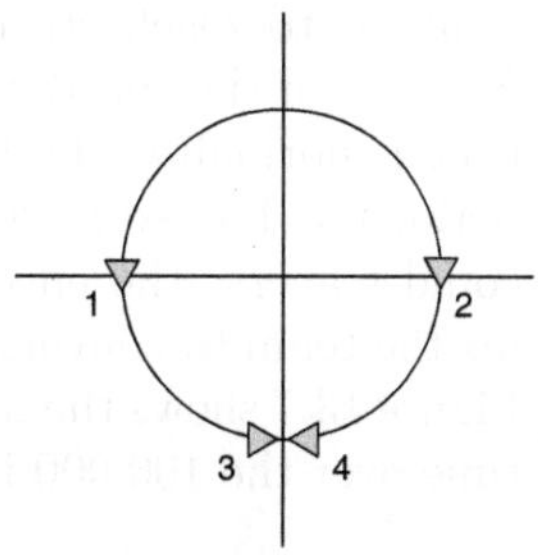

(f) UAVs 3 and 4 return to their previous rendezvous point; UAVs 1 and 2 begin loitering

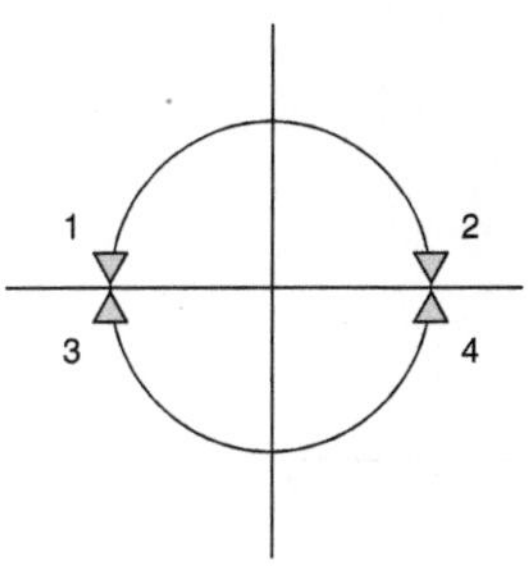

(g) $\ell_1 = \ell_3 = P/2$, so neither will loiter

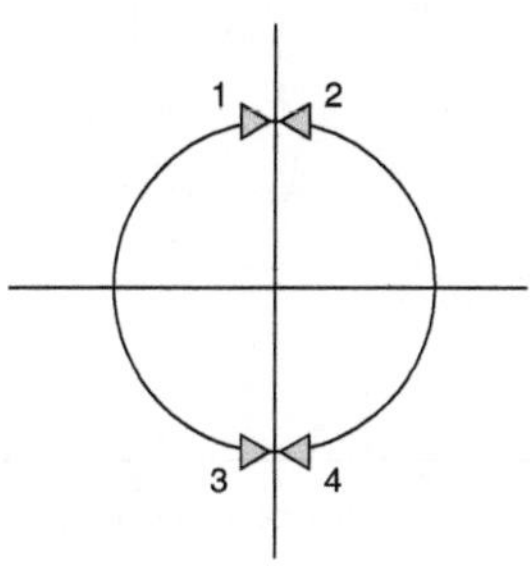

(h) $\ell_1 = \ell_2 = P/2$, so neither will loiter

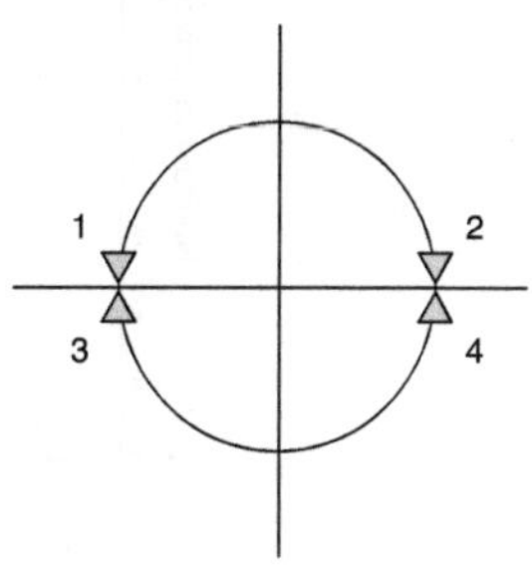

(i) Minimum latency configuration has been achieved

Fig. 13.6. Example of fixed perimeter length $2P$ when four UAVs implement the load balancing algorithm

4. At least one UAV in a rendezvous pair (*e.g.*, the one with the larger travel distance) must not loiter *en route* to the next anticipated rendezvous of this pair. This ensures that pairs of UAVs will always meet again, independent of the change in the perimeter.

We will show through Monte Carlo simulation that the load balancing algorithm converges to the minimum latency configuration for arbitrary initial conditions. A simulation instance consists of launching n pairs of UAVs from the base station at random times around a fixed length circular perimeter. Each member of the team continuously balances the load shared with each of its two neighbors. The simulation continues until all agents are within ϵ of the minimum latency configuration, or the maximum time is reached.

For each $n \in \{2, \ldots, 7\}$, 100,000 simulations were performed and the time required to reach steady state recorded. Because time to convergence is a function of the speed of the UAVs and the size of the perimeter, convergence time is normalized by the time required for information to travel around the perimeter. For example, if the convergence time is listed as T, then one UAV could traverse the entire perimeter T times in the amount of time required for the team to converge (to within ϵ) to the minimum latency configuration. Figure 13.7 shows the mean and standard deviation in normalized convergence time over the 100,000 iterations for each n with normalized $\epsilon = 0.0003$.

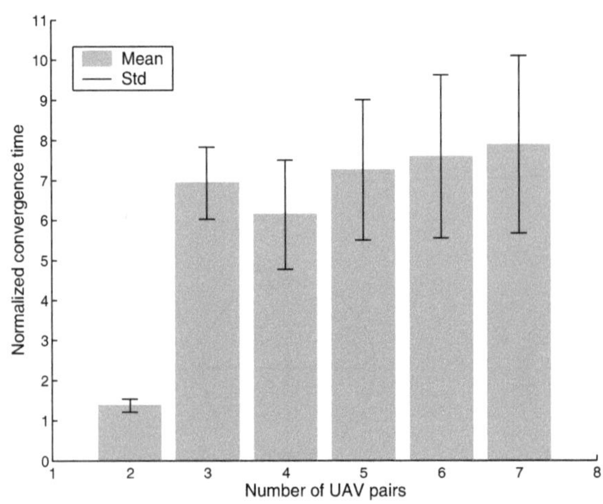

Fig. 13.7. Monte Carlo results for 100,000 iterations for $n = 2, \ldots, 7$ pairs of UAVs

Each simulation instance converged to the minimum latency configuration for every $n \in \{2, \ldots, 7\}$. An insertion or deletion of a pair of UAVs can be represented by a set of initial conditions given by the state of the system before the change with a new value of n. Monte Carlo simulations show that under any initial conditions for any n, the stability of the algorithm will not be affected.

The load balancing algorithm tracks the expansion/contraction of a perimeter by construction: the actual perimeter length is effectively sampled by the team continuously. In addition, Monte Carlo simulations have shown that insertion/deletion of UAV pairs will not affect the stability of the algorithm. We conclude that the load balancing algorithm will converge to the minimum latency configuration in the presence of team member insertion/deletion and finite changes in perimeter length.

13.4 Simulation Results

In this section, we present simulation results that highlight the effectiveness of the algorithms developed in this chapter. In Section 13.4.1, we describe the fire model that is used in all of our simulations. In Section 13.4.2, we present simulation results for the perimeter tracking algorithm described in Section 13.2. Section 13.4.3 contains the main cooperative fire tracking results.

13.4.1 Fire Model

To test the perimeter tracking and cooperative control algorithms effectively, a realistic, time-varying fire simulator was developed using the Ecological Model for Burning in the Yellowstone Region (EMBYR) (see [79, 86]). EMBYR divides the region of interest into a grid of cells, each with user specifiable properties that affect the spread of the fire. These properties include the type of foliage, the moisture level, and the elevation. At a given time step, the fire will spread from a burning cell to a nonburning cell, according to an independent stochastic event that is a function of the properties of the respective cells. By running EMBYR multiple times and averaging the result, we achieve realistic fire simulations like the one shown in Figs. 13.8–13.10.

13.4.2 Perimeter Tracking

Step (i) in the fire tracking algorithm, described in Section 13.2, scans through an image from an infrared camera and labels each pixel burned or unburned. In our simulation, we use the EMBYR fire model to generate an *a priori* model of the fire at each time step. During the execution of the simulation, the current simulation time and the state of the UAV are used to generate a binary image where 0 represents an unburned element and 1 represents a burned element. Steps (ii)–(vi) are implemented, as described in Section 13.2. Simulation results of a single UAV tracking a fire perimeter are shown in Fig. 13.11. The commanded offset distance from the boundary of the fire is 100 m.

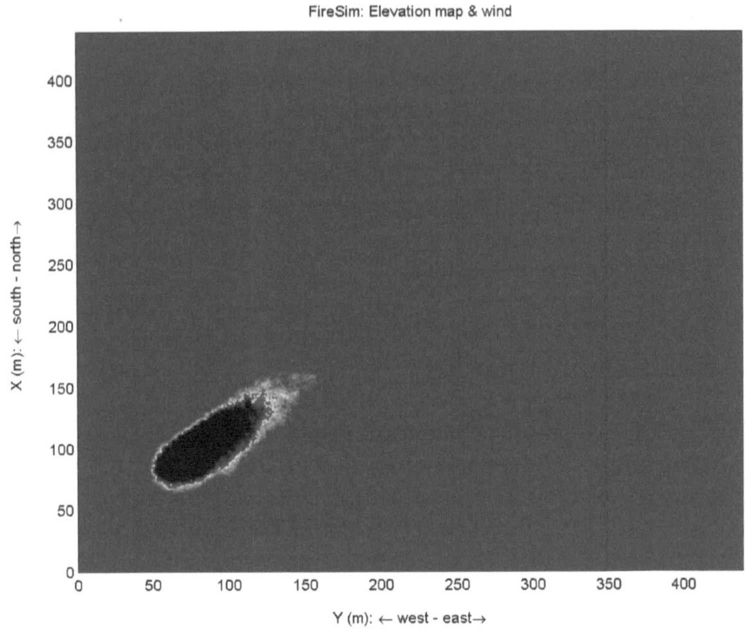

Fig. 13.8. Fire simulation of high wind conditions with an elevation gradient at $t = 2800$ s. The fire is spreading in the direction of the wind.

In reality, the fire's perimeter will not be contiguous due to noncombustible regions such as lakes or boulder fields. If the field of view of the camera is larger than gaps in the perimeter, than the algorithm in Section 13.2 will function properly because the linear classifier will find the best linear fit to the available data which will fit a line through the noncontiguous regions.

13.4.3 Cooperative Tracking

In this section, we describe the results of using multiple UAVs to monitor the perimeter of a forest fire simulated using the EMBYR model. The maximum communication range for the UAVs was set at 100 m (approximately 9 pixels). The velocity of the UAVs is 18 m/s. The fire perimeter is growing at an average of 2.8 m/s. Figure 13.12a shows four UAVs monitoring the fire with two more UAVs (bottom left) approaching the fire from the base station. A short time later, as shown in Fig. 13.12b, the two new UAVs (also at the bottom left) are loitering while waiting for their next rendezvous. After approximately 600 s have passed, the UAVs are in the steady-state configuration shown in Figs. 13.12c and 13.12d.

In the fire monitoring simulation, the UAV configuration does not precisely converge to the equilibrium predicted by the algorithm due to the dynamics of the UAVs and the fire. Specifically, the steady-state configuration error

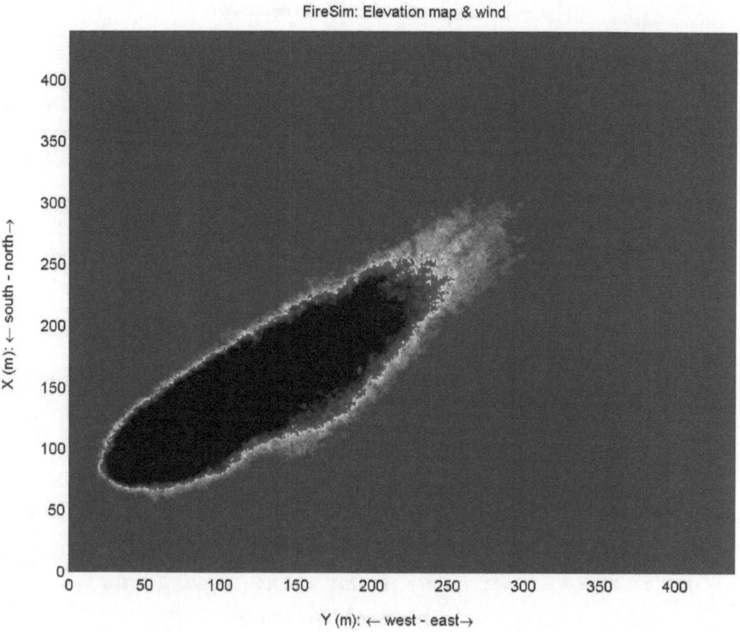

Fig. 13.9. Fire simulation of high wind conditions with an elevation gradient at $t = 6800$ s. The fire is spreading in the direction of the wind.

depends on the turning radius of the UAVs and the growth rate of the fire perimeter.

Despite these factors, the lengths $\ell_i(k)$ converge as time progresses. In Fig. 13.13, it can be seen that these lengths are increasing as time progresses due to the growth of the fire. Figures 13.14 and 13.15 also show the convergence of $\ell_i(k)$ for a static fire with perimeter of length 7.2 km, which is shown in Fig. 13.1. In Fig. 13.14, there are four UAVs initially monitoring the fire, and at $t = 1000$ s, two more UAVs are introduced. In Fig. 13.15, there are six UAVs monitoring the fire.

The checkpoints in Fig. 13.1 were used to show the latency of information at the base station. A UAV gathers information about a checkpoint when it is within 50 m from that point. This information is passed to its neighbor when the UAVs communicate. When a UAV passes the base station (halfway between the left and bottom checkpoints), the information concerning all the checkpoints known by that UAV is passed to the base station. The static perimeter is 7.2 km long, which would take a little less than 7 minutes to traverse, given that the UAVs were traveling at 18 m/s (a typical speed for small UAVs). The maximum latency for one UAV monitoring the fire would be 6.6 minutes. However, when more UAVs are used, this latency can be cut in half as well as reducing the latency for information near the base station. In

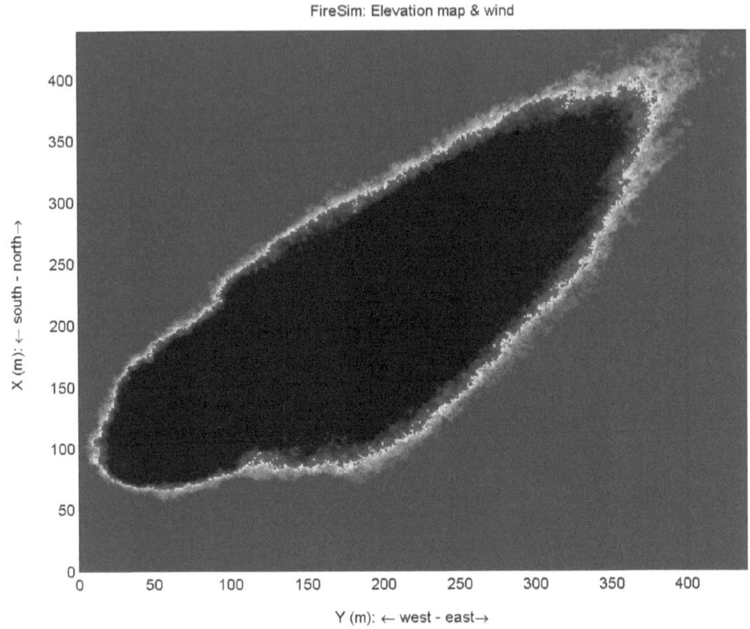

Fig. 13.10. Fire simulation of high wind conditions with an elevation gradient at $t = 10800$ s. The fire is spreading in the direction of the wind.

the event that firefighters can communicate with the UAVs, this would enable timely updates about the perimeter of the fire.

The minimum latency for each checkpoint is equivalent to a UAV flying directly from the checkpoint to the base station. The minimum latency of information for checkpoints 1 and 2 is about 45 s and for checkpoints 3 and 4 is about 125 s. The latencies for checkpoints 1 and 4 are plotted in Fig. 13.16 which demonstrates that the minimum latencies are achieved.

13.5 Notes

The results in this chapter are based mainly on [38].

Acknowledgment is given to

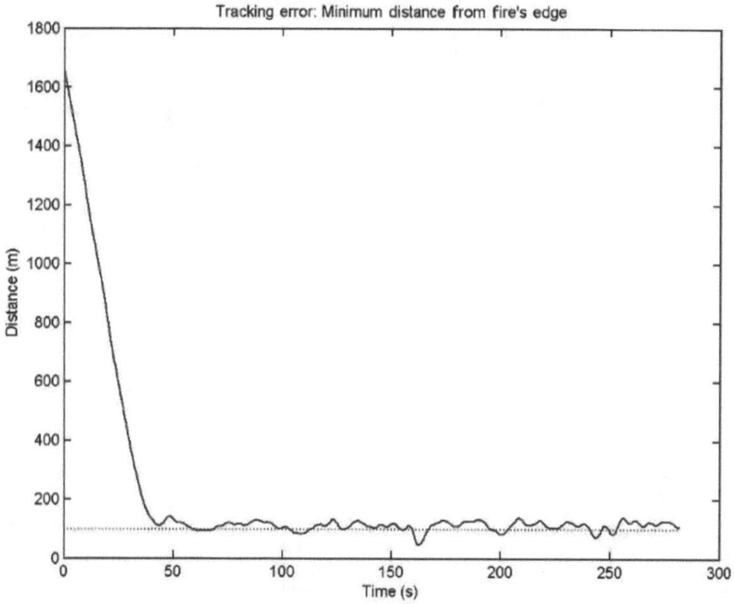

Fig. 13.11. Fire perimeter tracking error for a single UAV. The UAV was commanded to track the fire with an offset of 100 meters.

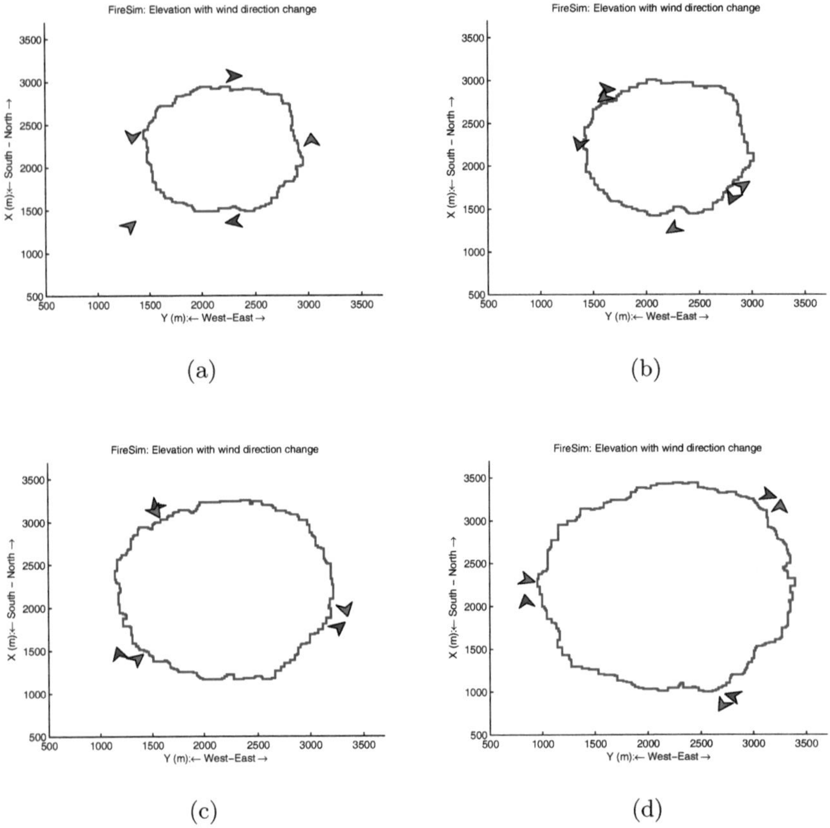

Fig. 13.12. Six UAVs are shown monitoring a growing fire

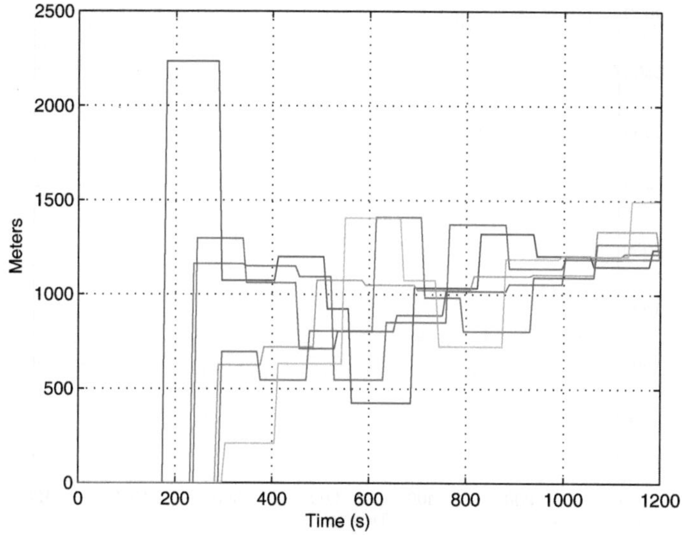

Fig. 13.13. The time histories of $\ell_i(k)$ are for each UAV

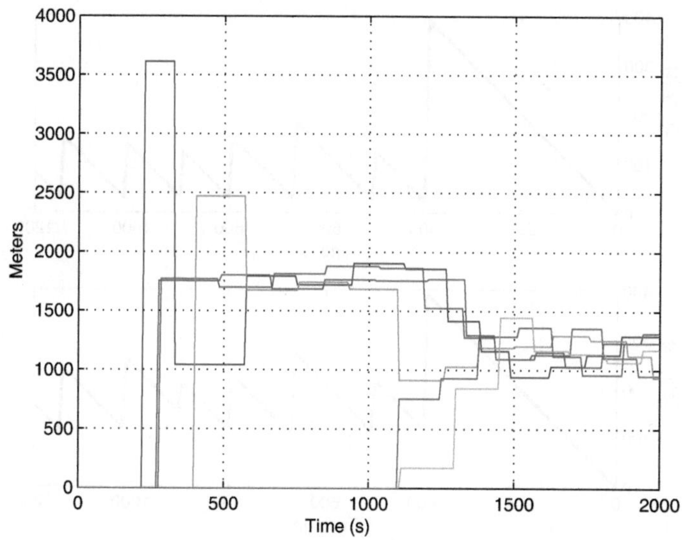

Fig. 13.14. Convergence of $\ell_i(k)$ for four UAVs monitoring the fire with two additional UAVs introduced at $t = 1000$ s

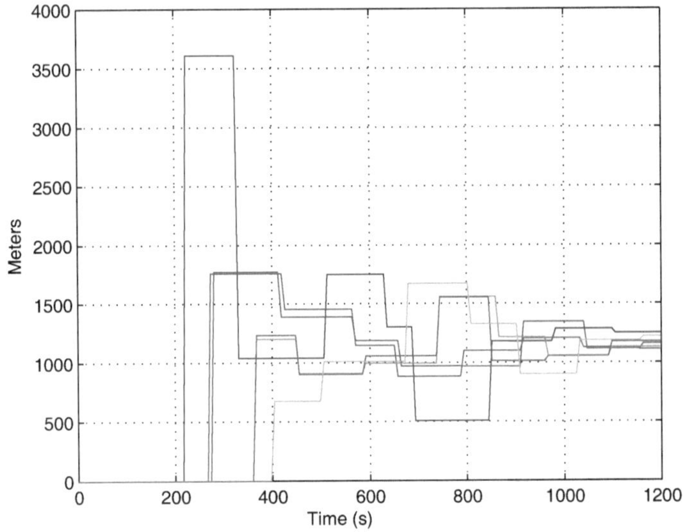

Fig. 13.15. Convergence of $\ell_i(k)$ for six UAVs

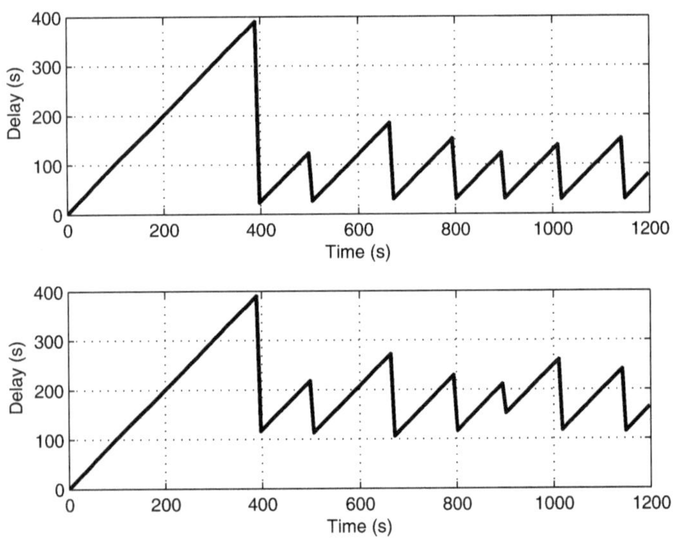

Fig. 13.16. Latencies for checkpoints 1 and 4 in Fig. 13.1

14

Cooperative Surveillance with Multiple UAVs

Contributed partly by Tim McLain

Numerous applications require aerial surveillance. Civilian applications include monitoring forest fires, oil fields and pipelines, and tracking wildlife. Applications to homeland security include border patrol and monitoring the perimeter of nuclear power plants. Military applications are numerous. The current approach to these applications is to use a single manned vehicle for surveillance. However, manned vehicles are typically large and expensive. In addition, hazardous environments and operator fatigue can potentially threaten the life of the pilot. Therefore, there is a critical need for automating aerial surveillance using UAVs. This chapter describes a cooperative control strategy for aerial surveillance that has been successfully flight tested on small (48 inch wingspan) UAVs. The cooperative control strategy follows the design methodology described in Section 8.3.

14.1 Experimental Test Bed

Over the past several years, BYU has developed a reliable and robust test bed for autonomous miniature air vehicles [21, 23]. Figure 14.1 shows the key elements of our test bed. The first frame shows BYU's Kestrel autopilot which is equipped with a Rabbit 3000 29 MHz processor, rate gyros, accelerometers, absolute and differential pressure sensors, and GPS. The autopilot measures only $1.5 \times 2.0 \times 0.75$ inches and weighs 18 grams. The second frame in Fig. 14.1 show the airframes used for the flight tests reported in this chapter. The airframes are 48 inch wingspan Zagi XS EPP foam flying wings that were selected for their durability and adaptability to different mission scenarios. Embedded in the airframe are the autopilot, batteries, a 1000 mW, 900 MHz radio modem, a GPS receiver, video transmitter, and a small analog camera. The third frame in Fig. 14.1 shows the ground station components. A laptop runs the virtual cockpit software that interfaces through a communication box to the UAVs. An RC transmitter is used as a stand-by fail-safe mechanism to facilitate safe operations.

Fig. 14.1. Hardware platform used to obtain experimental results. The first frame shows the Kestrel autopilot designed at BYU. The second frame shows the airframes used for this chapter. The third frame shows the ground station components for our test bed.

Cooperative flight tests with multiple UAVs are challenging to perform, and development of this capability has required several years of experimentation and refinement. One significant challenge in working with air vehicles is that the design/test cycle is considerably longer than that for ground robots. Flight experiments also entail a greater level of risk because minor errors can lead to catastrophic crashes. In doing experiments with UAVs, crashes are inevitable. Choosing an airframe that survives crashes greatly shortens the design and testing cycle. Experiments must be conducted in locations and at speeds where possible crashes do not pose a threat to life or property.

We have taken several steps to mitigate the logistical challenges of UAV flight experiments. These steps include the following:

Airframe Selection

The airframes that we use are designed to withstand crashes. Our airframes are constructed of EPP foam, Kevlar, and carbon fiber. Our workhorse airframe is the commercial off-the-shelf Zagi XS flying wing constructed of EPP foam [2]. The Zagi airframe was designed for midair RC dog fights and is extremely durable. It is inexpensive, hand-launchable, belly-landable, and easy to construct, modify, and repair. The EPP foam exterior protects the electronic components during crashes.

Preflight Protocols

Our standard preflight protocol includes a standard checklist to test for the most common causes of crashes. Items on the list include testing the telemetry link, calibrating the sensors, and verifying proper deflection of control surfaces.

In-flight Parameter Changes

We have developed a software architecture that allows any autopilot parameter to be modified in-flight. Parameters that can be modified in-flight include control loop gains and state estimation gains. Allowing parameters to be changed in-flight facilitates a rapid tune and debug cycle.

Switchable Software Modules

We have the capability to switch between different software modules in-flight. This has recently been used to test adaptive control algorithms on our UAVs. PID control was used during auto take-off and landing, but switched with an adaptive control module during flight. This capability allows rapid prototyping of algorithms without extensive validation and verification efforts.

Data Logging and Display

We can monitor and log any variable on the autopilot and display that data during the flight. Again, this capability allows algorithms to be debugged and tuned without requiring the UAV to be landed.

Hardware-in-the-Loop Simulator

We use a medium fidelity hardware-in-the-loop simulator that interfaces directly to the autopilot and runs in real time, allowing the autopilot code to be tested and debugged prior to flight experimentation.

In addition to the experimental capabilities described above, we developed two critical technologies before successfully demonstrating cooperative timing missions on UAVs: reliable intervehicle communication and robust trajectory tracking in wind.

Multiple Vehicle Communication. One of the challenges with miniature UAVs is that size, weight, and power constraints make high-speed, wide-bandwidth, digital communication links impractical. As an alternative, we use inexpensive low-power, half-duplex 9600-baud radio-frequency modems, which require all vehicles to share the same communication channel and the use of a turn-taking scheme to avoid packet collisions.

Cooperative control experiments have two competing communication demands: First, telemetry data from each of the vehicles must be communicated to the ground station on a regular basis for data logging and system monitoring. Second, coordination information must be sent among vehicles and between the ground station and vehicles. From an individual vehicle communication perspective, there are two important requirements: low latency of high priority data transmission and confirmation of properly receiving the transmission. To meet these communication challenges, the ground station acts as a server and controls which vehicles can communicate what information at what time. The transmissions can be divided into three categories: vehicle telemetry data, commands for the vehicle, and coordination information among vehicles. The ground station attempts to request telemetry data from each of the vehicles at a rate based on the number of vehicles (typically 3 to 4 Hz), and all other data are transmitted on a needs basis. The ground station dynamically assigns a priority to each type of transmission based on urgency, time in queue, and previous frequency of transmission. This allows high priority messages to be delivered with low latency and also prevents any type of transmission from flooding the communication channel. Although its bandwidth is suboptimal, this communication system meets the demanding requirements posed by cooperative UAVs and is very robust to packet loss.

Robust Trajectory Tracking in Wind. Airspeeds for the small UAVs that we fly are typically between 10 and 18 meters per second, or equivalently 20 to 40 miles per hour. A 10 mph wind, which is typical at 300 feet altitude on calm days, is a substantial disturbance for a small UAV. For the cooperative timing experiments presented in this chapter, the UAV is directed to be at a certain location at a specific time. Critical to completing the mission successfully are location and timing, which are significantly affected by wind. Experimental demonstration of cooperative timing requires high-fidelity trajectory tracking algorithms that mitigate the effects of wind.

For timing missions, preserving path length is critical to be able to satisfy the timing constraints. Flying extra distance due to poor path following adds unwanted time to the mission. To address these issues, a vector field path-following method has been implemented to enable the UAVs to accurately follow waypoint paths. It has been shown in [148, 149] that creating vector fields of desired heading to direct the UAV onto its desired path will result in asymptotically decaying tracking error in the presence of wind, provided ground track heading and ground speed are used in the control law. Use of this method has significantly decreased path-following errors, and consequently overall path length errors, thus enabling successful cooperative timing experiments.

14.2 Decentralized Cooperative Surveillance

The objective of this section is to apply the design methodology introduced in Section 8.3 to a cooperative aerial surveillance example and to demonstrate the effectiveness of the approach using high-fidelity simulation and flight tests of fixed-wing miniature air vehicles. We will consider two related variants of this problem. The first variant is persistent imaging, depicted in Fig. 14.2, where a team of n UAVs equipped with imaging sensors is tasked to persistently image a known target. If the field of view of the sensor is small with respect to the turning radius of the UAVs, the solution of this problem will require the team of UAVs to fly over the target at regular intervals. The second variant of the cooperative aerial surveillance problem is cooperative identification where a team of UAVs is required to fly over a target simultaneously, but along different approach angles. In the taxonomy introduced in Section 8.2, decentralized aerial surveillance problems are objectively coupled.

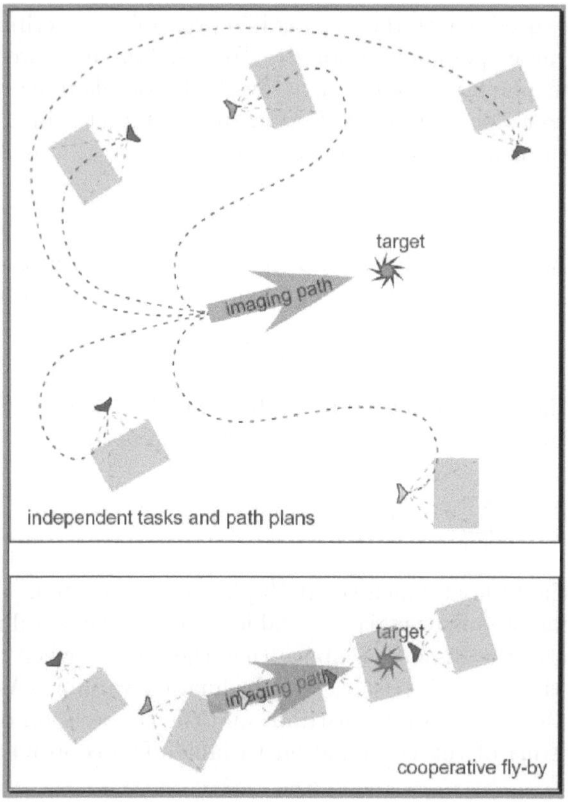

Fig. 14.2. Persistent aerial surveillance. The UAVs are initially performing an auxiliary task. Upon command, they coordinate their action to fly over the target at fixed time intervals.

14.2.1 Solution Methodology

The first step in addressing the aerial surveillance problem is to identify the cooperation constraint. Let $\hat{z}$ represent the location of the target which we assume is known to every vehicle, and let $z_i(t)$ denote the location of the ith vehicle at time t. The cooperation constraint is given by

$$J_{\text{constraint}}(\theta^*) = \sum_{i=1}^{n} \|z_i(\theta^* + \gamma_i \Delta) - \hat{z}\|^2, \tag{14.1}$$

where Δ is the desired spacing and $\gamma_i = (k-1)$ when the ith vehicle is assigned the kth position in the surveillance sequence. Note that if $J_{\text{constraint}} = 0$, then the vehicles fly by the target location at equally spaced intervals Δ seconds apart.

In (14.1), θ^* is the time at which the first vehicle passes over the target. It is clear that by increasing the loiter time, θ^* can be made arbitrarily large.

Therefore, we need to introduce an auxiliary optimization criterion that selects among the many possibilities for θ^*. Toward that end, we assume that a suitable path planning algorithm is available for planning waypoint paths from the current location of the UAV z_i to the target $\hat{z}$ in a certain time T. The algorithm will be denoted by the notation

$$\mathcal{W} = \texttt{planPath}(z_i, \hat{z}, T).$$

In addition, we assume a function $\texttt{Length}(\mathcal{W})$ that returns the path length of $\mathcal{W}$. The fuel expended in traversing a path is approximated by

$$\text{fuel} = c_f v_i \texttt{Length}\left[\texttt{planPath}(z_i, \hat{z}, T)\right],$$

where v_i is the airspeed along the path and c_f is a constant. Therefore, fuel will be minimized by selecting the cooperation objective as

$$J_{\text{objective}}(\theta^*) = \sum_{i=1}^{n} c_f v_i \texttt{Length}\{\texttt{planPath}[z_i(t_0), \hat{z}, t_0 + \theta^* + \gamma_i \Delta]\}, \quad (14.2)$$

where t_0 is the time at which the path planner is executed.

The second step is to make a suitable choice of the coordination variable. It is clear from the discussion above that the instant when the first vehicle passes over the target is a suitable coordination variable. As seen in (14.1) and (14.2), the cooperation constraint and the cooperation objective can be written in terms of the coordination variable. The coordination function is given by

$$J_{\text{cf},i}(\theta^*) = c_f v_i \texttt{Length}\{\texttt{planPath}[z_i(t_0), \hat{z}, t_0 + \theta^* + \gamma_i \Delta]\}.$$

The third step is to devise a centralized cooperation algorithm that solves (8.3). The formulation of the path planning problem ensures that the cooperation constraint is trivially satisfied. The objective function can be optimized with a Mixed Integer Linear Program (MILP) solver where γ_i are integers and θ^* is real. As it turns out, this particular problem has sufficient structure that it admits an analytic solution [139]. Once θ^* has been determined (by a centralized unit) then, according to (8.4), the ith vehicle implements the path given by

$$u_i = \texttt{planPath}[z_i(t_0), \hat{z}, t_0 + \theta^* + \gamma_i \Delta]. \quad (14.3)$$

The final step is to decentralize the algorithm using a consensus algorithm. To do so, let θ_i denote the ith vehicles instantiation of the coordination variable θ^*. Instead of (14.3), the ith vehicle implements

$$u_i = \texttt{planPath}[z_i(t_0), \hat{z}, t_0 + \theta_i + \gamma_i \Delta].$$

Given communication with the other vehicles, the coordination variable θ_i is then adjusted according to the consensus dynamics given in (2.2).

In the cooperative identification variant of this problem, we want all vehicles to arrive at the target location simultaneously and at different approach angles. For this problem, we set $\Delta = 0$ and pass the approach angle as an additional input to the path planning algorithm. In the next two sections, we will present simulation results for the decentralized cooperative identification problem and flight test results for the centralized persistent imaging and cooperative identification problem.

14.2.2 Simulation Results

To enable rapid prototyping of cooperative control algorithms, we have developed a medium-fidelity simulation environment for autonomous miniature air vehicles. The simulation environment consists of two components. The first is a six degree-of-freedom flight simulator with Digital Elevation Model (DEM) terrain data [3] and realistic wind models. The second component is an autopilot module that executes the same code that is implemented on the physical autopilot. The autopilot module connects to the same ground station software that is used to fly the miniature air vehicles. A screen shot of the flight simulator and the virtual cockpit are shown in Fig. 14.3.

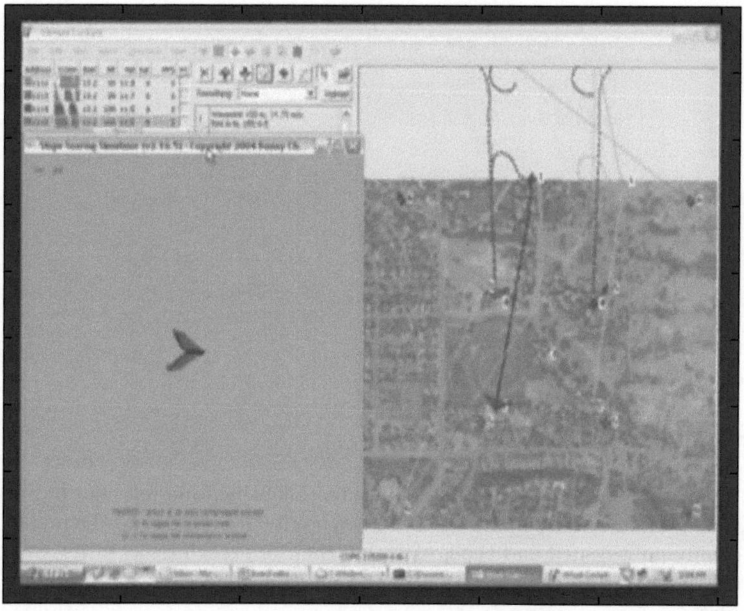

Fig. 14.3. Screen shot of the flight simulator (bottom left window in blue), and the virtual cockpit (background). The simulation environment enables rapid prototyping of cooperative control problems for autonomous miniature air vehicles.

The cooperative identification problem was simulated using four UAVs that were tasked to arrive at the target simultaneously with arrival angles differing by 90°. The average time to reach consensus to within 0.02 units was 6.2 seconds where one communication packet per second was allowed to be sent by each UAV to another UAV selected randomly from the team. Simulation results are shown in Fig. 14.4. Subplots (a) and (b) show the four UAVs loitering until the mission execution command is issued at the end of Subplot (b). The UAVs are all flying at distinct, preassigned altitudes to avoid collision as they pass over the target. Subplots (c) and (d) show the UAVs executing the cooperative identification mission.

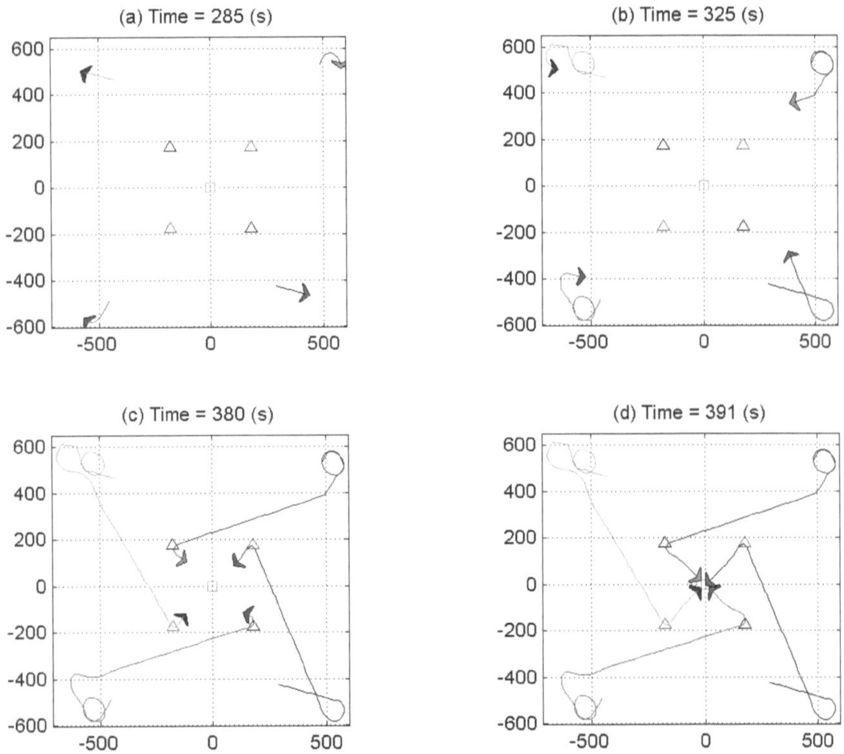

Fig. 14.4. Simulation results of a cooperative identification mission. In Subplot (a) the UAV are loitering around a specified waypoint. In Subplot (b) the mission execution command is issued and the UAVs plan their approach trajectories. Subplots (c) and (d) demonstrate the execution of the mission. The UAV are flying at distinct altitudes.

Figure 14.5 plots the average distance from the target versus time for each UAV and demonstrates the effectiveness of the decentralized coordination algorithm.

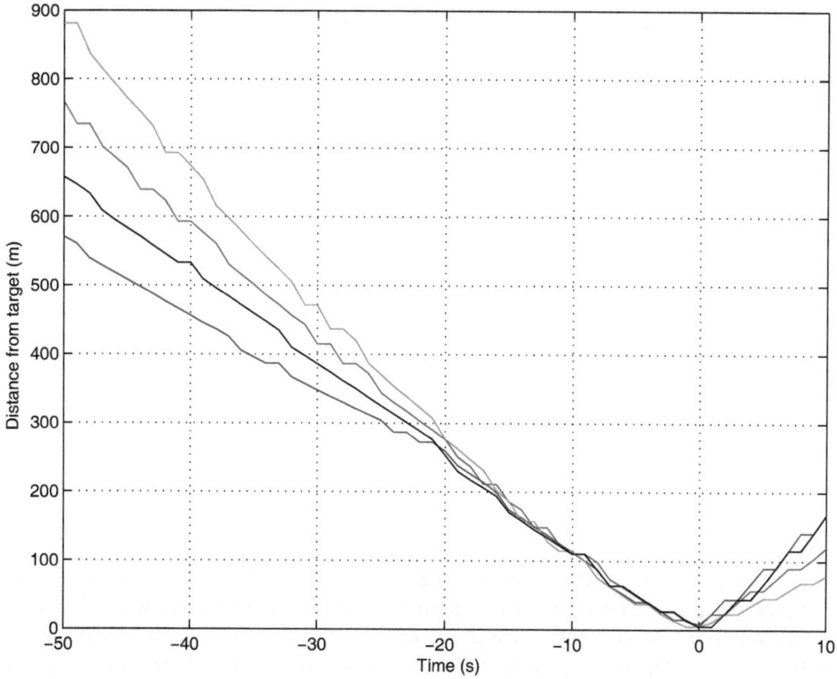

Fig. 14.5. A plot of the average distance from the target versus time for the simulated cooperative identification problem

14.2.3 Flight Tests

Flight tests were performed using three UAVs that were commanded to persistently image a target at fixed intervals of $\Delta = 3$ seconds. The three UAVs were initially commanded to loiter at specified GPS coordinates. The UAVs were then commanded to perform a persistent imaging mission and then return to their loitering coordinates. The mission was repeated six times during a 30-minute flight. The wind speeds during the flight were between 30 and 60% percent of the UAV airspeed and were from the southwest. The average error in arrival time for the six runs, in spite of the high wind conditions, was approximately 0.6 seconds.

Plots of the telemetry data are shown in Fig. 14.6. The imaging target is depicted by a circle, whereas the transition waypoints produced by the path planner are depicted by diamonds.

Figure 14.7 shows the case where one of the UAVs starts the mission from an initial target distance significantly different from that of the other vehicles. Note that the vehicle that is closest to the imaging point must plan a path that increases its distance, thus allowing the rest of the team to catch up with it.

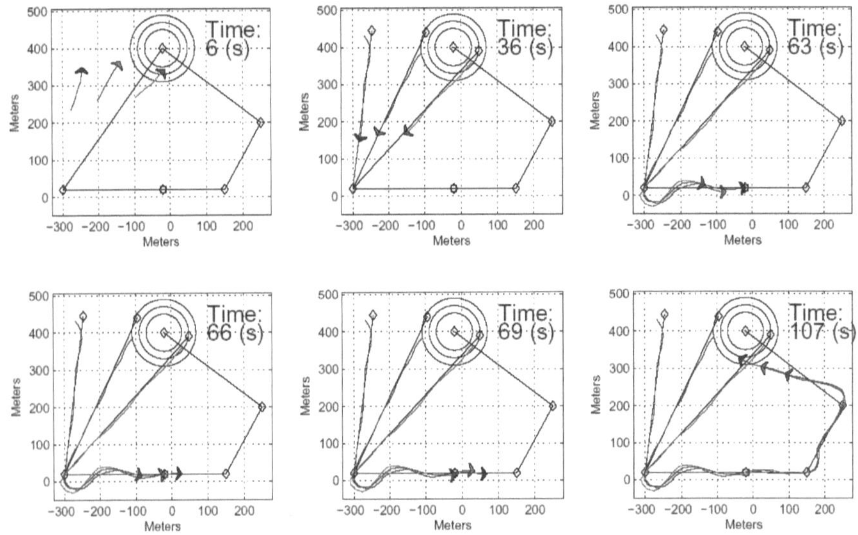

Fig. 14.6. Telemetry data for the flight tests of the persistent imaging mission. Subplot (a) shows the UAVs approaching a loiter waypoint. Subplot (b) shows the UAVs en route to the entry waypoint after the fly-by command has been issued. Subplots (c), (d), and (e) show the UAVs transition through the imaging target. Subplot (f) shows the UAVs returning to their loiter points.

In addition to the persistent imaging scenario, we have also flight tested the simultaneous arrival mission using three UAVs. Still shots of an arrival mission are shown in Fig. 14.8. Subplots (a) and (b) show the UAVs as they approach the cooperative rendezvous point. Subplot (c) shows the simultaneous arrival to within 0.3 seconds, and Subplot (d) shows the UAV departing from the rendezvous point. Telemetry plots for this mission are shown in Fig. 14.9.

14.3 Notes

The results in this chapter are based mainly on [26].

Acknowledgment is given to

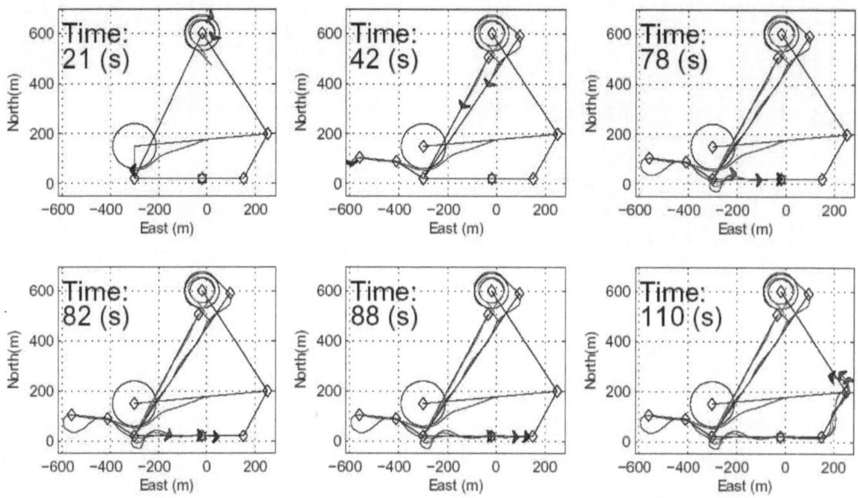

Fig. 14.7. Telemetry data for persistent imaging when one of the UAVs is significantly closer to the image point than the rest of the team.

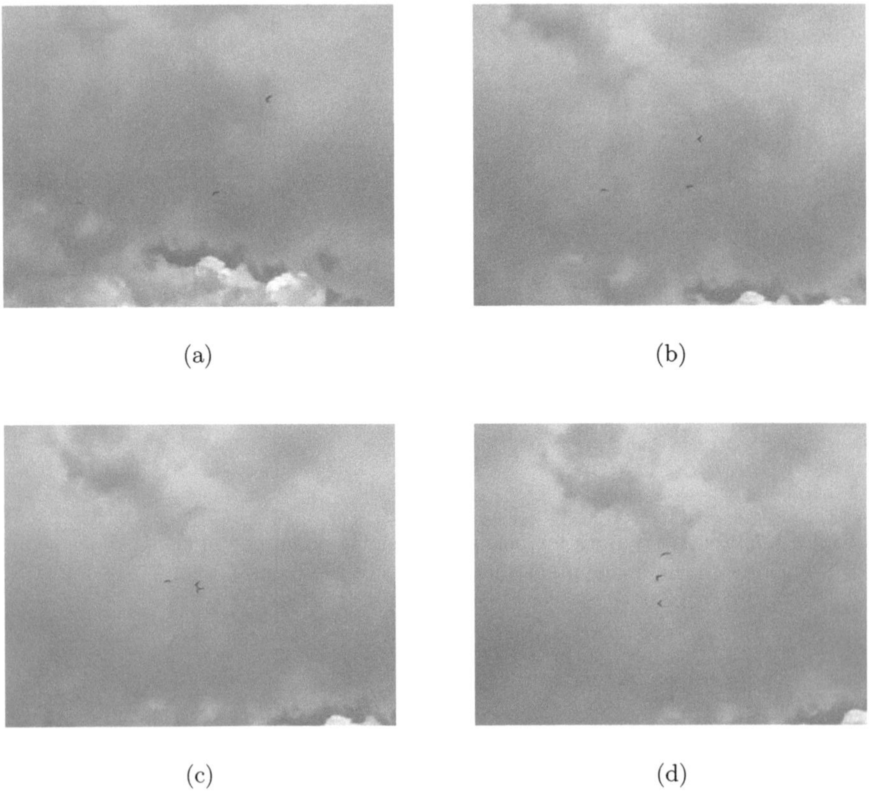

Fig. 14.8. Still shots of a cooperative timing scenario. Subplots (a) and (b) show three UAVs approaching the rendezvous point. Subplot (c) shows the rendezvous to within 0.3 seconds. Subplot (d) shows the UAVs immediately after the rendezvous.

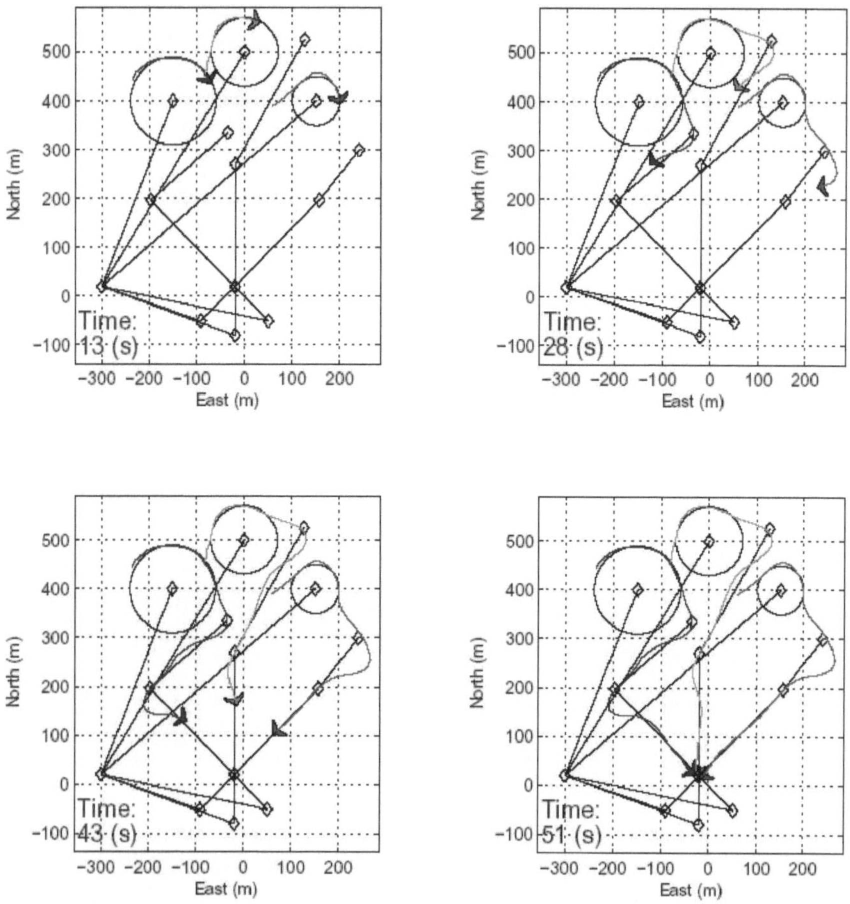

Fig. 14.9. Telemetry data for the simultaneous arrival mission. In the top right subplot, the UAVs are orbiting prescribed loitering points. In the top left subplot, the UAVs have just received the cooperative rendezvous signal. The bottom left subplot shows the UAVs *en route* to the rendezvous point, and the bottom right subplot shows the UAVs at the rendezvous point.

A

Selected Notations and Abbreviations

$\equiv$	identically equal
$\triangleq$	defined as
$\forall$	for all
$\in$	belongs to
$\subset$	a strict subset of
$\subseteq$	a subset of
$\rightarrow$	tends to
$\Longrightarrow$	implies
$\Longleftrightarrow$	equivalent to
$\sum$	summation
$\prod$	product
$\otimes$	Kronecker product
max	maximum
min	minimum
sup	supremum, the least upper bound
inf	infimum, the greatest lower bound
$\mathbb{R}^p$	set of $p \times 1$ real vectors
$\mathbb{R}^{m \times n}$	set of $m \times n$ real matrices
$\mathbb{C}$	set of complex numbers
$A > 0$	a positive matrix A
$A \geq 0$	a nonnegative matrix A
$\rho(A)$	spectral radius of matrix A
$\sigma(A)$	spectrum of matrix A
$\lambda_i(A)$	the ith eigenvalue of matrix A
$\lambda_{max}(A)$	the maximum eigenvalue of matrix A
$\lambda_{min}(A)$	the minimum eigenvalue of matrix A
$\det(A)$	determinant of matrix A
$\Gamma(A)$	directed graph of matrix A
$\mathrm{diag}(a_1, \ldots, a_p)$	a diagonal matrix with diagonal entries a_1 to a_p
$\mathrm{diag}(A_1, \ldots, A_p)$	a block diagonal matrix with diagonal blocks A_1 to A_p
log	logarithm function

tanh	tangent hyperbolic function
cosh	cosine hyperbolic function
$\|x\|$	norm of vector x
$\|x\|_p$	p-norm of vector x
$\|A\|_p$	induced p-norm of matrix A
$\mathrm{Re}(z)$	real part of number z
$\mathrm{Im}(z)$	imaginary part of number z
$\mathbf{1}_p$	$p \times 1$ column vector of all ones
$\mathbf{0}_p$	$p \times 1$ column vector of all zeros
$\mathbf{q}_I$	multiplicative identity quaternion $[0, 0, 0, 1]^T$
I_m	$m \times m$ identity matrix
$0_{m \times n}$	$m \times n$ zero matrix
$\mathcal{G}_p$	graph
$\mathcal{V}_p$	node set of a graph
$\mathcal{E}_p$	edge set of a graph
$\mathcal{A}_p$	$p \times p$ adjacency matrix
$\mathcal{L}_p$	$p \times p$ (nonsymmetrical) Laplacian matrix
AUV	autonomous underwater vehicle
CCD	charge-coupled device
GPS	global positioning system
HALE	high altitude long endurance
ISS	input-to-state stable
LASE	low altitude short endurance
LTI	linear time invariant
LTV	linear time varying
MAV	micro air vehicle
MRP	modified Rodriguez parameters
PID	proportional integral derivative
PWM	pulse-width modulation
RC	radio control
SIA	stochastic indecomposable aperiodic
UAV	unmanned air vehicle

B

Graph Theory Notations

It is natural to model information exchange among vehicles by directed or undirected graphs.[1] Suppose that a team consists of p vehicles. A *directed graph* is a pair $(\mathcal{V}_p, \mathcal{E}_p)$, where $\mathcal{V}_p = \{1, \ldots, p\}$ is a finite nonempty *node* set and $\mathcal{E}_p \subseteq \mathcal{V}_p \times \mathcal{V}_p$ is an *edge* set of ordered pairs of nodes, called *edges*. The edge (i, j) in the edge set of a directed graph denotes that vehicle j can obtain information from vehicle i, but not necessarily *vice versa*. Self-edges (i, i) are not allowed unless otherwise indicated. For the edge (i, j), i is the *parent node* and j is the *child node*. In contrast to a directed graph, the pairs of nodes in an *undirected graph* are unordered, where the edge (i, j) denotes that vehicles i and j can obtain information from each other. Note that an undirected graph can be viewed as a special case of a directed graph, where an edge (i, j) in the undirected graph corresponds to edges (i, j) and (j, i) in the directed graph. If an edge $(i, j) \in \mathcal{E}_p$, then node i is a *neighbor* of node j. The set of neighbors of node i is denoted as $\mathcal{N}_i$. A *weighted graph* associates a weight with every edge in the graph. In this book, all graphs are weighted. The *union* of a collection of graphs is a graph whose node and edge sets are the unions of the node and edge sets of the graphs in the collection.

A *directed path* is a sequence of edges in a directed graph of the form $(i_1, i_2), (i_2, i_3), \ldots$. An *undirected path* in an undirected graph is defined analogously. In a directed graph, a *cycle* is a directed path that starts and ends at the same node. A directed graph is *strongly connected* if there is a directed path from every node to every other node. An undirected graph is *connected* if there is an undirected path between every pair of distinct nodes. A *directed tree* is a directed graph in which every node has exactly one parent except for one node, called the *root*, which has no parent and which has a directed path to every other node. Note that a directed tree has no cycle because every edge is oriented away from the root. In undirected graphs, a *tree* is a graph in which every pair of nodes is connected by exactly one undirected path.

[1] This appendix is based mainly on [191].

A *subgraph* $(\mathcal{V}_p^s, \mathcal{E}_p^s)$ of $(\mathcal{V}_p, \mathcal{E}_p)$ is a graph such that $\mathcal{V}_p^s \subseteq \mathcal{V}_p$ and $\mathcal{E}_p^s \subseteq \mathcal{E}_p \bigcap (\mathcal{V}_p^s \times \mathcal{V}_p^s)$. A *(rooted) directed spanning tree* $(\mathcal{V}_p^s, \mathcal{E}_p^s)$ of the directed graph $(\mathcal{V}_p, \mathcal{E}_p)$ is a subgraph of $(\mathcal{V}_p, \mathcal{E}_p)$ such that $(\mathcal{V}_p^s, \mathcal{E}_p^s)$ is a directed tree and $\mathcal{V}_p^s = \mathcal{V}_p$. An *undirected spanning tree* of an undirected graph is defined analogously. The graph $(\mathcal{V}_p, \mathcal{E}_p)$ *has* or *contains* a directed spanning tree if a directed spanning tree is a subgraph of $(\mathcal{V}_p, \mathcal{E}_p)$. Note that the directed graph $(\mathcal{V}_p, \mathcal{E}_p)$ has a directed spanning tree if and only if $(\mathcal{V}_p, \mathcal{E}_p)$ has at least one node with a directed path to all other nodes. In undirected graphs, the existence of an undirected spanning tree is equivalent to being connected. However, in directed graphs, the existence of a directed spanning tree is a weaker condition than being strongly connected. Figure B.1 shows a directed graph that contains more than one directed spanning tree but is not strongly connected. Nodes 1 and 2 are both roots of directed spanning trees because both have directed paths to all other nodes. However, the graph is not strongly connected because nodes 3, 4, 5, and 6 do not have directed paths to all other nodes.

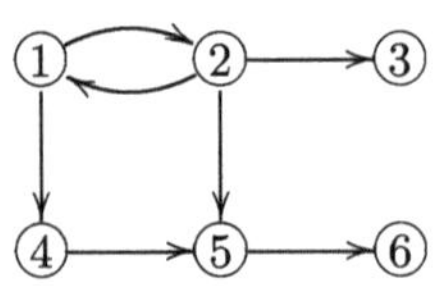

Fig. B.1. Directed graph among six vehicles. An arrow from node i to node j indicates that vehicle j receives information from vehicle i. This directed graph contains two directed spanning trees with root nodes 1 and 2 but is not strongly connected because nodes 3, 4, 5, and 6 do not have directed paths to all other nodes.

The *adjacency* matrix $\mathcal{A}_p = [a_{ij}] \in \mathbb{R}^{p \times p}$ of a directed graph $(\mathcal{V}_p, \mathcal{E}_p)$ is defined such that a_{ij} is a positive weight if $(j, i) \in \mathcal{E}_p$, and $a_{ij} = 0$ if $(j, i) \notin \mathcal{E}_p$. Self-edges are not allowed (*i.e.*, $a_{ii} = 0$) unless otherwise indicated. The adjacency matrix of an undirected graph is defined analogously except that $a_{ij} = a_{ji}$ for all $i \neq j$ because $(j, i) \in \mathcal{E}_p$ implies $(i, j) \in \mathcal{E}_p$. Note that a_{ij} denotes the weight for the edge $(j, i) \in \mathcal{E}_p$. If the weight is not relevant, then a_{ij} is set equal to 1 if $(j, i) \in \mathcal{E}_p$. A graph is *balanced* if $\sum_{j=1}^p a_{ij} = \sum_{j=1}^p a_{ji}$, for all i. For an undirected graph, $\mathcal{A}_p$ is symmetrical, and thus every undirected graph is balanced.

Define the matrix $\mathcal{L}_p = [\ell_{ij}] \in \mathbb{R}^{p \times p}$ as

$$\ell_{ii} = \sum_{j=1, j \neq i}^{n} a_{ij}, \qquad \ell_{ij} = -a_{ij}, \; i \neq j. \tag{B.1}$$

Note that if $(j, i) \notin \mathcal{E}_p$ then $\ell_{ij} = -a_{ij} = 0$. Matrix $\mathcal{L}_p$ satisfies

$$\ell_{ij} \leq 0, \; i \neq j, \qquad \sum_{j=1}^{p} \ell_{ij} = 0, \; i = 1, \ldots, p. \tag{B.2}$$

For an undirected graph, $\mathcal{L}_p$ is symmetrical and is called the *Laplacian matrix*. However, for a directed graph, $\mathcal{L}_p$ is not necessarily symmetrical and is sometimes called the *nonsymmetrical Laplacian matrix* [4] or *directed Laplacian matrix* [115].

Remark B.1. Note that $\mathcal{L}_p$ in (B.1) can be equivalently defined as $\mathcal{L}_p \triangleq D - \mathcal{A}_p$, where $D = [d_{ij}] \in \mathbb{R}^{p \times p}$ is the *in-degree matrix* given as $d_{ij} = 0$, $i \neq j$, and $d_{ii} = \sum_{j=1}^{p} a_{ij}$, $i = 1, \ldots, p$. Also note that for directed graphs, the definition of the nonsymmetrical Laplacian matrix given by (B.1) is different from the common definition of a Laplacian matrix for a directed graph in the graph theory literature (*e.g.*, [204]). However, we adopt the definition given by (B.1) for directed graphs due to its relevance to consensus algorithms. □

In both the undirected and directed cases, because $\mathcal{L}_p$ has zero row sums, 0 is an eigenvalue of $\mathcal{L}_p$ with the associated eigenvector $\mathbf{1}_p$, the $p \times 1$ column vector of ones. Note that $\mathcal{L}_p$ is diagonally dominant and has nonnegative diagonal entries. It follows from Theorem C.1 that, for an undirected graph, all nonzero eigenvalues of $\mathcal{L}_p$ are positive ($\mathcal{L}_p$ is positive semidefinite), whereas, for a directed graph, all nonzero eigenvalues of $\mathcal{L}_p$ have positive real parts. Therefore, all nonzero eigenvalues of $-\mathcal{L}_p$ have negative real parts. For an undirected graph, 0 is a simple eigenvalue of $\mathcal{L}_p$ if and only if the undirected graph is connected [141, p. 147]. For a directed graph, 0 is a simple eigenvalue of $\mathcal{L}_p$ if the directed graph is strongly connected [69, Proposition 3], although the converse does not hold. For an undirected graph, let $\lambda_i(\mathcal{L}_p)$ be the ith eigenvalue of $\mathcal{L}_p$ with $\lambda_1(\mathcal{L}_p) \leq \lambda_2(\mathcal{L}_p) \leq \cdots \leq \lambda_p(\mathcal{L}_p)$, so that $\lambda_1(\mathcal{L}_p) = 0$. For an undirected graph, $\lambda_2(\mathcal{L}_p)$ is the *algebraic connectivity*, which is positive if and only if the undirected graph is connected [141, p. 147]. The algebraic connectivity quantifies the convergence rate of consensus algorithms [106].

Given a matrix $S = [s_{ij}] \in \mathbb{R}^{p \times p}$, the *directed graph of* S, denoted by $\Gamma(S)$, is the directed graph with node set $\mathcal{V}_p = \{1, \ldots, p\}$ such that there is an edge in $\Gamma(S)$ from j to i if and only if $s_{ij} \neq 0$ [91, p. 357]. In other words, the entries of the adjacency matrix satisfy $a_{ij} > 0$ if $s_{ij} \neq 0$ and $a_{ij} = 0$ if $s_{ij} = 0$.

C

Matrix Theory Notations

We need the following definitions, lemmas, and theorems from matrix theory.

Theorem C.1. *[91, Theorem 6.1.1 (Gershgorin's disc theorem), p. 344] Let* $A = [a_{ij}] \in \mathbb{R}^{n \times n}$*, and let*

$$R_i'(A) \equiv \sum_{j=1, j \neq i}^{n} |a_{ij}|, \quad i = 1, \ldots, n,$$

denote the deleted absolute row sums of A*. Then all eigenvalues of* A *are located in the union of* n *discs*

$$\bigcup_{i=1}^{n} \{z \in \mathbb{C} : |z - a_{ii}| \leq R_i'(A)\} \equiv G(A).$$

Furthermore, if a union of k *of these* n *discs forms a connected region that is disjoint from all of the remaining* $n - k$ *discs, then there are precisely* k *eigenvalues of* A *in this region.*

Lemma C.2. *[91, Lemma 8.2.7 part(i), p. 498] Let* $A \in \mathbb{R}^{n \times n}$ *be given, let* $\lambda \in \mathbb{C}$ *be given, and suppose* x *and* y *are vectors such that (i)* $Ax = \lambda x$*, (ii)* $A^T y = \lambda y$*, and (iii)* $x^T y = 1$*. If* $|\lambda| = \rho(A) > 0$*, where* $\rho(A)$ *denotes the spectral radius of* A*, and* λ *is the only eigenvalue of* A *with modulus* $\rho(A)$*, then* $\lim_{m \to \infty} (\lambda^{-1} A)^m \to xy^T$.

The matrix $A \in \mathbb{R}^{n \times n}$ is *reducible* if either (i) $n = 1$ and $A = 0$, or (ii) $n \geq 2$ and there exists a permutation matrix $P \in \mathbb{R}^{n \times n}$ such that $P^T AP$ is in block upper triangular form. A matrix is *irreducible* if it is not reducible.

Theorem C.3. *[91, Theorem 6.2.24, p. 362] The matrix* $A \in \mathbb{R}^{n \times n}$ *is irreducible if and only if* $\Gamma(A)$ *is strongly connected.*

A vector x or matrix A is *nonnegative* (respectively, *positive*), denoted as $x \geq 0$ (respectively, $x > 0$) or $A \geq 0$ (respectively, $A > 0$), if all of its entries are nonnegative (respectively, positive). For nonnegative matrices, $A \geq B$ (respectively, $A > B$) implies that $A - B$ is a nonnegative (respectively, positive) matrix. Two $n \times n$ nonnegative matrices P and Q are of the same *type* if they have zero entries and positive entries in the same locations (see [253]). We use the notation $P \sim Q$ to denote that P and Q are of the same type.

Theorem C.4. *[91, Theorem 8.3.1, p. 503] If $A \in \mathbb{R}^{n \times n}$ is nonnegative, then $\rho(A)$ is an eigenvalue of A and there is a nonnegative vector $x \geq 0$, $x \neq 0$, such that $Ax = \rho(A)x$.*

Theorem C.5. *[91, Perron-Frobenius theorem, p. 508] If $A \in \mathbb{R}^{n \times n}$ is irreducible and nonnegative, then (i) $\rho(A) > 0$, (ii) $\rho(A)$ is an eigenvalue of A, (iii) there is a positive vector x such that $Ax = \rho(A)x$, and (iv) $\rho(A)$ is an algebraically (and hence geometrically) simple eigenvalue of A.*

A square nonnegative matrix is *primitive* if it is irreducible and has exactly one eigenvalue of maximum modulus, which is necessarily positive. A square nonnegative matrix is *row stochastic* if all of its row sums are 1 [91, p. 526]. Every row-stochastic matrix has 1 as an eigenvalue with an associated eigenvector $\mathbf{1}_n$. The spectral radius of a row-stochastic matrix is 1 because 1 is an eigenvalue and Theorem C.1 implies that all of the eigenvalues are contained in a closed unit disc. The row-stochastic matrix A is *indecomposable and aperiodic* (SIA) if $\lim_{k\to\infty} A^k = \mathbf{1}_n \nu^T$, where ν is a column vector [253]. Given a row-stochastic matrix $S = [s_{ij}] \in \mathbb{R}^{n \times n}$, define the matrix function

$$\chi(S) = 1 - \min_{i_1, i_2} \sum_j \min(s_{i_1 j}, s_{i_2 j}).$$

Note that $\chi(S) \leq 1$ for every row-stochastic matrix S. If $\chi(S) < 1$, S is a *scrambling* matrix. Also note that $\chi(S) = 0$ if and only if the rows of S are identical [253].

Theorem C.6. *[91, Theorem 8.5.1, p. 516] If $A \in \mathbb{R}^{n \times n}$ is nonnegative and primitive, then $\lim_{k\to\infty} [\rho(A)^{-1} A^k] \to w\nu^T$, where $Aw = \rho(A)w$, $A^T\nu = \rho(A)\nu$, $w > 0$, $\nu > 0$, and $w^T\nu = 1$.*

Theorem C.7. *[91, Theorem 8.5.2, p. 516] If $A \in \mathbb{R}^{n \times n}$ is nonnegative, then A is primitive if and only if $A^m > 0$ for some positive integer m.*

Because the spectral radius of a row-stochastic matrix is 1, Theorem C.6 implies that if A is row stochastic and primitive; then $\lim_{k\to\infty} A^k = \mathbf{1}_n \nu^T$, where $A^T\nu = \nu$, $\nu > 0$, and $\mathbf{1}_n^T \nu = 1$.

Lemma C.8. *[84] Suppose that $U \in \mathbb{R}^{p \times p}$, $V \in \mathbb{R}^{q \times q}$, $X \in \mathbb{R}^{p \times p}$, and $Y \in \mathbb{R}^{q \times q}$. The following arguments are valid.*

(i) $(U \otimes V)(X \otimes Y) = UX \otimes VY$.
(ii) $(U \otimes V)^T = U^T \otimes V^T$.
(iii) Suppose that U *and* V *are invertible. Then* $(U \otimes V)^{-1} = U^{-1} \otimes V^{-1}$.
(iv) If U *and* V *are symmetrical, so is* $U \otimes V$.
(v) If U *and* V *are symmetrical positive definite (respectively, positive semi-definite), so is* $U \otimes V$.

D

Rigid Body Attitude Dynamics

Given a vector v with coordinate representation $[v_1, v_2, v_3]^T$, the cross-product operator is denoted by [249]

$$v^\times = \begin{bmatrix} 0 & -v_3 & v_2 \\ v_3 & 0 & -v_1 \\ -v_2 & v_1 & 0 \end{bmatrix},$$

which represents the fact that $v \times w = v^\times w$, where $w = [w_1, w_2, w_3]^T$.

A *quaternion* is defined as $q = [\hat{q}^T, \bar{q}]^T \in \mathbb{R}^4$, where $\hat{q} \in \mathbb{R}^3$ denotes the vector part and $\bar{q} \in \mathbb{R}$ denotes the scalar part. The product of two quaternions p and q is defined by

$$qp = \begin{bmatrix} \bar{q}\hat{p} + \bar{p}\hat{q} + \hat{q} \times \hat{p} \\ \bar{q}\bar{p} - \hat{q}^T \hat{p} \end{bmatrix}. \tag{D.1}$$

Quaternions are not commutative, *i.e.*, $qp \neq pq$, but they are associative, *i.e.*, $(pq)s = p(qs)$, and distributive, *i.e.*, $(p + q)s = ps + qs$. The conjugate of quaternion q is defined by $q^* = [-\hat{q}^T, \bar{q}]^T$. The conjugate of qp is given by $(qp)^* = p^* q^*$.

Rigid body attitudes can be represented by *Euler parameters*, *i.e.*, *unit quaternions* [93,250]. A unit quaternion is defined as $q = [\hat{q}^T, \bar{q}]^T \in \mathbb{R}^4$, where $\hat{q} = a\sin(\frac{\phi}{2}) \in \mathbb{R}^3$ denotes the vector part and $\bar{q} = \cos(\frac{\phi}{2}) \in \mathbb{R}$ denotes the scalar part of the unit quaternion. In this notation, $a \in \mathbb{R}^3$ is a unit vector in the direction of rotation with a coordinate representation $[a_1, a_2, a_3]^T$, called the *Euler axis*, and ϕ is the rotational angle about a, called the *Euler angle*. Note that for the unit quaternion q, $q^T q = 1$ by definition. The product of two unit quaternions is also a unit quaternion. The conjugate of a unit quaternion is also a unit quaternion. The *multiplicative identity quaternion* is denoted by

$$\mathbf{q}_I \triangleq [0, 0, 0, 1]^T,$$

where $qq^* = q^* q = \mathbf{q}_I$ and $q\mathbf{q}_I = \mathbf{q}_I q = q$. A unit quaternion is not unique because q and $-q$ represent the same attitude. However, uniqueness can be achieved by restricting ϕ to the range $0 \leq \phi \leq \pi$ so that $\bar{q} \geq 0$ [93].

Suppose that q^d and q represent, respectively, the desired and actual attitude, then the attitude error is given by $q_e = q^{d*}q = [\hat{q}_e^T, \bar{q}_e]^T$, which represents the attitude of the actual reference frame $\mathcal{F}$ with respect to the desired reference frame $\mathcal{F}^d$. The relationship between the rotational matrix C_{ab} and the unit quaternion q is given by

$$C_{ab} = (2\bar{q}^2 - 1)I + 2\hat{q}\hat{q}^T - 2\bar{q}\hat{q}^{\times},$$

where q represents the attitude of $\mathcal{F}_a$ with respect to $\mathcal{F}_b$ [250].

Attitude dynamics for rigid bodies using Euler parameters are given by

$$\dot{\widehat{q}_i} = -\frac{1}{2}\omega_i \times \widehat{q}_i + \frac{1}{2}\overline{q_i}\omega_i, \qquad \dot{\overline{q_i}} = -\frac{1}{2}\omega_i \cdot \widehat{q}_i, \tag{D.2a}$$

$$J_i\dot{\omega}_i = -\omega_i \times (J_i\omega_i) + \tau_i, \quad i = 1, \cdots, n, \tag{D.2b}$$

where $\widehat{q}_i \in \mathbb{R}^3$ and $\overline{q_i} \in \mathbb{R}$ are vector and scalar parts of the unit quaternion of the ith rigid body, $\omega_i \in \mathbb{R}^3$ is the angular velocity, and $J_i \in \mathbb{R}^{3\times 3}$ and $\tau_i \in \mathbb{R}^3$ are, respectively, the inertia and control torque [93]. Note that (D.2a) can be written as

$$\dot{q}_i = \frac{1}{2}q_i\breve{\omega}_i, \tag{D.3}$$

where $\breve{\omega}_i = [\omega_i^T, 0]^T \in \mathbb{R}^4$ is a quaternion and $q_i\breve{\omega}_i$ is computed by following (D.1) with q and p replaced by q_i and $\breve{\omega}_i$, respectively.

Lemma D.1. *Suppose that the unit quaternion and angular velocity pairs* (q_k, ω_k) *and* (q_ℓ, ω_ℓ) *satisfy* (D.2) *with the same inertia* J *and with the control torques* τ_k *and* τ_ℓ*, respectively. Then the unit quaternion and angular velocity pair* $(q_\ell^* q_k, \omega_k - \omega_\ell)$ *also satisfies* (D.2a)*. In addition, if* $V_q = \|q_\ell^* q_k - \mathbf{q}_I\|^2$*, then* $\dot{V}_q = (\omega_k - \omega_\ell)^T \widehat{q_\ell^* q_k}$*, where* $\widehat{q_\ell^* q_k}$ *denotes the vector part of the unit quaternion* $q_\ell^* q_k$*. Furthermore, if* $V_q = \|q_k - q_\ell\|^2$*, then* $\dot{V}_q = (\omega_k - \omega_\ell)^T \widehat{q_\ell^* q_k}$*. Moreover, if* $V_\omega = \frac{1}{2}(\omega_k - \omega_\ell) \cdot J(\omega_k - \omega_\ell)$*, then* $\dot{V}_\omega = (\omega_k - \omega_\ell) \cdot \{\tau_k - \tau_\ell - \frac{1}{2}[\omega_k \times J(\omega_k - \omega_\ell)]\}$*.*

Proof: See [244, 249]. ∎

Lemma D.2. *Let* $\tau_i = \omega_i \times (J_i\omega_i) - k_{qi}\hat{q}_i - K_{\omega i}\omega_i$*, where* k_{qi} *is a positive scalar and* $K_{\omega i}$ *is a positive-definite matrix, in* (D.2b)*. Then* $\hat{q}_i(t) \to 0$ *and* $\omega_i(t) \to 0$*, as* $t \to \infty$*.*

Proof: The proof follows from [252] by using the Lyapunov function $V = k_{qi}\|q_i - \mathbf{q}_I\|^2 + \frac{1}{2}\omega_i^T J_i \omega_i$. ∎

Modified Rodriguez parameters (MRPs) [235] can also be used to represent the attitude of a rigid body with respect to an inertial frame. Let $\sigma_i = a_i \tan(\frac{\phi_i}{4}) \in \mathbb{R}^3$ represent the MRPs for the ith rigid body, where a_i is the Euler axis and ϕ_i is the Euler angle.

Attitude dynamics for rigid bodies using MRPs are given by

$$\dot{\sigma}_i = F(\sigma_i)\omega_i \tag{D.4a}$$

$$J_i\dot{\omega}_i = -\omega_i \times J_i\omega_i + \tau_i \quad i = 1, \ldots, n, \tag{D.4b}$$

where $\omega_i \in \mathbb{R}^3$ denotes the angular velocity, $J_i \in \mathbb{R}^{3\times 3}$ is the inertia, $\tau_i \in \mathbb{R}^3$ is the control torque, and

$$F(\sigma_i) \triangleq \frac{1}{2}[(\frac{1 - \sigma_i^T\sigma_i}{2})I_3 + \sigma_i^\times + \sigma_i\sigma_i^T].$$

Following [217, 254], (D.4) can be written as

$$H_i^*(\sigma_i)\ddot{\sigma}_i + C_i^*(\sigma_i, \dot{\sigma}_i)\dot{\sigma}_i = F^{-T}(\sigma_i)\tau_i, \tag{D.5}$$

where

$$H_i^*(\sigma_i) \triangleq F^{-T}(\sigma_i)J_iF^{-1}(\sigma_i),$$

and

$$\begin{aligned} C_i^*(\sigma_i, \dot{\sigma}_i) \triangleq & -F^{-T}(\sigma_i)J_iF^{-1}(\sigma_i)\dot{F}(\sigma_i)F^{-1}(\sigma_i) \\ & - F^{-T}(\sigma_i)(J_iF^{-1}(\sigma_i)\dot{\sigma}_i)^\times F^{-1}(\sigma_i). \end{aligned}$$

Note that $H_i^*(\sigma_i)$ is a symmetrical positive-definite matrix and $\dot{H}_i^*(\sigma_i) - 2C_i^*(\sigma_i, \dot{\sigma}_i)$ is a skew-symmetrical matrix [217].

E

Linear System Theory Background

Consider the linear state equation

$$\dot{x}(t) = A(t)x(t) + B(t)u(t), \quad x(t_0) = x_0, \tag{E.1a}$$

$$y(t) = C(t)x(t) + D(t)u(t), \tag{E.1b}$$

where $x(t) \in \mathbb{R}^n$ is the state vector, $u(t) \in \mathbb{R}^m$ is the input signal, $y(t) \in \mathbb{R}^p$ is the output signal, $A(t) \in \mathbb{R}^{n\times n}$, $B(t) \in \mathbb{R}^{n\times m}$, $C(t) \in \mathbb{R}^{p\times n}$, and $D(t) \in \mathbb{R}^{p\times m}$. The linear state equation (E.1) is *linear time invariant* (LTI) if all of the matrices A, B, C, and D are constant and is *linear time varying* (LTV) otherwise.

The solution of the state is given by

$$x(t) = \Phi(t, t_0)x_0 + \int_{t_0}^{t} \Phi(t, \sigma)B(\sigma)u(\sigma)d\sigma, \quad t \geq t_0,$$

where $\Phi(t, t_0)$ is *transition matrix* defined as

$$\Phi(t, t_0) \triangleq I_n + \int_{t_0}^{t} A(\sigma_1)\, d\sigma_1 + \int_{t_0}^{t} A(\sigma_1) \int_{t_0}^{\sigma_1} A(\sigma_2)\, d\sigma_2\, d\sigma_1 + \cdots .$$

For an LTI system, $\Phi(t, t_0) = e^{A(t-t_0)}$. The solution of the output is given by

$$y(t) = C(t)\Phi(t, t_0)x_0 + \int_{t_0}^{t} C(t)\Phi(t, \sigma)B(\sigma)u(\sigma)d\sigma + D(t)u(t), \quad t \geq t_0.$$

Internal stability deals with the stability of the zero-input response of (E.1a), *i.e.*, the linear state equation

$$\dot{x} = A(t)x, \quad x(t_0) = x_0. \tag{E.2}$$

The following definitions and theorem characterize internal stability.

Definition E.1. *[42, p. 138] The linear state equation* (E.2) *is marginally stable or stable if every finite initial state excites a bounded response. The linear state equation* (E.2) *is asymptotically stable if the response excited by every finite initial state is bounded and approaches zero, as* $t \to \infty$.

Definition E.2. *[205, Definitions 6.1, 6.5, and 6.12] The linear state equation* (E.2) *is uniformly stable if there exists a finite positive constant* γ *such that for any* t_0 *and* x_0*, the corresponding solution satisfies* $\|x(t)\| \leq \gamma \|x_0\|$, $\forall t \geq t_0$. *The linear state equation* (E.2) *is uniformly exponentially stable if there exist finite positive constants* γ *and* λ *such that for any* t_0 *and* x_0*, the corresponding solution satisfies* $\|x(t)\| \leq \gamma e^{-\lambda(t-t_0)} \|x_0\|$, $t \geq t_0$. *The linear state equation* (E.2) *is uniformly asymptotically stable if it is uniformly stable, and if given any positive constant* δ*, there exists a positive* T *such that for any* t_0 *and* x_0*, the corresponding solution satisfies* $\|x(t)\| \leq \delta \|x_0\|$, $t \geq t_0 + T$.

Theorem E.3. *[205, Theorem 6.13] The linear state equation* (E.2) *is uniformly asymptotically stable if and only if it is uniformly exponentially stable.*

Input-output stability deals with the stability of (E.1) with zero initial state (*i.e.*, $x_0 = 0$). The following definition characterizes input-output stability.

Definition E.4. *[205, Definition 6.1] The linear state equation* (E.1) *is uniformly bounded-input, bounded-output stable (BIBO) if there exists a finite constant* η *such that for any* t_0 *and any input signal* $u(t)$*, the corresponding zero-state response satisfies* $\sup_{t \geq t_0} \|y(t)\| \leq \eta \sup_{t \geq t_0} \|u(t)\|$.

Remark E.5. The definition of uniformly bounded-input, bounded-state stability follows Definition E.4 with $y(t)$ replaced by $x(t)$. □

Note that for an LTI system, asymptotical stability always implies uniformly bounded-input, bounded-state stability and uniformly bounded-input, bounded-output stability. However, for an LTV system, asymptotical stability does not necessarily imply uniformly bounded-input, bounded-state and bounded-input, bounded-output stability [42, p. 140]. The following lemma shows their relationship for an LTV system.

Lemma E.6. *[205, Lemma 12.4] Suppose that the linear state equation* (E.2) *is uniformly exponentially stable. Then the linear state equation* (E.1) *is uniformly bounded-input, bounded-state stable. Furthermore, if there exist finite constants* β, μ, *and* γ *such that for all* t, $\|B(t)\| \leq \beta$, $\|C(t)\| \leq \mu$, *and* $\|D(t)\| \leq \gamma$, *then* (E.1) *is uniformly bounded-input, bounded-output stable.*

F

Nonlinear System Theory Background

We need the following definitions, lemmas, and theorems from nonlinear system theory:

Consider the autonomous system

$$\dot{x} = f(x), \tag{F.1}$$

where $f : D \to \mathbb{R}^n$ is a locally Lipschitz map from a domain $D \subseteq \mathbb{R}^n$ into $\mathbb{R}^n$.

Theorem F.1. *[105, Theorem 4.2] Let $x = 0$ be an equilibrium point for* (F.1)*. Let $V : \mathbb{R}^n \to \mathbb{R}$ be a continuously differentiable function such that*

$$\begin{aligned} V(0) = 0 \text{ and } V(x) > 0, \quad \forall x \neq 0, \\ \|x\| \to \infty \Rightarrow V(x) \to \infty, \\ \dot{V}(x) < 0, \quad \forall x \neq 0. \end{aligned}$$

Then $x = 0$ is globally asymptotically stable.

Definition F.2. *[105, p. 127] A set M is said to be an invariant set with respect to* (F.1) *if $x(0) \in M$ implies $x(t) \in M$, $\forall t \in \mathbb{R}$. A set M is said to be a positively invariant set if $x(0) \in M$ implies $x(t) \in M$, $\forall t \geq 0$.*

Theorem F.3. *[218, Theorem 3.4 (Local Invariance Set Theorem)] Consider the autonomous system* (F.1)*. Let $V : \mathbb{R}^n \to \mathbb{R}$ be a continuously differentiable function. Assume that*

- *for some $c > 0$, the region Ω_c defined by $V(x) < c$ is bounded;*
- *$\dot{V}(x) \leq 0$, $\forall x \in \Omega_c$.*

Let E be the set of all points in Ω_c where $\dot{V}(x) = 0$, and let M be the largest invariant set in E. Then every solution $x(t)$ starting in Ω_c approaches M, as $t \to \infty$.

Theorem F.4. *[218, Theorem 3.5 (Global Invariance Set Theorem)] Consider the autonomous system* (F.1). *Let $V : \mathbb{R}^n \to \mathbb{R}$ be a continuously differentiable function. Assume that*

- $\dot{V}(x) \leq 0, \forall x \in \mathbb{R}^n$;
- $V(x) \to \infty$ *as* $\|x\| \to \infty$.[1]

Let E be the set of all points where $\dot{V}(x) = 0$, and let M be the largest invariant set in E. Then all solutions globally asymptotically converge to M, as $t \to \infty$.

Lemma F.5. *[218, Lemma 4.2 (Barbalat)] If the differentiable function $f(t)$ has a finite limit, as $t \to \infty$, and if $\dot{f}(t)$ is uniformly continuous,*[2] *then $\dot{f}(t) \to 0$, as $t \to \infty$.*

Lemma F.6. *[218, Lemma 4.3 (Lyapunov-like Lemma)] If a scalar function $V(x,t)$ satisfies the following conditions:*

- *$V(x,t)$ is lower bounded;*
- *$\dot{V}(x,t)$ is negative semidefinite;*
- *$\dot{V}(x,t)$ is uniformly continuous in t;*

then $\dot{V}(x,t) \to 0$, as $t \to \infty$.[3]

Definition F.7. *[105, Defnition 4.2, p. 144] A continuous function $\alpha : [0,a) \to [0,\infty)$ is said to belong to class $\mathcal{K}$ if it is strictly increasing and $\alpha(0) = 0$. It is said to belong to class $\mathcal{K}_\infty$ if $a = \infty$ and $\alpha(r) \to \infty$, as $r \to \infty$.*

Definition F.8. *[105, Defnition 4.3, p. 144] A continuous function $\beta : [0,a) \times [0,\infty) \to [0,\infty)$ is said to belong to class $\mathcal{KL}$ if, for each fixed s, the mapping $\beta(r,s)$ belongs to class $\mathcal{K}$ with respect to r and, for each fixed r, the mapping $\beta(r,s)$ is decreasing with respect to s and $\beta(r,s) \to 0$, as $s \to \infty$.*

Consider the system

$$\dot{x} = f(t,x,u) \tag{F.2}$$

where $f : [0,\infty) \times \mathbb{R}^n \times \mathbb{R}^m \to \mathbb{R}^n$ is piecewise continuous in t and locally Lipschitz in x and u. The input $u(t)$ is a piecewise continuous, bounded function of t for all $t \geq 0$.

Definition F.9. *[105, Defnition 4.7, p. 175] The system* (F.2) *is said to be input-to-state stable (ISS) if there exist a class $\mathcal{KL}$ function β and a class $\mathcal{K}$ function γ such that for any initial state $x(t_0)$ and any bounded input $u(t)$, the solution $x(t)$ exists for all $t \geq t_0$ and satisfies*

$$\|x(t)\| \leq \beta[\|x(t_0)\|, t - t_0] + \gamma\left[\sup_{t_0 \leq \tau \leq t} \|u(\tau)\|\right]. \tag{F.3}$$

[1] A function satisfying this condition is said to be *radially unbounded*

[2] A sufficient condition for a differentiable function to be uniformly continuous is that its derivative is bounded.

[3] Lemma F.6 follows from Lemma F.5.

As stated in [105, p. 175], (F.3) implies that for any bounded $u(t)$, the state $x(t)$ will be bounded. In addition, if $u(t)$ converges to zero, as $t \to \infty$, so does $x(t)$.

Theorem F.10. *[132, Theorem 7.9, p. 196] Consider the cascade interconnection of the systems*

$$\dot{x} = f(x, z) \tag{F.4a}$$

$$\dot{z} = g(z, u). \tag{F.4b}$$

If both (F.4a) *and* (F.4b) *are ISS, so is* (F.4).

References

1. http://www.crossbow.com.
2. http://zagi.com.
3. USGS digital elevation models. http://data.geocomm.com/dem/.
4. R. Agaev and P. Chebotarev. On the spectra of nonsymmetric Laplacian matrices. *Linear Algebra and its Applications*, 399:157–178, 2005.
5. S. Akella and S. Hutchinson. Coordinating the motions of multiple robots with specific trajectories. In *Proceedings of the IEEE International Conference on Robotics and Automation*, pages 624–631, Washington DC, May 2002.
6. M. Alighanbari and J. How. Decentralized task assignment for unmanned air vehicles. In *Proceedings of the IEEE Conference on Decision and Control, and the European Control Conference*, pages 5668–5673, Seville, Spain, December 2005.
7. M. Alighanbari, Y. Kuwata, and J. How. Coordination and control of multiple UAVs with timing constraints and loitering. In *Proceedings of the American Control Conference*, pages 5311–5316, Denver, CO, June 2003.
8. V. Ambrosia, S. Wegener, J. Brass, and S. Schoenung. The UAV western states fire mission: Concepts, plans, and developmental advancements. In *Proceedings of the AIAA 3rd Unmanned Unlimited Technical Conference*, Chicago, IL, September 2004. Paper no. AIAA-2004-6415.
9. E. P. Anderson and R. W. Beard. An algorithmic implementation of constrained extremal control for UAVs. In *Proceedings of the AIAA Guidance, Navigation, and Control Conference*, Monterey, CA, August 2002. Paper no. AIAA-2002-4470.
10. M. Anderson, M. Mckay, and B. Richardson. Multirobot automated indoor floor characterization team. In *Proceedings of the IEEE International Conference on Robotics and Automation*, pages 1750–1753, Minneapolis, April 1996.
11. M. R. Anderson and A. C. Robbins. Formation flight as a cooperative game. In *Proceedings of the AIAA Guidance, Navigation, and Control Conference*, pages 244–251, Boston, MA, August 1998. Paper no. AIAA-98-4124.
12. D. Angeli and P.-A. Bliman. Extension of a result by Moreau on stability of leaderless multi-agent systems. In *Proceedings of the IEEE Conference on Decision and Control, and the European Control Conference*, pages 759–764, Seville, Spain, December 2005.

13. M. Arcak. Passivity as a design tool for group coordination. *IEEE Transactions on Automatic Control*, 52(8):1380–1390, August 2007.
14. A. Astolfi. Exponential stabilization of a wheeled mobile robot via discontinuous control. *Journal of Dynamic Systems, Measurement, and Control*, 121(1):121–126, March 1999.
15. T. Balch and R. C. Arkin. Behavior-based formation control for multirobot teams. *IEEE Transactions on Robotics and Automation*, 14(6):926–939, December 1998.
16. T. Balch and L. E. Parker, editors. *Robot Teams: From Diversity to Polymorphism.* A. K. Peters, Natick, MA, 2002.
17. D. Bauso, L. Giarre, and R. Pesenti. Non-linear protocols for optimal distributed consensus in networks of dynamic agents. *Systems and Control Letters*, 55(11):918–928, 2006.
18. R. Beard and T. McLain. Multiple UAV cooperative search under collision avoidance and limited range communication constraints. In *Proceedings of the IEEE Conference on Decision and Control*, pages 25–30, Maui, HI, December 2003.
19. R. W. Beard and F. Y. Hadaegh. Constellation templates: An approach to autonomous formation flying. In *World Automation Congress*, pages 177.1–177.6, Anchorage, AK, May 1998. ISIAC.
20. R. W. Beard and F. Y. Hadaegh. Fuel optimization for unconstrained rotation of spacecraft formations. *Journal of the Astronautical Sciences*, 43(3):259–273, July–December 1999.
21. R. W. Beard, D. Kingston, M. Quigley, D. Snyder, R. Christiansen, W. Johnson, T. McLain, and M. Goodrich. Autonomous vehicle technologies for small fixed-wing UAVs. *Journal of Aerospace Computing, Information, and Communication*, 2(1):92–108, 2005.
22. R. W. Beard, J. R. Lawton, and F. Y. Hadaegh. A coordination architecture for spacecraft formation control. *IEEE Transactions on Control Systems Technology*, 9(6):777–790, November 2001.
23. R. W. Beard, D. Lee, M. Quigley, S. Thakoor, and S. Zornetzer. A new approach to observation of descent and landing of future Mars mission using bioinspired technology innovations. *AIAA Journal of Aerospace Computing, Information, and Communication*, 2(1):65–91, January 2005.
24. R. W. Beard, T. W. McLain, M. Goodrich, and E. P. Anderson. Coordinated target assignment and intercept for unmanned air vehicles. *IEEE Transactions on Robotics and Automation*, 18(6):911–922, 2002.
25. R. W. Beard, T. W. McLain, and F. Y. Hadaegh. Fuel optimization for constrained rotation of spacecraft formations. *Journal of Guidance, Control, and Dynamics*, 23(2):339–346, March–April 2000.
26. R. W. Beard, T. W. McLain, D. Nelson, D. Kingston, and D. Johanson. Decentralized cooperative aerial surveillance using fixed-wing miniature UAVs. *Proceedings of the IEEE*, 94(7):1306–1324, 2006.
27. J. Bellingham, M. Tillerson, A. Richards, and J. P. How. Multi-task allocation and path planning for cooperating UAVs. In S. Butenko, R. Murphey, and P. M. Pardalos, editors, *Cooperative Control: Models, Applications and Algorithms.* Kluwer, Norwell, MA, 2003.
28. C. Belta and V. Kumar. Abstraction and control for groups of robots. *IEEE Transactions on Robotics and Automation*, 20(5):865–875, October 2004.

29. W. Blake and D. Multhopp. Design, performance and modeling considerations for close formation flight. In *Proceedings of the AIAA Guidance, Navigation, and Control Conference*, pages 476–486, Boston, MA, August 1998. Paper no. AIAA-98-4343.
30. P.-A. Bliman and G. Ferrari-Trecate. Average consensus problems in networks of agents with delayed communications. In *Proceedings of the IEEE Conference on Decision and Control, and the European Control Conference*, pages 7066–7071, Seville, Spain, December 2005.
31. V. D. Blondel, J. M. Hendrickx, A. Olshevsky, and J. N. Tsisiklis. Convergence in multiagent coordination, consensus, and flocking. In *Proceedings of the IEEE Conference on Decision and Control, and the European Control Conference*, pages 2996–3000, Seville, Spain, December 2005.
32. V. Borkar and P. Varaiya. Asymptotic agreement in distributed estimation. *IEEE Transactions on Automatic Control*, 27(3):650–655, 1982.
33. R. W. Brockett. Asymptotic stability and feedback stabilization. In R. W. Brockett, R. S. Millman, and H. J. Sussmann, editors, *Differential Geometric Control Theory*, pages 181–191, Boston, 1983. Birkhaüser.
34. M. Campbell, R. D'Andrea, J. Lee, and E. Scholte. Experimental demonstrations of semi-autonomous control. In *Proceedings of the American Control Conference*, pages 5338–5343, Boston, MA, July 2004.
35. M. Cao, A. S. Morse, and B. D. O. Anderson. Coordination of an asynchronous multi-agent system via averaging. In *IFAC World Congress*, Prague, Czech Republic, 2005. Paper code Mo-A09-TO/5.
36. M. Cao, D. A. Spielman, and A. S. Morse. A lower bound on convergence of a distributed network consensus algorithm. In *Proceedings of the IEEE Conference on Decision and Control, and the European Control Conference*, pages 2356–2361, Seville, Spain, December 2005.
37. J. R. Carpenter. Decentralized control of satellite formations. *International Journal of Robust and Nonlinear Control*, 12:141–161, 2002.
38. D. W. Casbeer, D. B. Kingston, R. W. Beard, T. W. McLain, S.-M. Li, and R. Mehra. Cooperative forest fire surveillance using a team of small unmanned air vehicles. *International Journal of Systems Sciences*, 37(6):351–360, 2006.
39. S. Chakravorty. *Design and Optimal Control of Multi-Spacecraft Interferometric Imaging Systems.* PhD thesis, University of Michigan, Ann Arbor, MI, 2004.
40. P. Chandler, S. Rasumussen, and M. Pachter. UAV cooperative path planning. In *Proceedings of the AIAA Guidance, Navigation, and Control Conference*, Denver, CO, August 2000. Paper no. AIAA-2000-4370.
41. S. Chatterjee and E. Seneta. Towards consensus: Some convergence theorems on repeated averaging. *Journal of Applied Probability*, 14(1):89–97, March 1977.
42. C.-T. Chen. *Linear System Theory and Design.* Oxford University Press, New York, 3rd edition, 1999.
43. P. Chen, Z. Song, Z. Wang, and Y. Chen. Pattern formation experiments in mobile actuator and sensor network (Mas-net). In *Proceedings of the IEEE/RSJ International Conference on Intelligent Robots and Systems*, pages 3658–3663, Edmonton, Alberta, Can, August 2005.
44. Q. Chen and J. Y. S. Luh. Coordination and control of a group of small mobile robots. In *Proceedings of the IEEE International Conference on Robotics and Automation*, pages 2315–2320, San Diego, 1994.

45. D. F. Chichka and J. L. Speyer. Solar-powered, formation-enhanced aerial vehicle systems for sustained endurance. In *Proceedings of the American Control Conference*, pages 684–688, Philadelphia, PA, June 1998.
46. N. Chopra and M. W. Spong. On synchronization of Kuramoto oscillators. In *Proceedings of the IEEE Conference on Decision and Control, and the European Control Conference*, pages 3916–3922, Seville, Spain, December 2005.
47. J. Cortes. Analysis and design of distributed algorithms for chi-consensus. In *Proceedings of the IEEE Conference on Decision and Control*, pages 3363–3368, San Diego, 2006.
48. J. Cortes. Finite-time convergent gradient flows with applications to network consensus. *Automatica*, 42(11):1993–2000, November 2006.
49. J. Cortés, S. Martínez, T. Karatas, and F. Bullo. Coverage control for mobile sensing networks. *IEEE Transactions on Robotics and Automation*, 20:243–255, 2004.
50. F. Cucker and S. Smale. Emergent behavior in flocks. *IEEE Transactions on Automatic Control*, 52(5):852–862, 2007.
51. A. K. Das, R. Fierro, V. Kumar, J. P. Ostrowski, J. Spletzer, and C. J. Taylor. A vision-based formation control framework. *IEEE Transactions on Robotics and Automation*, 18(5):813–825, October 2002.
52. G. A. de Castro and F. Paganini. Convex synthesis of controllers for consensus. In *Proceedings of the American Control Conference*, pages 4933–4938, Boston, MA, 2004.
53. M. S. de Queiroz, V. Kapila, and Q. Yan. Adaptive nonlinear control of multiple spacecraft formation flying. *Journal of Guidance, Control, and Dynamics*, 23(3):385–390, May–June 2000.
54. C. C. de Wit and O. Sordalen. Exponential stabilization of mobile robots with nonholonomic constraints. *IEEE Transactions on Automatic Control*, 37(11):1791–1797, November 1992.
55. M. H. DeGroot. Reaching a consensus. *Journal of American Statistical Association*, 69(345):118–121, 1974.
56. J. P. Desai, J. Ostrowski, and V. Kumar. Controlling formations of multiple mobile robots. In *Proceedings of the IEEE International Conference on Robotics and Automation*, pages 2864–2869, Leuven, Belgium, May 1998.
57. W. C. Dickson, R. H. Cannon, and S. M. Rock. Symbolic dynamic modeling and analysis of object/robot-team systems with experiments. In *Proceedings of the IEEE Conference on Robotics and Automation*, pages 1413–1420, Minneapolis, April 1996.
58. D. V. Dimarogonas and K. J. Kyriakopoulos. On the rendezvous problem for multiple nonholonomic agents. *IEEE Transactions on Automatic Control*, 52(5):916–922, 2007.
59. D. V. Dimarogonas, S. G. Loizou, K. J. Kyriakopoulos, and M. M. Zavlanos. A feedback stabilization and collision avoidance scheme for multiple independent non-point agents. *Automatica*, 42(2):229–243, 2006.
60. D. V. Dimarogonas, P. Tsiotras, and K. J. Kyriakopoulos. Laplacian cooperative attitude control of multiple rigid bodies. In *Proceedings of the IEEE International Symposium on Intelligent Control*, pages 3064–3069, Munich, Germany, October 2006.
61. K. D. Do. Bounded controllers for formation stabilization of mobile agents with limited sensing ranges. *IEEE Transactions on Automatic Control*, 52(3):569–576, March 2007.

62. M. Egerstedt, X. Hu, and A. Stotsky. Control of mobile platforms using a virtual vehicle approach. *IEEE Transactions on Automatic Control*, 46(11):1777–1782, November 2001.
63. R. Emery, K. Sikorski, and T. Balch. Protocols for collaboration, coordination and dynamic role assignment in a robot team. In *Proceedings of the IEEE International Conference on Robotics and Automation*, pages 3008–3015, Washington DC, May 2002.
64. T. Eren, P. N. Belhumeur, and A. S. Morse. Closing ranks in vehicle formations based on rigidity. In *Proceedings of the IEEE Conference on Decision and Control*, pages 2959–2964, Las Vegas, NV, December 2002.
65. F. Fahimi. Sliding-mode formation control for underactuated surface vessels. *IEEE Transactions on Robotics*, 23(3):617–622, 2007.
66. N. Faiz, S. K. Agrawal, and R. M. Murray. Trajectory planning of differentially flat systems with dynamics and inequalities. *Journal of Guidance, Control, and Dynamics*, 24(2):219–227, March–April 2001.
67. L. Fang and P. J. Antsaklis. Information consensus of asynchronous discrete-time multi-agent systems. In *Proceedings of the American Control Conference*, pages 1883–1888, Portland, OR, June 2005.
68. L. Fang and P. J. Antsaklis. On communication requirements for muli-agent consensus seeking. In P. Antsaklis and P. Tabuada, editors, *Networked Embedded Sensing and Control*, volume 331 of *Lecture Notes in Control and Information Sciences*, pages 53–68, Berlin, 2006. Springer.
69. J. A. Fax and R. M. Murray. Information flow and cooperative control of vehicle formations. *IEEE Transactions on Automatic Control*, 49(9):1465–1476, September 2004.
70. J. T. Feddema, C. Lewis, and D. A. Schoenwald. Decentralized control of cooperative robotic vehicles: Theory and application. *IEEE Transactions on Robotics and Automation*, 18(5):852–864, 2002.
71. G. Ferrari-Trecate, A. Buffa, and M. Gati. Analysis of coordination in multi-agent systems through partial difference equations. *IEEE Transactions on Automatic Control*, 51(6):1058–1063, September 2006.
72. R. Fierro, A. K. Das, V. Kumar, and J. P. Ostrowski. Hybrid control of formations of robots. In *Proceedings of the IEEE International Conference on Robotics and Automation*, pages 157–162, Seoul, Korea, May 2001.
73. D. Folta, F. Bordi, and C. Scolese. Considerations on formation flying separations for earth observing satellite missions. *Advances in the Astronautical Sciences*, 79(2):803–822, 1992.
74. D. Folta and D. Quinn. A universal 3-D method for controlling the relative motion of multiple spacecraft in any orbit. In *Proceedings of the AIAA Guidance, Navigation, and Control Conference*, Boston, MA, August 1998. Paper no. AIAA-98-4193.
75. E. Frazzoli, M. A. Dahleh, and E. Feron. Real-time motion planning for agile autonomous vehicles. *Journal of Guidance, Control, and Dynamics*, 25(1):116–129, January–February 2002.
76. R. A. Freeman, P. Yang, and K. M. Lynch. Distributed estimation and control of swarm formation statistics. In *Proceedings of the American Control Conference*, pages 749–755, Minneapolis, June 2006.
77. J. Fujiwara, K.; Kudoh. Forest fire detection in 2001 using three-dimensional histogram. In *Proceedings of the IEEE International Geoscience and Remote Sensing Symposium*, volume 4, pages 2057 – 2059, Toronto, Can, June 2002.

78. L. Gao and D. Cheng. Comments on coordination of groups of mobile agents using nearest neighbor rules. *IEEE Transactions on Automatic Control*, 50(11):1913–1916, November 2005.
79. R. Gardner, W. Hargrove, M. Turner, and W. Romme. *Climate change, disturbances, and landscape dynamics*, pages 149–172. Global Change and Terrestrial Ecosystems. International Geosphere-Biosphere Programme Book Series - Book #2. Cambridge University Press, Cambridge, UK, 1996.
80. V. Gazi and K. M. Passino. Stability analysis of swarms. *IEEE Transactions on Automatic Control*, 48(4):692–697, 2003.
81. G. L. Gilardoni and M. K. Clayton. On reaching a consensus using degroot's iterative pooling. *The Annals of Statistics*, 21(1):391–401, 1993.
82. F. Giulietti, L. Pollini, and M. Innocenti. Autonomous formation flight. *IEEE Control Systems Magazine*, 20(6):34–44, December 2000.
83. R. E. Glickman. TIDE: The timed-destination approach to constellation formation keeping. *Advances in the Astronautical Sciences*, 87(2):725–743, 1994.
84. A. Graham. *Kronecker Products and Matrix Calculus With Applications*. Halsted, New York, 1981.
85. F. Y. Hadaegh, W.-M. Lu, and P. K. C. Wang. Adaptive control of formation flying spacecraft for interferometry. In *Proceedings of the IFAC Symposium on Large Scale Systems: Theory and Applications*, pages 97–102, Patras, Greece, 1998.
86. W. Hargrove, R. Gardner, M. Turner, W. Romme, and D. Despain. Simulating fire patterns in heterogeneous landscapes. *Ecological Modelling*, 135:243–263, 2000.
87. Y. Hatano, A. K. Das, and M. Mesbahi. Agreement in presence of noise: Pseudogradients on random geometric networks. In *Proceedings of the IEEE Conference on Decision and Control, and the European Control Conference*, pages 6382–6387, Seville, Spain, December 2005.
88. Y. Hatano and M. Mesbahi. Agreement over random networks. *IEEE Transactions on Automatic Control*, 50(11):1867–1872, November 2005.
89. Y. Hong, L. Gao, D. Cheng, and J. Hu. Lyapunov-based approach to multiagent systems with switching jointly connected interconnection. *IEEE Transactions on Automatic Control*, 52(5):943–948, May 2007.
90. Y. Hong, J. Hu, and L. Gao. Tracking control for multi-agent consensus with an active leader and variable topology. *Automatica*, 42(7):1177–1182, July 2006.
91. R. A. Horn and C. R. Johnson. *Matrix Analysis*. Cambridge University Press, Cambridge, UK, 1985.
92. J. K. Howlett. Path planning for sensing multiple targets from an aircraft. Master's thesis, Mechanical Engineering, Brigham Young University, Provo, UT, 2002.
93. P. C. Hughes. *Spacecraft Attitude Dynamics*. Wiley, New York, 1986.
94. I. I. Hussein. *Motion Planning for Multi-Spacecraft Interferometric Imaging Systems*. PhD thesis, University of Michigan, Ann Arbor, MI, 2005.
95. G. Inalhan, D. M. Stipanovic, and C. J. Tomlin. Decentralized optimization, with application to multiple aircraft coordination. In *Proceedings of the IEEE Conference on Decision and Control*, pages 1147–1155, Las Vegas, NV, December 2002.
96. A. Isidori. *Nonlinear Control Systems*. Communication and Control Engineering. Springer-Verlag, New York, 2nd edition, 1989.

97. A. Jadbabaie, J. Lin, and A. S. Morse. Coordination of groups of mobile autonomous agents using nearest neighbor rules. *IEEE Transactions on Automatic Control*, 48(6):988–1001, June 2003.
98. A. Jadbabaie, N. Motee, and M. Barahona. On the stability of the Kuramoto model of coupled nonlinear oscillators. In *Proceedings of the American Control Conference*, pages 4296–4301, Boston, MA, 2004.
99. Z. Jin and R. M. Murray. Consensus controllability for coordinated multiple vehicle control. In *The 6th International Conference on Cooperative Control and Optimization*, Gainesville, FL, February 2006.
100. E. W. Justh and P. S. Krishnaprasad. Equilibria and steering laws for planar formations. *Systems and Control Letters*, 52:25–38, 2004.
101. W. Kang, N. Xi, and A. Sparks. Formation control of autonomous agents in 3D workspace. In *Proceedings of the IEEE International Conference on Robotics and Automation*, pages 1755–1760, San Francisco, CA, April 2000.
102. W. Kang and H.-H. Yeh. Co-ordinated attitude control of multi-satellite systems. *International Journal of Robust and Nonlinear Control*, 12:185–205, 2002.
103. V. Kapila, A. G. Sparks, J. M. Buffington, and Q. Yan. Spacecraft formation flying: Dynamics and control. *Journal of Guidance, Control, and Dynamics*, 23(3):561–564, May-June 2000.
104. A. Kashyap, T. Basar, and R. Srikant. Quantized consensus. *Automatica*, 43:1192–1203, 2007.
105. H. K. Khalil. *Nonlinear Systems.* Prentice Hall, Upper Saddle River, NJ, 3rd edition, 2002.
106. Y. Kim and M. Mesbahi. On maximizing the second smallest eigenvalue of a state-dependent graph Laplacian. *IEEE Transactions on Automatic Control*, 51(1):116–120, 2006.
107. E. King, Y. Kuwata, M. Alighangari, L. Bertucelli, and J. How. Coordination and control experiments on a multi-vehicle testbed. In *Proceedings of the American Control Conference*, pages 5315–5320, Boston, MA, July 2004.
108. D. Kingston, R. Beard, T. McLain, M. Larsen, and W. Ren. Autonomous vehicle technologies for small fixed wing UAVs. In *Proceedings of the AIAA 2nd Unmanned Unlimited Systems, Technologies, and Operations–Aerospace, Land, and Sea Conference*, San Diego, September 2003. Paper no. AIAA-2003-6559.
109. D. B. Kingston, R. S. Holt, R. W. Beard, T. W. McLain, and D. W. Casbeer. Decentralized perimeter surveillance using a team of UAVs. In *Proceedings of the AIAA Guidance, Navigation, and Control Conference*, San Francisco, August 2005. Paper no. AIAA-2005-5831.
110. D. B. Kingston, W. Ren, and R. W. Beard. Consensus algorithms are input-to-state stable. In *Proceedings of the American Control Conference*, pages 1686–1690, Portland, OR, June 2005.
111. C. R. Kube and H. Zhang. The use of perceptual cues in multi-robot box-pushing. In *Proceedings of the IEEE International Conference on Robotics and Automation*, pages 2085–2090, Minneapolis, April 1996.
112. J.-I. Kudoh and K. Hosoi. Two dimensional forest fire detection method by using NOAA AVHRR images. In *Proceedings of the IEEE International Geoscience and Remote Sensing Symposium*, volume 4, pages 2494 – 2495, Toulouse, France, July 2003.

113. D. Kurabayashi, J. Ota, T. Arai, and E. Yoshida. Cooperative sweeping by multiple mobile robots. In *Proceedings of the IEEE International Conference on Robotics and Automation*, pages 1744–1749, Minneapolis, April 1996.
114. Y. Kuramoto. *Chemical Oscillations, Waves and Turbulence.* Springer, Berlin, 1984.
115. G. Lafferriere, A. Williams, J. Caughman, and J. J. P. Veerman. Decentralized control of vehicle formations. *Systems and Control Letters*, 54(9):899–910, 2005.
116. J. R. Lawton and R. W. Beard. A projection approach to spacecraft formation attitude control. In *23rd Annual AAS Guidance and Control Conference*, Breckenridge, CO, February 2000. Paper no. AAS 00-011.
117. J. R. Lawton and R. W. Beard. Synchronized multiple spacecraft rotations. *Automatica*, 38(8):1359–1364, 2002.
118. J. R. Lawton, R. W. Beard, and B. Young. A decentralized approach to formation maneuvers. *IEEE Transactions on Robotics and Automation*, 19(6):933–941, December 2003.
119. D. Lee and M. W. Spong. Agreement with non-uniform information delays. In *Proceedings of the American Control Conference*, pages 756–761, Minneapolis, June 2006.
120. D. Lee and M. W. Spong. Stable flocking of multiple inertial agents on balanced graphs. *IEEE Transactions on Automatic Control*, 52(8):1469–1475, August 2007.
121. N. E. Leonard and E. Fiorelli. Virtual leaders, artificial potentials and coordinated control of groups. In *Proceedings of the IEEE Conference on Decision and Control*, pages 2968–2973, Orlando, FL, December 2001.
122. F. L. Lewis. *Optimal Control.* Wiley, New York, 1986.
123. M. A. Lewis and K.-H. Tan. High precision formation control of mobile robots using virtual structures. *Autonomous Robots*, 4:387–403, 1997.
124. J. Lin, A. S. Morse, and B. D. O. Anderson. The multi-agent rendezvous problem. In *Proceedings of the IEEE Conference on Decision and Control*, pages 1508–1513, Maui, HI, 2003.
125. J. Lin, A. S. Morse, and B. D. O. Anderson. The multi-agent rendezvous problem - the asynchronous case. In *Proceedings of the IEEE Conference on Decision and Control*, pages 1926–1931, Paradise Island, Bahamas, December 2004.
126. Z. Lin, M. Broucke, and B. Francis. Local control strategies for groups of mobile autonomous agents. *IEEE Transactions on Automatic Control*, 49(4):622–629, April 2004.
127. Z. Lin, B. Francis, and M. Maggiore. Necessary and sufficient graphical conditions for formation control of unicycles. *IEEE Transactions on Automatic Control*, 50(1):121–127, 2005.
128. Z. Lin, B. Francis, and M. Maggiore. State agreement for continuous-time coupled nonlinear systems. *SIAM Journal on Control and Optimization*, 46(1):288–307, 2007.
129. F. Lizarralde and J. Wen. Attitude control without angular velocity measurement: A passivity approach. *IEEE Transactions on Automatic Control*, 41(3):468–472, March 1996.
130. A. Loria, R. Kelly, R. Ortega, and V. Santibanez. On global output feedback regulation of Euler-Lagrange systems with bounded inputs. *IEEE Transactions on Automatic Control*, 42(8):1138–1143, August 1997.

131. N. A. Lynch. *Distributed Algorithms.* Morgan Kaufmann, San Francisco, 1996.
132. H. J. Marquez. *Nonlinear Control Systems: Analysis and Design.* Wiley, Hoboken, NJ, 2003.
133. J. A. Marshall, M. E. Broucke, and B. A. Francis. Formations of vehicles in cyclic pursuit. *IEEE Transactions on Automatic Control*, 49(11):1963–1974, 2004.
134. J. A. Marshall, T. Fung, M. E. Broucke, G. M. T. D'Eleuterio, and B. A. Francis. Experiments in multirobot coordination. *Robotics and Autonomous Systems*, 54(3):265–275, 2006.
135. S. Martinez, J. Cortes, and F. Bullo. On robust rendezvous for mobile autonomous agents. In *IFAC World Congress*, Prague, Czech Republic, 2005. Paper code We-A18-TO/3.
136. C. R. McInnes. Autonomous ring formation for a planar constellation of satellites. *Journal of Guidance, Control, and Dynamics*, 18(5):1215–1217, 1995.
137. T. W. McLain and R. W. Beard. Cooperative rendezvous of multiple unmanned air vehicles. In *Proceedings of the AIAA Guidance, Navigation, and Control Conference*, Denver, CO, August 2000. Paper no. AIAA-2000-4369.
138. T. W. McLain and R. W. Beard. Coordination variables, coordination functions, and cooperative timing missions. In *Proceedings of the American Control Conference*, pages 296–301, Denver, CO, June 2003.
139. T. W. McLain and R. W. Beard. Coordination variables, coordination functions, and cooperative timing missions. *Journal of Guidance, Control, and Dynamics*, 28(1):150–161, January 2005.
140. M. Mehyar, D. Spanos, J. Pongsajapan, S. H. Low, and R. M. Murray. Distributed averaging on asynchronous communication networks. In *Proceedings of the IEEE Conference on Decision and Control, and the European Control Conference*, pages 7446–7451, Seville, Spain, December 2005.
141. R. Merris. Laplacian matrices of graphs: A survey. *Linear Algebra and its Applications*, 197-198:143–176, 1994.
142. M. Mesbahi and F. Y. Hadaegh. Formation flying control of multiple spacecraft via graphs, matrix inequalities, and switching. *Journal of Guidance, Control, and Dynamics*, 24(2):369–377, March–April 2000.
143. K. Moore and D. Lucarelli. Decentralized adaptive scheduling using consensus variables. *International Journal of Robust and Nonlinear Control*, 17(10–11):921–940, July 2007.
144. L. Moreau. Stability of continuous-time distributed consensus algorithms. In *Proceedings of the IEEE Conference on Decision and Control*, pages 3998–4003, Paradise Island, Bahamas, December 2004.
145. L. Moreau. Stability of multi-agent systems with time-dependent communication links. *IEEE Transactions on Automatic Control*, 50(2):169–182, February 2005.
146. A. S. Morse. Supervisory control of families of linear set-point controllers-part 1: Exact matching. *IEEE Transactions on Automatic Control*, 41(10):1413–1431, 1996.
147. N. Moshtagh and A. Jadbabaie. Distributed geodesic control laws for flocking of nonholonomic agents. *IEEE Transactions on Automatic Control*, 52(4):681–686, April 2007.
148. D. R. Nelson. Cooperative control of miniature air vehicles. Master's thesis, Brigham Young University, Provo, UT, December 2005.

149. D. R. Nelson, D. B. Barber, T. W. McLain, and R. W. Beard. Vector field path following for small unmanned air vehicles. In *Proceedings of the American Control Conference*, pages 5788–5794, Minneapolis, June 2006.
150. D. R. Nelson, T. W. McLain, R. S. Christiansen, R. W. Beard, and D. Johansen. Initial experiements in cooperative control of unmanned air vehicles. In *Proceedings of the AIAA 3rd Unmanned Unlimited Systems Conference*, Chicago, IL, 2004. Paper no. AIAA-2004-6533.
151. P. Ogren, M. Egerstedt, and X. Hu. A control Lyapunov function approach to multiagent coordination. *IEEE Transactions on Robotics and Automation*, 18(5):847–851, October 2002.
152. P. Ogren, E. Fiorelli, and N. E. Leonard. Formations with a mission: Stable coordination of vehicle group maneuvers. In *Proceedings of the 15th International Symposium on Mathematical Theory of Networks and Systems*, Notre Dame, IN, 2002.
153. P. Ogren, E. Fiorelli, and N. E. Leonard. Cooperative control of mobile sensor networks: Adaptive gradient climbing in a distributed environment. *IEEE Transactions on Automatic Control*, 49(8):1292–1302, August 2004.
154. R. Olfati-Saber. Distributed Kalman filter with embedded consensus filters. In *Proceedings of the IEEE Conference on Decision and Control, and the European Control Conference*, pages 8179–8184, Seville, Spain, December 2005.
155. R. Olfati-Saber. Flocking for multi-agent dynamic systems: Algorithms and theory. *IEEE Transactions on Automatic Control*, 51(3):401–420, March 2006.
156. R. Olfati-Saber, J. A. Fax, and R. M. Murray. Consensus and cooperation in networked multi-agent systems. *Proceedings of the IEEE*, 95(1):215–233, January 2007.
157. R. Olfati-Saber and R. M. Murray. Distributed cooperative control of multiple vehicle formations using structural potential functions. In *IFAC World Congress*, Barcelona, Spain, July 2002.
158. R. Olfati-Saber and R. M. Murray. Consensus problems in networks of agents with switching topology and time-delays. *IEEE Transactions on Automatic Control*, 49(9):1520–1533, September 2004.
159. R. Olfati-Saber and J. S. Shamma. Consensus filters for sensor networks and distributed sensor fusion. In *Proceedings of the IEEE Conference on Decision and Control, and the European Control Conference*, pages 6698–6703, Seville, Spain, December 2005.
160. R. Ortega, A. Loria, R. Kelly, and L. Praly. On passivity-based output feedback global stabilization of Euler-Lagrange systems. *International Journal of Robust and Nonlinear Control*, 5:313–323, 1995.
161. M. Pachter, J. J. D'Azzo, and A. W. Proud. Tight formation flight control. *Journal of Guidance, Control, and Dynamics*, 24(2):246–254, March–April 2001.
162. A. Papachristodoulou and A. Jadbabaie. Synchronization in oscillator networks: Switching topologies and non-homogeneous delays. In *Proceedings of the IEEE Conference on Decision and Control, and the European Control Conference*, pages 5692–5697, Seville, Spain, December 2005.
163. S. T. Pledgie, Y. Hao, A. M. Ferreira, S. K. Agrawal, and R. Murphey. Groups of unmanned vehicles: Differential flatness, trajectory planning, and control. In *Proceedings of the IEEE International Conference on Robotics and Automation*, pages 3461–3466, Washington DC, May 2002.

164. J. P. Pomet. Explicit design of time-varying stabilizing control laws for a class of controllable systems without drift. *Systems and Control Letters*, 18:147–158, 1992.
165. M. Porfiri, D. G. Roberson, and D. J. Stilwell. Tracking and formation control of multiple autonomous agents: A two-level consensus approach. *Automatica*, 43(8):1318–1328, 2007.
166. V. M. Preciado and G. C. Verghese. Synchronization in generalized Erdos-Renyi networks of nonlinear oscillators. In *Proceedings of the IEEE Conference on Decision and Control, and the European Control Conference*, pages 4628–4633, Seville, Spain, December 2005.
167. A. W. Proud, M. Pachter, and J. J. D'Azzo. Close formation flight control. In *Proceedings of the AIAA Guidance, Navigation, and Control Conference*, pages 1231–1246, Portland, OR, August 1999. Paper no. AIAA-99-4207.
168. Z. Qu, J. Wang, and R. A. Hull. Products of row stochastic matrices and their applications to cooperative control for autonomous mobile robots. In *Proceedings of the American Control Conference*, pages 1066–1071, Portland, OR, June 2005.
169. A. Regmi, R. Sandoval, R. Byrne, H. Tanner, and C. Abdallah. Experimental implementation of flocking algorithms in wheeled mobile robots. In *Proceedings of the American Control Conference*, pages 4917–4922, Portland, OR, 2005.
170. W. Ren. Consensus based formation control strategies for multi-vehicle systems. In *Proceedings of the American Control Conference*, pages 4237–4242, Minneapolis, June 2006.
171. W. Ren. Cooperative control design strategies with local interactions. In *Proceedings of IEEE Conference on Networking, Sensing, and Control*, Ft. Lauderdale, FL, April 2006.
172. W. Ren. Distributed attitude consensus among multiple networked spacecraft. In *Proceedings of the American Control Conference*, pages 1760–1765, Minneapolis, June 2006.
173. W. Ren. Consensus seeking in multi-vehicle systems with a time-varying reference state. In *Proceedings of the American Control Conference*, pages 717–722, New York, July 2007.
174. W. Ren. Consensus strategies for cooperative control of vehicle formations. *IET Control Theory & Applications*, 1(2):505–512, March 2007.
175. W. Ren. Consensus tracking under directed interaction topologies: Algorithms and experiments. *IEEE Transactions on Control Systems Technology*, 2007. (in review).
176. W. Ren. Distributed attitude alignment in spacecraft formation flying. *International Journal of Adaptive Control and Signal Processing*, 21(2–3):95–113, March–April 2007.
177. W. Ren. Distributed attitude synchronization for multiple rigid bodies with Euler-Lagrange equations of motion. In *Proceedings of the IEEE Conference on Decision and Control*, New Orleans, December 2007. (to appear).
178. W. Ren. Distributed cooperative attitude synchronization and tracking for multiple rigid bodies. *IEEE Transactions on Control Systems Technology*, 2007. (in review).
179. W. Ren. Formation keeping and attitude alignment for multiple spacecraft through local interactions. *Journal of Guidance, Control, and Dynamics*, 30(2):633–638, March–April 2007.

180. W. Ren. Multi-vehicle consensus with a time-varying reference state. *Systems and Control Letters*, 56(7–8):474–483, July 2007.
181. W. Ren. On consensus algorithms for double-integrator dynamics. *IEEE Transactions on Automatic Control*, 2007. (in review).
182. W. Ren. On consensus algorithms for double-integrator dynamics. In *Proceedings of the IEEE Conference on Decision and Control*, New Orleans, December 2007. (to appear).
183. W. Ren. Second-order consensus algorithm with extensions to switching topologies and reference models. In *Proceedings of the American Control Conference*, pages 1431–1436, New York, July 2007.
184. W. Ren. Synchronized multiple spacecraft rotations: A revisit in the context of consensus building. In *Proceedings of the American Control Conference*, pages 3174–3179, New York, July 2007.
185. W. Ren and E. M. Atkins. Second-order consensus protocols in multiple vehicle systems with local interactions. In *Proceedings of the AIAA Guidance, Navigation, and Control Conference*, San Francisco, August 2005. Paper no. AIAA-2005-6238.
186. W. Ren and E. M. Atkins. Distributed multi-vehicle coordinated control via local information exchange. *International Journal of Robust and Nonlinear Control*, 17(10–11):1002–1033, July 2007.
187. W. Ren and R. W. Beard. Consensus of information under dynamically changing interaction topologies. In *Proceedings of the American Control Conference*, pages 4939–4944, Boston, MA, June 2004.
188. W. Ren and R. W. Beard. Decentralized scheme for spacecraft formation flying via the virtual structure approach. *Journal of Guidance, Control, and Dynamics*, 27(1):73–82, January–February 2004.
189. W. Ren and R. W. Beard. Formation feedback control for multiple spacecraft via virtual structures. *IEE Proceedings - Control Theory and Applications*, 151(3):357–368, May 2004.
190. W. Ren and R. W. Beard. Consensus seeking in multiagent systems under dynamically changing interaction topologies. *IEEE Transactions on Automatic Control*, 50(5):655–661, May 2005.
191. W. Ren, R. W. Beard, and E. M. Atkins. Information consensus in multivehicle cooperative control: Collective group behavior through local interaction. *IEEE Control Systems Magzine*, 27(2):71–82, April 2007.
192. W. Ren, R. W. Beard, and D. B. Kingston. Multi-agent Kalman consensus with relative uncertainty. In *Proceedings of the American Control Conference*, pages 1865–1870, Portland, OR, June 2005.
193. W. Ren, R. W. Beard, and T. W. McLain. Coordination variables and consensus building in multiple vehicle systems. In V. Kumar, N. E. Leonard, and A. S. Morse, editors, *Cooperative Control*, volume 309 of *Lecture Notes in Control and Information Sciences*, pages 171–188, Berlin, 2005. Springer-Verlag.
194. W. Ren, H. Chao, W. Bourgeous, N. Sorensen, and Y. Chen. Experimental implementation and validation of consensus algorithms on a mobile actuator and sensor network platform. In *Proceedings of the IEEE International Conference on Systems, Man, and Cybernetics*, Montreal, Can, October 2007. (to appear).
195. W. Ren, H. Chao, W. Bourgeous, N. Sorensen, and Y. Chen. Experimental validation of consensus algorithms for multi-vehicle cooperative control. *IEEE Transactions on Control Systems Technology*, 2007. (to appear).

196. W. Ren, K. L. Moore, and Y. Chen. High-order and model reference consensus algorithms in cooperative control of multi-vehicle systems. *Journal of Dynamic Systems, Measurement, and Control*, 129(5):678–688, September 2007.
197. W. Ren and N. Sorensen. Distributed coordination architecture for multi-robot formation control. *Robotics and Autonomous Systems*, 2007. (to appear).
198. C. W. Reynolds. Flocks, herds, and schools: A distributed behavioral model. *Computer Graphics*, 21:25–34, 1987.
199. A. Richards, J. Bellingham, M. Tillerson, and J. How. Coordination and control of UAVs. In *Proceedings of the AIAA Guidance, Navigation, and Control Conference*, Monterey, CA, August 2002. Paper no. AIAA-2002-4588.
200. A. Robertson, G. Inalhan, and J. P. How. Formation control strategies for a separated spacecraft interferometer. In *Proceedings of the American Control Conference*, pages 4142–4147, San Diego, June 1999.
201. A. Rodriguez-Angeles and H. Nijmeijer. Mutual synchronization of robots via estimated state feedback: A cooperative approach. *IEEE Transactions on Control Systems Technology*, 9(4):380–386, 2004.
202. S. Roy, A. Saberi, and K. Herlugson. A control-theoretic perspective on the design of distributed agreement protocols. *International Journal of Robust and Nonlinear Control*, 17(10–11):1002–1033, July 2007.
203. S. Roy, A. Saberi, and K. Herlugson. Formation and alignment of distributed sensing agents with double-integrator dynamics and actuator saturation. In S. Phoha, T. LaPorta, and C. Griffin, editors, *Sensor Network Applications*, Piscataway, NJ, 2007. IEEE Press.
204. G. Royle and C. Godsil. *Algebraic Graph Theory*. Springer Graduate Texts in Mathematics #207, New York, 2001.
205. W. J. Rugh. *Linear System Theory*. Prentice Hall, Englewood Cliffs, NJ, 2nd edition, 1996.
206. C. Samson. Time-varying feedback stabilization of car-like wheeled mobile robots. *International Journal of Robotics Research*, 12(1):55–64, February 1993.
207. S. Sastry, G. Meyer, C. Tomlin, J. Lygeros, D. Godbole, and G. Pappas. Hybrid control in air traffic management systems. In *Proceedings of the IEEE Conference on Decision and Control*, pages 1478–1483, New Orleans, December 1995.
208. A. V. Savkin. Coordinated collective motion of groups of autonomous mobile robots: Analysis of Vicsek's model. *IEEE Transactions on Automatic Control*, 49(6):981–983, 2004.
209. C. Schumacher, P. R. Chandler, and S. J. Rasmussen. Task allocation for wide area search munitions via network flow optimization. In *Proceedings of the AIAA Guidance, Navigation, and Control Conference*, Montreal, Can, 2001.
210. C. Schumacher and S. N. Singh. Nonlinear control of multiple UAVs in close-coupled formation flight. In *Proceedings of the AIAA Guidance, Navigation, and Control Conference*, Denver, CO, August 2000. Paper no. AIAA-2000-4373.
211. C. Scolese, D. Folta, and F. Bordi. Field of view location and formation flying for polar orbiting missions. *Advances in the Astronautical Sciences*, 75(2):949–966, 1991.
212. B. Seanor, G. Campa, Y. Gu, M. Napolitano, L. Rowe, and M. Perhinschi. Formation flight test results for UAV research aircraft models. In *Proceedings of the AIAA 1st Intelligent Systems Technical Conference*, Chicago, IL, September 2004. Paper no. AIAA-2004-6251.

213. R. Sepulchre, D. Paley, and N. Leonard. Stabilization of planar collective motion: All-to-all communication. *IEEE Transactions on Automatic Control*, 52(5):811–824, 2007.
214. S. Sheikholeslam and C. A. Desoer. Control of interconnected nonlinear dynamical systems: The platoon problem. *IEEE Transactions on Automatic Control*, 37(6):806–810, June 1992.
215. D. D. Siljak. *Decentralized control of complex systems.* Academic, Boston, 1991.
216. A. Sinha and D. Ghose. Generalization of linear cyclic pursuit with application to rendezvous of multiple autonomous agents. *IEEE Transactions on Automatic Control*, 51(11):1819–1824, November 2006.
217. J.-J. E. Slotine and M. D. D. Benedetto. Hamiltonian adaptive control of spacecraft. *IEEE Transactions on Automatic Control*, 35(7):848–852, July 1990.
218. J.-J. E. Slotine and W. Li. *Applied Nonlinear Control.* Prentice Hall, Englewood Cliffs, NJ, 1991.
219. J.-J. E. Slotine and W. Wang. A study of synchronization and group cooperation using partial contraction theory. In V. Kumar, N. E. Leonard, and A. S. Morse, editors, *Cooperative Control*, volume 309 of *Lecture Notes in Control and Information Sciences*, pages 207–228, Berlin, 2005. Springer-Verlag.
220. S. L. Smith, M. E. Broucke, and B. A. Francis. A hierarchical cyclic pursuit scheme for vehicle networks. *Automatica*, 41:1045–1053, 2005.
221. S. L. Smith, M. E. Broucke, and B. A. Francis. Curve shortening and the rendezvous problem for mobile autonomous robots. *IEEE Transactions on Automatic Control*, 52(6):1154–1159, 2007.
222. N. Sorensen and W. Ren. A unified formation control scheme with a single or multiple leaders. In *Proceedings of the American Control Conference*, pages 5412–5418, New York, July 2007.
223. D. P. Spanos and R. M. Murray. Distributed sensor fusion using dynamic consensus. In *IFAC World Congress*, Prague, Czech Republic, 2005. Paper code We-E04-TP/3.
224. D. P. Spanos, R. Olfati-Saber, and R. M. Murray. Dynamic consensus on mobile networks. In *IFAC World Congress*, Prague, Czech Republic, 2005. Paper code We-A18-TO/1.
225. D. J. Stilwell and B. E. Bishop. Platoons of underwater vehicles. *IEEE Control Systems Magazine*, 20(6):45–52, December 2000.
226. D. M. Stipanovic, G. Inalhan, R. Teo, and C. J. Tomlin. Decentralized overlapping control of a formation of unmanned aerial vehicles. *Automatica*, 40(8):1285–1296, 2004.
227. T. Sugar and V. Kumar. Decentralized control of cooperating mobile manipulators. In *Proceedings of the IEEE International Conference on Robotics and Automation*, pages 2916–2921, Leuven, Belgium, May 1998.
228. J. Tan, N. Xi, W. Sheng, and J. Xiao. Modeling multiple robot systems for area coverage and cooperation. In *Proceedings of the IEEE International Conference on Robotics and Automation*, pages 2568–2573, New Orleans, 2004.
229. H. G. Tanner. On the controllability of nearest neighbor interconnections. In *Proceedings of the IEEE Conference on Decision and Control*, pages 2467–2472, Paradise Island, Bahamas, December 2004.
230. H. G. Tanner and D. K. Christodoulakis. State synchronization in local-interaction networks is robust with respect to time delays. In *Proceedings*

of the IEEE Conference on Decision and Control, and the European Control Conference, pages 4945–4950, Seville, Spain, December 2005.
231. H. G. Tanner, A. Jadbabaie, and G. J. Pappas. Stable flocking of mobile agents, part ii: Dynamic topology. In *Proceedings of the IEEE Conference on Decision and Control*, pages 2016–2021, Maui, HI, December 2003.
232. H. G. Tanner, A. Jadbabaie, and G. J. Pappas. Flocking in fixed and switching networks. *IEEE Transactions on Automatic Control*, 52(5):863–868, May 2007.
233. H. G. Tanner, G. J. Pappas, and V. Kumar. Input-to-state stability on formation graphs. In *Proceedings of the IEEE Conference on Decision and Control*, pages 2439–2444, Las Vegas, NV, December 2002.
234. H. G. Tanner, G. J. Pappas, and V. Kumar. Leader-to-formation stability. *IEEE Transactions on Robotics and Automation*, 20(3):433–455, June 2004.
235. P. Tsiotras. Further passivity results for the attitude control problem. *IEEE Transactions on Automatic Control*, 43(11):1597–1600, 1998.
236. J. N. Tsitsiklis, D. P. Bertsekas, and M. Athans. Distributed asynchronous deterministic and stochastic gradient optimization algorithms. *IEEE Transactions on Automatic Control*, 31(9):803–812, 1986.
237. Y. Ulybyshev. Long-term formation keeping of satellite constellation using linear-quadratic controller. *Journal of Guidance, Control, and Dynamics*, 21(1):109–115, January–February 1998.
238. J. J. P. Veerman, G. Lafferriere, J. S. Caughman, and A. Williams. Flocks and formations. *Journal of Statistical Physics*, 121:901–936, December 2005.
239. T. Vicsek, A. Czirok, E. B. Jacob, I. Cohen, and O. Schochet. Novel type of phase transitions in a system of self-driven particles. *Physical Review Letters*, 75(6):1226–1229, 1995.
240. T. Vidal, M. Ghallab, and R. Alami. Incremental mission allocation to a large team of robots. In *Proceedings of the IEEE International Conference on Robotics and Automation*, pages 1620–1625, Minneapolis, April 1996.
241. M. A. Vincent. Design of the TOPSAT mission. *Advances in the Astronautical Sciences*, 85(2):1137–1146, 1993.
242. P. Wang and F. Hadaegh. Minimum-fuel formation reconfiguration of multiple free-flying spacecraft. ENG 97–187, UCLA, Los Angeles, December 1997.
243. P. K. C. Wang. Navigation strategies for multiple autonomous mobile robots moving in formation. *Journal of Robotic Systems*, 8(2):177–195, 1991.
244. P. K. C. Wang and F. Y. Hadaegh. Coordination and control of multiple microspacecraft moving in formation. *Journal of the Astronautical Sciences*, 44(3):315–355, 1996.
245. P. K. C. Wang, F. Y. Hadaegh, and K. Lau. Synchronized formation rotation and attitude control of multiple free-flying spacecraft. *Journal of Guidance, Control, and Dynamics*, 22(1):28–35, January 1999.
246. W. Wang and J.-J. E. Slotine. Contraction analysis of time-delayed communications and group cooperation. *IEEE Transactions on Automatic Control*, 51(4):712–717, 2006.
247. W. Wang and J.-J. E. Slotine. A theoretical study of different leader roles in networks. *IEEE Transactions on Automatic Control*, 51(7):1156–1161, 2006.
248. S. Wegener. White paper on UAV over-the-horizon disaster management demonstration projects, Feb 2000. `http://geo.arc.nasa.gov/sge/UAVFiRE/whitepaper.html`.
249. J. T.-Y. Wen and K. Kreutz-Delgado. The attitude control problem. *IEEE Transactions on Automatic Control*, 36(10):1148–1162, October 1991.

250. J. R. Wertz, editor. *Spacecraft Attitude Determination and Control.* Kluwer, Norwell, MA, 1978.
251. B. Wie. *Space Vehicle Dynamics and Control.* AIAA Education Series, Reston, VA, 1998.
252. B. Wie, H. Weiss, and A. Apapostathis. Quaternion feedback regulator for spacecraft eigenaxis rotations. *Journal of Guidance, Control, and Dynamics*, 12(3):375–380, May 1989.
253. J. Wolfowitz. Products of indecomposable, aperiodic, stochastic matrices. *Proceedings of the American Mathematical Society*, 15:733–736, 1963.
254. H. Wong, M. S. de Queiroz, and V. Kapila. Adaptive tracking control using synthesized velocity from attitude measurements. *Automatica*, 37(6):947–953, 2001.
255. C. W. Wu. Agreement and consensus problems in groups of autonomous agents with linear dynamics. In *Proceedings of the IEEE International Symposium on Circuits and Systems*, pages 292–295, Kobe, Japan, 2005.
256. C. W. Wu. Synchronization and convergence of linear dynamics in random directed networks. *IEEE Transactions on Automatic Control*, 51(7):1207–1210, July 2006.
257. W. Xi, X. Tan, and J. S. Baras. Gibbs sampler-based coordination of autonomous swarms. *Automatica*, 42(7):1107–1119, 2006.
258. F. Xiao and L. Wang. State consensus for multi-agent systems with switching topologies and time-varying delays. *International Journal of Control*, 79(10):1277–1284, October 2006.
259. L. Xiao and S. Boyd. Fast linear iterations for distributed averaging. *Systems and Control Letters*, 53(1):65–78, 2004.
260. L. Xiao, S. Boyd, and S. Lall. A scheme for robust distributed sensor fusion based on average consensus. In *Proceedings of the International Conference on Information Processing in Sensor Networks*, pages 63–70, Los Angeles, April 2005.
261. G. Xie and L. Wang. Consensus control for a class of networks of dynamic agents. *International Journal of Robust and Nonlinear Control*, 17(10–11):941–959, July 2007.
262. O. A. Yakimenko. Direct method for rapid prototyping of near-optimal aircraft trajectories. *Journal of Guidance, Control, and Dynamics*, 23(5):865–875, September–October 2000.
263. H. Yamaguchi. Adaptive formation control for distributed autonomous mobile robot groups. In *Proceedings of the IEEE International Conference on Robotics and Automation*, pages 2300–2305, Albuquerque, NM, April 1997.
264. H. Yamaguchi and J. W. Burdick. Asymptotic stabilization of multiple nonholonomic mobile robots forming group formations. In *Proceedings of the IEEE International Conference on Robotics and Automation*, pages 3573–3580, Leuven, Belgium, May 1998.
265. Q. Yan, G. Yang, V. Kapila, and M. S. de Queiroz. Nonlinear dyanamics and adaptive control of multiple spacecraft in periodic relative orbits. In *Proceedings of the 23rd Annual AAS Guidance and Control Conference*, pages AAS 00–013, Breckenridge, CO, February 2000.
266. G. Yang and V. Kapila. Optimal path planning for unmanned air vehicles with kinematic and tactical constraints. In *Proceedings of the IEEE Conference on Decision and Control*, pages 1301–1306, Las Vegas, NV, 2002.

267. B. Young, R. W. Beard, and J. Kelsey. A control scheme for improving multi-vehicle formation maneuvers. In *Proceedings of the American Control Conference*, Arlington, VA, June 2001.
268. C. Yu, J. M. Hendrickx, B. Fidan, B. D. O. Anderson, and V. D. Blondel. Three and higher dimensional autonomous formations: Rigidity, persistence and structural persistence. *Automatica*, 43(3):387–402, 2007.
269. X. Yun, G. Alptekin, and O. Albayrak. Line and circle formation of distributed physical mobile robots. *Journal of Robotic Systems*, 14(2):63–76, 1997.

Index

MIX
Papier aus verantwortungsvollen Quellen
Paper from responsible sources
FSC® C105338

If you have any concerns about our products,
you can contact us on
ProductSafety@springernature.com

In case Publisher is established outside the EU,
the EU authorized representative is:
Springer Nature Customer Service Center GmbH
Europaplatz 3, 69115 Heidelberg, Germany

Printed by Libri Plureos GmbH
in Hamburg, Germany